SEXUAL BEHAVIOR IN
THE HUMAN FEMALE

By the Staff of the Institute for
Sex Research, Indiana University

ALFRED C. KINSEY   WARDELL B. POMEROY

CLYDE E. MARTIN   PAUL H. GEBHARD
Research Associates

JEAN M. BROWN	Research Assistant, Library
CORNELIA V. CHRISTENSON	Research Assistant, in charge of Reference Research
DOROTHY COLLINS	Research Assistant, Statistical Calculator
RITCHIE G. DAVIS	Research Associate, Legal Studies
WILLIAM DELLENBACK	Research Assistant, Photographic Studies
ALICE W. FIELD	Research Associate, Legal Studies
HEDWIG G. LESER	Research Assistant, Translator
HENRY H. REMAK	Special Translator
ELEANOR L. ROEHR	Research Assistant, Secretary

Volume Two
ISHI PRESS
INTERNATIONAL

Sexual Behavior in the Human Female
Volume Two

by Alfred C. Kinsey
Wardell B. Pomeroy
Clyde E. Martin
and Paul H. Gebhard

First Published in 1953 by W. B. Saunders Company

This Printing divided into two volumes in
April, 2010 by Ishi Press in New York and Tokyo

with a new foreword by Sam Sloan

ISBN 4-87187-705-1
978-4-87187-705-3

Ishi Press International
1664 Davidson Avenue, Suite 1B
Bronx NY 10453-7877
1-917-507-7226
Printed in the United States of America

SEXUAL BEHAVIOR IN
THE HUMAN FEMALE

CONSULTING EDITORS

To the nearly 8000 females
who contributed the data on
which this book is based

Sexual Behavior in the Human Female

by Alfred C. Kinsey

Foreword by Sam Sloan

There are two Kinsey reports, one for the human male and one for the human female. This is the Second Part of "**The Kinsey Report**", the part that deals with the Human Female. This Report on the human female addresses the hot topics of when do girls start having sex and what percentage of all married women commit adultery. (The surprising answer is: About the same as the percentage of married men who commit adultery.)

Because the Kinsey reports are lengthy, they have each been divided into two volumes here. Volume One of this book, "**Sexual Behavior in the Human Female**", is about the scope of the study, the problems of sampling, the sources of the data, and the types of sexual activity including pre-adolescent behavior, masturbation, nocturnal sex dreams, petting, pre-marital coitus and marital coitus.

Volume Two deals with Extra-marital coitus, homosexual contacts, sex with animals, physiology of sexual response, psychology of sexual response and hormonal factors in sexual response, including studies on the castration of human males.

Although this book was published in 1953 and it is assumed that our sexual mores have vastly changed since then, attempts to repeat the study by Dr. Kinsey have come up with about the same results.

Here, it is important to note that although this part of the Kinsey Report was published in 1953, it shows statistics derived from interviews taken years earlier of females describing events of their youth. For example, it shows what percentage of females born before 1900 had engaged in sexual intercourse or in petting to orgasm by age 20. This basically shows the sexual practices of the World War I era. It does show the expected increase, that there was more sex during the Roaring Twenties than during the World War I era, but the increase was not dramatic and just a few percentage point.

During the eras covered by the Kinsey Report, birth control pills were not yet developed or just in the process of being developed and the first effective drugs to cure venereal diseases were not available or were just becoming available. Also, abortions were illegal. With the changes in the modern era, it seems surprising that the sex rate and the adultery rates have not soared, although of course nobody really knows for sure.

It is also important to remember that Kinsey was a zoologist. He constantly compares sex in humans to sex in animals and finds them to be about the same. For example he finds that cows, horses, dogs and pigs engage in petting sort of activities just as do humans.

Kinsey explodes many myths. For example, it was widely believed that all boys masturbate but few girls masturbate. Kinsey found that 65% of all females masturbate. Among married females, the frequency rate is 0.2 to 0.3 per week and among unmarried females the rate is between 0.3 to 0.4 times per week.

Kinsey found that 58% of all females had learned to masturbate through self-discovery. However, only 25% of all boys had learned about masturbation in this way. The rest of the boys had heard about it from other boys.

One thing that may have changed is the methods of sex. Dr. Kinsey has a long chapter on "petting", how many females engage in petting, how often do they do it, at what age and how far do they go.

Petting is hugging and kissing and other sex play that does not lead to intercourse. It is assumed that nowadays there is little petting of the type practiced in 1953. Nowadays couples are believed to go to straight to intercourse, without bothering with the preliminaries.

In 1953, couples would typically go to a drive-in movie or a "Lovers Lane" and hug and kiss in a car. Perhaps the female would pull up her blouse or unfasten her bra and let the male caress or play with her breasts. The movie "The Last Picture Show" in the opening scene has a typical example of petting as practiced at that time.

But that was usually as far as it went.

Kinsey has statistics on how many females have experienced petting to orgasm. This seems surprising because one would not imagine that petting would ever lead to orgasm. However, Kinsey reports that 23% of all females have experienced petting to orgasm by their late teens.

Perhaps the most famous statistic in this Kinsey Report is that 26% of all married women have committed adultery. One would imagine that the figure would have gone up since them, but modern statistics seem to indicate that the rate is about the same. There are definite reasons why it would be easier for a woman to commit adultery nowadays. Cell Phones, beepers, email and the Internet make it much easier for women to contact their lovers or to find new lovers, with almost no chance of being caught by their husbands. Also, the vast amounts of porn now available on the Internet make such activities seem more acceptable. Also, the risks involved in adultery are greatly reduced from what they were in 1953.

We are relieved to learn from Dr. Kinsey that sex with animals is less common among human females than among human males, Donkey shows in Tijuana notwithstanding.

Kinsey found only 6 females in his study who were achieveing orgasm on a regular basis through sexual relations with her own dog, whereas among farm boys in some regions of the country, nearly half of all boys had had orgasms through sex with their farm animals.

A conclusion reached by Kinsey is that among humans things have not changed much. The ancient Greeks and Romans did it pretty much the same way, concludes Kinsey.

Other sources have reported that only 2% of females are lesbians. Kinsey shows that the true figure is much higher. He shows that by age 40, 19% of all females have had sexual contact with other females and 13% of all females have experienced orgasm with other females. However, among married females, only 3% have experienced orgasm with other females. Kinsey believes that marriage simply stops homosexual activity from taking place, but it probably does not limit the inclination towards such activity.

In Kinsey's time, females were more secretive and much less willing to admit to lesbian tendencies. Nowadays, with even popular movie stars admitting to being lesbians it is much more acceptable.

The Kinsey Reports together sold three-quarters of a million copies and were translated in thirteen languages. They may be considered as part of the most successful and influential scientific books of the 20th century. However, almost all sales were in the first year of publication. Since then, the books have been reprinted only once, that being by **Indiana University Press** in **1998**.

In this study, Dr. Kinsey, as a zoologist, applies the same scientific methods that he used to study the lower animals, which is in some ways both the strongest and the weakest part of his study.

This study was first published in 1948 and was based on interviews of only 8,000 men and 8,000 women. In spite of the small number of participants, subsequent studies by others have verified these results. Perhaps the most surprising of these subsequent studies is that studies of vastly different human groups, such as Chinese, Africans and Native American Indians, have produced similar results. This proves that all humans regardless of cultural grouping have about the same sexual habits. It also proves that such institutions as marriage and religion have no real impact on human behavior, as societies that do not have marriage and have vastly different religions still have about the same sexual habits.

Dr. Kinsey, a Professor of Zoology, goes even further than that. He finds that not only are all humans similar but we are similar to other primates including gorillas and chimpanzees, who have about the same sexual habits as we do.

Dr. Kinsey does not stop there. Delving into his knowledge of zoology, he compares us humans to moths and wasps. Did you know that a male moth can smell miles away when a female moth is ovulating? Did you know that different kinds of animals often try to have sex with each other? Dr. Kinsey tells us all about that.

This particular report, "**Sexual Behavior in the Human Female**", is almost impossible to obtain. I searched for one in good condition for months without success. I finally found one in Vienna, Austria of all places and used it to create this reprinting.

Sam Sloan
New York
April 10, 2010

Original Back Cover and Flap Blurb

Sexual Behavior in the Human Female

By Alfred Kinsey

Original Back Cover and Flap Blurb

This book is based on studies pursued during the past fifteen years by Dr. Kinsey and his associates at the Institute for Sex Research at Indiana University. The studies have been supported throughout the years by Indiana University, and by funds from the Rockefeller Foundation (administered by the Committee for Research on Problems of Sex of the Medical Division of the National Research Council). All incomes from the Institute's publications and other work have been used for the further advancement of the research.

The material presented in this book has been derived from personal interviews with nearly 8,000 women; from special research studies in sexual anatomy, physiology, psychology, and endocrinology; and from an exhaustive study of the literature.

The book presents data on the incidence and frequency with which women participate in various types of sexual activity. The authors show how such factors as age, decade of birth, and religious adherence are reflected in patterns of sexual behavior. Some measure of the social significance of the various types of sexual behavior is provided. Legal aspects are presented: The authors make comparisons of female and male sexual activities, and investigate the factors which account for the similarities and differences between female and male patterns of behavior.

The book analyzes the anatomic and physiologic background of sexual response. The findings indicate that there is need for revision of many current theories concerning the supposed difference in the sexual responses of females and males.

Much original information is offered on the significance of

Original Back Cover and Flap Blurb

psychological factors in sexual response with indications that this is the area in which very real and marked differences do exist between female and male. The authors examine the relation of the so-called sex hormones to sexual response, and they investigate the role played by the nervous system in sexual physiology.

It would be difficult to overemphasize the book's ultimate importance in brushing away the confusion that has for many centuries cobwebbed the subject of sex. Its impact will be felt immediately in such problems as sexual adjustment in marriage, sexual education of children, and the social control of sex offenders.

This is the only report on female sexual behavior written or authorized by Dr. Kinsey and his associates. Neither the authors nor the publisher of this volume has any connection with, or any control over, any other book on female sexual behavior.

a companion volume .. .
SEXUAL BEHAVIOR in the HUMAN MALE

Out of the storm of acclaim, condemnation and controversy which accompanied publication of this volume in 1948, an unmistakable fact has emerged: SEXUAL BEHAVIOR IN THE HUMAN MALE was a milestone on the road toward a truly scientific approach to human sexual behavior.

Since its publication, state legislators, students of social problems, and others in similar capacities have turned to this volume for data on which to base their thinking; the movement for intelligent sex education in home and school has gained tremendous momentum; and literally hundreds of articles, based on the book, have besought physicians, parents and teachers to look at sex as it is-rather than as they think it might be, or as they feel it should be.

Original Back Cover and Flap Blurb

Kinsey, Pomeroy and Martin's SEXUAL BEHAVIOR IN THE HUMAN MALE is-and will be for decades to come-indispensable in securing a factual understanding of the sexual outlets of men.

why and how these studies were made

Dr. Alfred C. Kinsey, the senior author, is Professor of Zoology at the University of Indiana and has been on the faculty there for 33 years.

In the course of his teaching, Dr. Kinsey was approached by his biology students with questions concerning sexual behavior which he was unable to answer, or to find answers for, in the previously published studies on human activities.

He thus became convinced of the need for a comprehensive survey of sexual behavior based on interviews with many thousands of persons. The National Research Council and Indiana University supported the project, which has thus far spanned fifteen years and produced the two books, SEXUAL BEHAVIOR IN THE HUMAN MALE and SEXUAL BEHAVIOR IN THE HUMAN FEMALE.

CONTENTS

PART III. COMPARISONS OF FEMALE AND MALE

LIST OF TABLES

PART I. HISTORY AND METHOD

CHAPTER 6. NOCTURNAL SEX DREAMS

CHAPTER 9. MARITAL COITUS

PART III. COMPARISONS OF FEMALE AND MALE

LIST OF FIGURES

PART III. COMPARISONS OF FEMALE AND MALE

CHAPTER 14. ANATOMY OF SEXUAL RESPONSE AND ORGASM

CHAPTER 15. PHYSIOLOGY OF SEXUAL RESPONSE AND ORGASM

CHAPTER 17. NEURAL MECHANISMS OF SEXUAL RESPONSE

Chapter 10

EXTRA-MARITAL COITUS

It is widely understood that many males fail to be satisfied with sexual relations that are confined to their wives and would like to make at least occasional contacts with females to whom they are not married. While it is generally realized that there are some females who similarly desire and actually engage in extra-marital coitus, public opinion is less certain about the inclination and behavior of the average female in this regard.

Most males can immediately understand why most males want extra-marital coitus. Although many of them refrain from engaging in such activity because they consider it morally unacceptable or socially undesirable, even such abstinent individuals can usually understand that sexual variety, new situations, and new partners might provide satisfactions which are no longer found in coitus which has been confined for some period of years to a single sexual partner. To most males the desire for variety in sexual activity seems as reasonable as the desire for variety in the books that one reads, the music that one hears, the recreations in which one engages, and the friends with whom one associates socially. On the other hand, many females find it difficult to understand why any male who is happily married should want to have coitus with any female other than his wife. The fact that there are females who ask such questions seems, to most males, the best sort of evidence that there are basic differences between the two sexes (Chapter 16).

As we have remarked in our volume on the male, the preoccupation of the world's biography and fiction, through all ages and in all human cultures, with the non-marital sexual activities of married females and males, is evidence of the universality of human desires in these matters, and of the universal failure of the existent social regulations to resolve the basic issues which are involved. The record of extra-marital coitus in our sample of American females, and our examination of the outcome of their experience, may contribute to an understanding of the nature and magnitude of the problem; and an examination of the anthropologic and more ancient mammalian backgrounds of this aspect of human behavior may show something of the origins of the conflict between the individual's personal desires and the social regulation of individual behavior.

409

MAMMALIAN ORIGINS

Since extra-marital coitus can exist only where there is an institution of marriage, there is nothing strictly comparable to it among man's mammalian ancestors, although a rather similar phenomenon may develop in any species of animal which establishes sexual partnerships that last through at least one full breeding season. Such more or less prolonged relationships exist among many of the primates (the monkeys and apes) and among some of the other larger mammals such as sea lions, elephants, some of the dogs, beaver, horses, deer, and still other species. These animals travel in family groups or packs, or herds which may include several or many adult females and their offspring but only a limited number of adult males. It is usually a single male, however, which dominates the whole group.

The dominant male in such a mammalian family claims exclusive sexual rights to all of the females in his group, and attempts to prevent other males from having access to them even when he himself is sexually satiated. The females are property which he has acquired, maintains, and can defend because of his physical strength and aggressiveness, and less aggressive males usually find it difficult or impossible to invade his domain. In the same way, many animals consider certain geographic locations as their own, and will fight to keep other animals from invading their territory.[1] Similarly, a gorged animal may still try to prevent other individuals from eating his food, even though he no longer wants or needs it. The maintenance of a sexual monopoly appears to emanate from a similar attempt to maintain the exclusive possession of one's property, and this appears to be as true in the human species as in the lower mammals.

Both the females and the males in these mammalian mateships are quite ready to accept coitus with individuals who are not their established mates; but females are deterred by their male mates, and males are limited in their promiscuity by other males who bar them from their groups. Sometimes males are also deterred from becoming promiscuous by the fact that they are already satiated by the coitus which they have had with their own mates.

In both the baboon and the rhesus monkey, females soliciting new sexual partners have been known to utilize a remarkably human procedure to escape the anger of their established mates. When the mates discover them in coitus with other males, or seem about to discover them, the females may cease their sexual activities and attack the new male partners. A high proportion of the human "rape" cases which we

---

[1] An excellent description of animal territories, and the possessive attitudes toward them, is in: Hediger 1950:7–18.

have had the opportunity to examine involve something of the same motifs. In the case of the infra-human primates, such action diverts the anger of the established mates who turn on the other males and drive them away, often with the help of the unfaithful females. After such an episode, the females may then present themselves to their own mates for coitus.[2]

There are occasional records of infra-human females objecting to other females having coitus with their mates, but this is not the rule.[3] Males among these other mammals, just as in the human species, are the ones who are most often and most violently disturbed at any sexual infidelity on the part of their mates. While cultural traditions may account for some of the human male's behavior, his jealousies so closely parallel those of the lower species that one is forced to conclude that his mammalian heritage may be partly responsible for his attitudes.

Among most species of mammals, sexual response is likely to become most vigorous when the animal encounters a new situation or meets a new sexual partner. Many species become psychologically fatigued when relationships with a single partner are maintained over any long period of time. The introduction of a new partner will then revive the sexual interest. Among cattle, for instance, a new bull brought into the corral may incite the other bulls to a renewal of the heterosexual or homosexual activities which had subsided before the new animal was introduced.[4] Among monkeys it has been noted that animals caged together gradually become less aroused by each other, the preliminary sex play must be more extended before they are stimulated enough to attempt coitus, and the subsequent copulation is less vigorous. With the introduction of new partners, both the female and the male may become more aroused and copulate with them more vigorously and with a minimum of foreplay.[5] Psychologic fatigue must be a prime source of the difficulty in keeping married human mates strictly monogamous.

But even among those mammals which have reasonably durable sexual partnerships, coitus outside of the partnership is still a definite part of the life pattern. The sexual history of the male in such a species (e.g., the seal, elephant, gibbon, or baboon) usually passes through three

[2] Escaping the anger of an offended mate by offering coitus is described by: Hamilton 1914:304–305. Zuckerman 1932:228.
[3] One of the few cases of females fighting over the possession of a male, is recorded for the monkey by: Hamilton 1914:304. Carpenter 1942:*passim* believes that the female gibbon drives other adult females away from her mate.
[4] That there may be a renewal of sexual activity among bulls following the introduction of a new bull is common knowledge among cattle breeders, and we have frequently observed such behavior.
[5] The increased vigor in copulation, and the reduction of foreplay which characterizes sexual activity with a new partner, are described for monkeys in: Hamilton 1914:303. Kempf 1917:143.

phases. First, as a young adolescent he is a bachelor, primarily because he is not yet dominant enough to keep a female from other male aspirants. He hunts for the occasional, unattached female, or hangs around a family unit (which may be his own parental family group or some other family group) and surreptitiously secures coitus from whatever females are available. Periodically he has to fight the older, more dominant males who discover him while he is trying to secure sexual relations with their mates.

But as the young male becomes more mature and physically and psychologically more dominant, he may succeed in taking some other male's mate, or in obtaining or keeping a just-adult female which has not been mated. This constitutes the second period of his career, and the one in which he is sexually most active. If he is very dominant he collects several females as mates; if he is less dominant he has fewer chances to accumulate female mates. At this stage he must periodically fight off the current group of bachelors, including his own sons who try to secure coitus from his mates. For some years he may succeed in dominating the situation, but finally, as an old and physically less powerful animal, he begins to lose his females to younger males and he himself eventually ends up as a bachelor again. Now he hangs around other family units trying to obtain sexual relations or, in many instances, he may have to go away and live alone for the rest of his life.[6]

There are obvious parallels between the situation in these other mammalian families, and the course of marital infidelity in the human species. Not all of the human problems are cultural developments or the product of particular social philosophies. It is evident that interest in a variety of sexual partners is of ancient standing in the mammalian stocks, and occurs among both females and males. The human male's interest in maintaining his property rights in his female mate, his objection to his wife's extra-marital coitus, and her lesser objection to his extra-marital activity, are mammalian heritages. These heritages, human females and males must accept, or rise above them if they intend to control their patterns of sexual behavior.

ANTHROPOLOGIC DATA

In the course of human history, and in various cultures, there has been some recognition of the human mammal's desire for coitus with individuals to whom he or she is not wedded, and various means have been devised to cope with these demands for non-marital sexual experience.[7] All cultures recognize the desirability of maintaining the

---

[6] The rise from youthful bachelorhood to adult mateship, and the subsequent return to bachelorhood, is well known for animals such as the elephant and stallion. It is described for the baboon by: Zuckerman 1932.

[7] The omnipresent desire for extra-marital coitus is noted, for instance, by Reichard

family as a stabilizing unit in the social organization, but it still remains to decide whether it is necessary to forbid and try to prevent all non-marital activity, or whether it is possible to accept and regulate such activity so it will do a minimum of damage to the institution of the family.

In no society, anywhere in the world, does there seem to have been any serious acceptance of complete sexual freedom as a substitute for the arrangements of a formal marriage. On the other hand, some cultures allow considerable freedom for both females and males in non-marital relationships. This is primarily true in groups which do not associate sex with social goals, with love, and with other emotional values. Thus, one anthropologist records for the Lepcha that "sexual activity is practically divorced from emotion; it is a pleasant and amusing experience, and as much a necessity as food and drink; and like food and drink it does not matter from whom you receive it, as long as you get it; although you are naturally grateful to the people who provide you with either regularly." [8] Another records for a second group that "intercourse is among all people whatsoever regarded essentially as a pleasure, and among the Arunta . . . there is no evidence that it is invested with any more meaning than that." [9]

The anthropologists find that most societies recognize the necessity for accepting at least some extra-marital coitus as an escape valve for the male, to relieve him from the pressures put on him by society's insistence on stable marital partnerships. These same societies, however, less often permit it for the female. But most societies have also recognized that some restraint on extra-marital activities is necessary if marriages and homes are to be maintained and if the social organization is to function effectively. As one anthropologist puts it, "unrestrained competition over food, drink, and sexual partners would soon involve the destruction of any society." [10] As another puts it, "Prohibitory regulations curb the socially more disruptive forms of social competition. Permissive regulations allow at least the minimum impulse gratification required for individual well-being." [11] And a third sums up the significance of a permissive attitude by pointing out that "Pre-nuptial license and the relaxation of the matrimonial bonds must not be regarded as a

---

in Boas 1938:435 ("There is probably no tribe in which formal marriage alone is sexually satisfactory"). The checking of this desire by various social devices is noted by Linton 1936:136 ("All societies inhibit the male's tendency to collect females to some degree, setting limits to the competition for them and, through marriage, assuring the male of the possession of those which he has already gathered").

[8] Gorer 1938:170.
[9] Ashley Montagu 1937:236.
[10] Ford 1945:93.
[11] Murdock 1949:261.

denial of marriage, as its abrogation, but rather as its complement. The function of license is not to upset but rather to maintain marriage." [12]

Most societies, in consequence, permit or condone extra-marital coitus for the male if he is reasonably circumspect about it, and if he does not carry it to extremes which would break up his home, lead to any neglect of his family, outrage his in-laws, stir up public scandal, or start difficulties with the husbands or other relatives of the women with whom he has his extra-marital relationships. Even in those societies which overtly forbid all non-marital coitus, there is a covert toleration of occasional lapses if social difficulties do not arise from such acts. There are few if any human societies in which the male's extra-marital coitus is very stringently suppressed or very severely punished. [13]

On the other hand, such extra-marital activity is much less frequently permitted or condoned for the female. Only 10 per cent of the cultures freely permit it. In another 40 per cent, the female may be allowed extra-marital experience on special occasions or with particular persons. [14] For instance, non-marital activity may be permitted at certain orgiastic ceremonies which are recurrent and usually seasonal events. It may be permitted or even required for the new bride as part of the marriage ceremony. [15] In a few instances it is the customary means of entertaining the husband's guests, but in this case it is the male who lends his wife to the guests. Occasionally non-marital coitus is allowed and sometimes required of siblings-in-law. Occasionally non-marital coitus has been allowed and even required when a marriage was barren.

In many of these societies, extra-marital coitus is overtly forbidden for the female, although it may be covertly condoned if it is not too

---

[12] Malinowski in Marcuse 1928:104.

[13] This lesser social concern in the majority of primitive societies over extra-marital coitus for the male is noted by: Ford 1945:29–30.

[14] Using the "Cross Cultural Survey" files, Ford 1945:29–30, 101, found under one-third of the societies permitting extra-marital coitus. In a later work, Ford and Beach 1951:113 estimated 39 per cent. Utilizing the same files, Murdock 1949:265 reports a smaller percentage but adds: "A substantial majority of all societies . . . permit extra-marital relations with certain affinal relatives." Considering these sources and still additional data, and defining a "permissive" society as one which permits female extra-marital coitus under *any* situation, including wife-lending and orgiastic ceremonies or festivities, we arrive at our figures of 10 per cent and 40 per cent. A broad definition is imperative if one is to avoid such complications as arise, for example, in the case of the Yakut who are listed as forbidding female extra-marital coitus (Ford and Beach 1951:115), and who also demand extra-marital coitus for females in the form of wife-lending (Ford 1945:30).

[15] The widespread practice of defloration or coitus with a bride by someone other than the groom, throughout European and derivative cultures, is the so-called "Jus primae noctis" or "Droit du seigneur." Its history is excellently covered by: Schmidt 1881. Foras 1886. Westermarck 1922(1):ch.5. The custom is also listed for almost a score of primitive groups, as in: Crawley 1927(2):66, 255.

flagrant and if the husband is not particularly disturbed.[16] As our later data may indicate, this seems to be the direction toward which American attitudes may be moving.

In about half of the known human societies, extra-marital coitus is completely prohibited for the female, and in many of them she may be rigorously and severely punished for such activity. In not a few cultures it is the husband's privilege and often his obligation to kill the offending wife. In many cultures, his failure to exact such a penalty from his wife and/or the offending suitor is taken as disgraceful evidence of his insufficient masculinity. He loses caste if he does not seek revenge from those who threaten his conjugal rights. Even in cultures where there is no established law or custom allowing such retribution, the male would, nevertheless, be supported by public opinion if he took the law into his own hands, and he would not be penalized for doing so. In our own European and American past, this attitude predominated; but the privilege of the husband to exact any severe penalty from either the unfaithful wife or her paramour has largely disappeared in all but a few parts of the United States.

Nevertheless, even in those cultures which most rigorously attempt to control the female's extra-marital coitus, it is perfectly clear that such activity does occur, and in many instances it occurs with considerable regularity. Even those males who disapprove of extra-marital coitus for their own wives may be interested in securing such contacts for themselves, and this in most instances means securing coitus with the wives of other males. Such inconsistencies would be unexplainable if we did not understand the mammalian backgrounds of human male behavior.

The reasons given in any particular society for the prohibition of the female's extra-marital coitus may appear logical. It is, for instance, pointed out that such activity represents a defiance of social convention and threatens a husband's right to sufficiently frequent and regular coitus with his spouse. It is said that the wife's involvement in such activity would spoil the prestige of her husband and his kin. It is pointed out that such non-marital activity may cause her to neglect her duties and obligations in her home, and social interests are clearly involved if the coitus leads to extra-marital pregnancies. It is generally believed that extra-marital coitus invariable leads to marital discord and/or divorce, with all of their consequent social implications. In many instances there is an insistence that any coitus outside of marriage is, intrinsically, in itself, morally wrong; and irrespective of the social

---

[16] Covert toleration of female extra-marital coitus in societies forbidding such coitus, is not uncommon. See: Ford 1945:30. Ford and Beach 1951:116.

consequences of such activity, it constitutes a profanation of the sacra-
ment of marriage and consequently a sin against God and against
society. These considerations have all been part of the rationale by
which various societies have supported their condemnation of extra-
marital activities for the female. The arguments would be more impressive if they led to any resolution of behavior which does not appear
in the unreasoned behavior of the males of the lower mammalian
groups.

In the light of these mammalian and historic backgrounds, it has
been significant to examine the status of extra-marital coitus among
the females in our sample.

RELATION TO AGE

Among the married females in the sample, about a quarter (26 per
cent) had had extra-marital coitus by age forty (Table 115, Figure 74).
Between the ages of twenty-six and fifty, something between one in six
and one in ten was having extra-marital coitus (Table 114, Figure 75).
Both the accumulative and active incidences of extra-marital coitus
were remarkably uniform for many of the subdivisions of the sample,
but they had varied in relation to the ages, the educational levels, the
decades of birth, and the religious backgrounds of the various groups.
The frequencies had increased somewhat with advancing age.

Since the cover-up on any socially disapproved sexual activity may
be greater than the cover-up on more accepted activities, it is possible
that the incidences and frequencies of extra-marital coitus in the sample
had been higher than our interviewing disclosed.

Accumulative Incidence. In their late teens, 7 per cent of the married females in the sample were having coitus with males other than
their husbands. The accumulative incidences did not materially increase in the next five years, but after age twenty-six they gradually
and steadily rose until they reached their maximum of 26 per cent by
forty years of age (Table 115, Figure 74). After that age only a few
females began for the first time to have extra-marital coitus.[17]

---

[17] American data on incidences of extra-marital activities among females are also
found in: Hamilton 1929:350. Dickinson and Beam 1931:313, 315, 394.
Strakosch 1934:77. Glueck and Glueck 1934:432. Landis et al. 1940:97. Dearborn in Fishbein and Burgess 1947:168. Locke 1951:152. The findings in these
studies range from the low figure of 1.2 per cent in Locke's happily married
sample to a 24 per cent figure in the studies of Hamilton and Glueck. The
European studies show the following: Schbankov acc. Weissenberg 1924a:13
(6 per cent of 53 Russian students). Golossowker acc. Weissenberg 1925:176
(30 per cent of 107 Russian students). Gurewitsch and Grosser 1929:535
(18 per cent of 166 Russian female students). Gurewitsch and Woroschbit
1931:91 (8 per cent of over 1500 Russian peasant women, but 12 per cent
in the 35–40 age group). Friedeburg 1950:13 (10 per cent of 517 German
women in a questionnaire survey). England acc. Rosenthal 1951:59 (18 per
cent of British middle-class women).

Active Incidence. The number of females in the sample who were having extra-marital coitus in any particular five-year period had been lowest in the youngest and in the oldest age groups (Table 114, Figure 75). The incidences had reached their maxima somewhere in the thirties and early forties. For the total sample the active incidences had begun at about 6 per cent in the late teens, increased to 14 per cent by the late twenties, and reached 17 per cent by the thirties. They began to decrease after the early forties. They had dropped to 6 per cent by the early fifties.

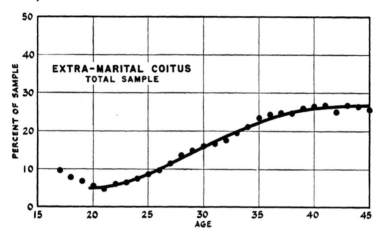

Figure 74. Accumulative incidence: experience in extra-marital coitus
Data from Table 115.

The younger married females had not so often engaged in extra-marital coitus, partly because they were still very much interested in their husbands and partly because the young husbands were particularly jealous of their marital rights. Moreover, at that age both the male and the female were more often concerned over the morality of non-marital sexual relationships. In time, however, many of these factors had seemed less important, and the middle-aged and older females had become more inclined to accept extra-marital coitus, and at least some of the husbands no longer objected if their wives engaged in such activities.

Although it is commonly believed that most males prefer sexual relations with distinctly younger partners, and although most males are attracted by the physical charms of younger females, data which we have on our histories show that many of them actually prefer to have coitus with middle-aged or older females. Many younger females become much disturbed over non-marital irregularities in which they may have engaged, and many males fear the social difficulties that may arise from such disturbances. Older females are not so likely to become disturbed, and often have a better knowledge of sexual techniques. In consequence many males find the older females more effective as sexual partners. All of these factors probably contributed to the fact that the peak of the extra-marital activities of the females in the sample had come in the mid-thirties and early forties.

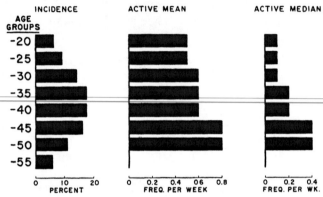

Figure 75. Active incidence, mean, median: experience in extra-marital coitus, by age

Data from Table 114.

Incidence of Orgasm. In the available sample, about 85 per cent (in most groups 78 to 100 per cent) of all those who were engaging in extra-marital activity were responding, at least on occasion, to orgasm (compare the incidences of experience and orgasm in Table 114). For most age groups the incidences of response were about the same as those in marital coitus.

On the other hand, comparisons of the median frequencies of experience and of experience to the point of orgasm indicate that orgasm in the extra-marital relationships had occurred in a high proportion of the contacts (Table 114). In some cases, this had been more often than those same females were reaching orgasm in their marital coitus. Some of the females had experienced multiple orgasms, and the total number of orgasms had actually exceeded the number of contacts in some of the groups. Selective factors may have been involved, and the more responsive females may have been the ones who had most often accepted extra-marital coitus; but the high rate of response appears to have depended also on the fact that the extra-marital experience had provided a new situation, a new partner, and a new type of relationship which had been as stimulating to some of the females as it would have been to most males. Some females who had never or rarely reached orgasm with their husbands had responded regularly in their extra-marital relationships (p. 432).

Frequency. In that segment of the sample which was having extra-marital coitus (the active sample), the frequencies had begun at the rate of once in ten weeks (0.1 per week) in the married teen-age and twenty-year-old groups (Table 114, Figure 75). They had steadily increased in the later age groups. By the forties, the extra-marital coitus was occurring once in two to three weeks for those who were having any experience at all. This means that the active median frequencies

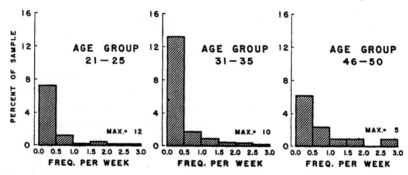

Figure 76. Individual variation: frequency of experience in extra-marital coitus

Each class interval includes the upper but not the lower frequency. For incidences of females not having extra-marital coitus, see Table 114.

of the extra-marital coitus were of about the same order as the active median frequencies of masturbation and twice as high as the frequencies of nocturnal dreams in marriage. The frequencies of the extra-marital coitus were in actuality second only to the frequencies of marital coitus in the sample of middle-aged and older married females.

Because there were some individuals in each age group who were having extra-marital experience much more frequently than the average female in the sample, the active mean frequencies were much higher than the active median frequencies. The active mean frequencies had begun at once in two weeks (0.5 per week) among the teen-aged, married females, and had risen to about once in eight or nine days (0.8 per week) among the females in their forties (Table 114, Figure 75).

However, since the incidences of the extra-marital coitus were relatively low, the mean frequencies for the total sample, including those who were having and those who were not having experience, were very low. They had not averaged more than one extra-marital contact in something like ten weeks (the total mean frequencies), even during their peak between the ages of thirty and forty (Table 114).

Sporadic Nature of Frequency. There are few types of sexual activity which occur more irregularly than extra-marital coitus. This is be-

cause the opportunities to make such contacts usually occur only sporadically, and it is often difficult to find the time and place where the
coitus may be had without the spouse or someone else becoming aware
of the activity. Married persons may find more difficulty in arranging
their non-marital activities than single persons find in arranging their
pre-marital activities. Moreover, many married persons sharply limit
their extra-marital activities in order to avoid emotional relationships
which might break up their marriages.

Consequently the average frequencies shown in our calculations are
misleading if they suggest that the extra-marital contacts had occurred
with any weekly or monthly regularity. It is a prime weakness of statistical averages that they suggest a regularity in the occurrence of
activities which do not actually occur with any regularity. A dozen
sexual contacts which are made in two weeks of a summer vacation
may show up as frequencies of once per month for the whole of a year,
and such an even distribution of activity does not often occur. It is
more usual to find several non-marital contacts occurring in the matter
of a few days or in a single week when the spouse is away on a trip,
or when the female is traveling and putting up at a hotel, or at a summer resort, or on an ocean voyage, or visiting at a friend's home; and
then there may be no further contacts for months or for a year or more.
Only a smaller proportion of the females in the sample had ever developed regular and long-time relationships with males who were not
their husbands.

Among the females in the available sample, the highest average frequencies of extra-marital coitus had occurred in the twenties, when
three individuals were averaging seven contacts, one was averaging
twelve, and one was averaging nearly thirty contacts per week over a
five-year period. The maximum frequencies for particular individuals
had dropped in the older groups. By fifty years of age, only one female
in the total sample of 261 was having extra-marital coitus with a frequency which averaged more than three times per week (Figure 76).
Just as with pre-marital coitus, the high frequencies were often attained
by assured and socially effective individuals who had not been emotionally disturbed by their departures from the social code and who,
therefore, had not gotten into difficulties because of their non-marital
sexual activities.

Percentage of Total Outlet. Because of the relatively low incidences
and low average frequencies of the extra-marital coitus, only a relatively small proportion of the sexual outlet of the total female sample
had been derived from that source. Such activity had accounted for
only 3 per cent of the orgasms of the females who were in their early
twenties, but an increasing proportion of the outlet in the later age

groups, until 13 per cent of the outlet had come from the non-marital activity by the late forties (Table 114). At that age in many of the marriages there had been some drop in the frequencies of the marital coitus, and the female who was still as responsive as or even more responsive than she had been in her earlier years had become more inclined to accept extra-marital coitus as a substitute for her reduced marital outlet.

RELATION TO EDUCATIONAL LEVEL

There had been only minor differences in the accumulative incidences of extra-marital coitus among the females of the different edu-

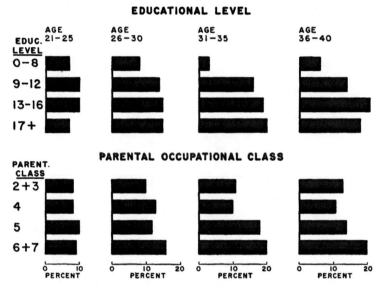

Figure 77. Active incidence: experience in extra-marital coitus, by educational level and parental occupational class

Data from Tables 116 and 117.

cational levels (Table 115). Some 31 per cent of the females in the college sample had had some extra-marital coitus by age forty. Some 27 per cent of those who had done graduate work, and about 24 per cent of those who had gone into but not beyond high school had had such experience. The differences had not been great.[18]

The active incidences had hardly differed among the females of the several educational levels during their late teens (Table 116), but after age twenty-five the limited grade school sample showed definitely lower incidences than the rest of the sample (Table 116, Figure 77).

[18] In the several Russian studies (see footnote 17), the incidence of extra-marital coitus among peasant women is lower than the incidence among students, as in: Gurewitsch and Woroschbit 1931:91.

The active incidences of the extra-marital coitus had steadily become higher in the older age groups, often because of the more deliberate acceptance of such activity among the older husbands and wives, especially in the better educated groups.

None of the differences in the average frequencies of the extra-marital coitus seemed to have been related in any significant way with the educational backgrounds of the females in the sample.

RELATION TO PARENTAL OCCUPATIONAL CLASS

In the late teens, the active incidences of extra-marital coital experience and orgasm had not significantly differed among the females in the sample who had come from laborers' homes, from the homes of skilled mechanics, and from the homes of lower and upper white collar and professional groups. But after age twenty-five more of the females who had come from upper white collar and professional homes were having extra-marital coitus, and extra-marital coitus in which they had reached orgasm (Table 117, Figure 77).

The average frequencies of the extra-marital coitus had not varied in any way which seemed significantly correlated with the occupational classes of the homes in which the females had been raised.

RELATION TO DECADE OF BIRTH

The number of married females who were ultimately involved in extra-marital coitus (the accumulative incidences), and the number who were involved in any particular five-year period (the active incidences), appear to have been more or less markedly affected by the increased acceptance of sexual activities which began in this country with the generation that was born immediately after the turn of the century and which was, therefore, sexually most active immediately after the first World War or in the 1920's (Tables 119, 120, Figures 78, 79).

The accumulative incidences of extra-marital coitus among the females in the sample who were born before 1900 had reached 22 per cent by forty years of age. The incidences among the females who were born in the first decade after 1900 had reached 30 per cent by that same age. The later generations seem to be maintaining that level of incidence.[19]

The lowest active incidences of extra-marital coitus had been among

---

[19] These changes in patterns of marital fidelity are commented on by: Kühn 1932: 228–229 (increase is product of freer life of women). Folsom 1937:720–723 (increase due to revaluation of importance of pleasure in sexual experience). Locke 1951:149–150 (reported no differences in incidences of extra-marital coitus among females born in four successive decades; but based his findings on an active sample of not more than 12 cases).

the females who were born before 1900. Most of the groups born after that showed a somewhat and, in many instances, a markedly increased incidence (Table 120, Figure 79). For instance, in the sample of females between twenty-one and twenty-five years of age, 4 per cent of

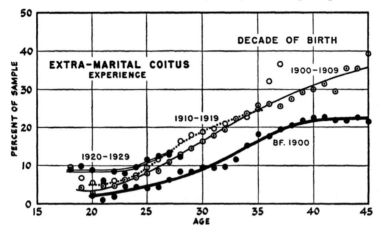

Figure 78. Accumulative incidence: experience in extra-marital coitus, by decade of birth

Data from Table 119.

Figure 79. Active incidence: experience in extra-marital coitus, by decade of birth

Data from Table 120. Incidence for ages 26–30 in youngest generation not shown because their marriages are of such short duration that data are incomplete.

the generation that was born before 1900, but 8 per cent of those who were born in the first decade after 1900, had been involved in extra-marital coitus. For the age group twenty-six to thirty, the figures were 9 per cent for the older generation and 16 per cent for the next generation. These increased active incidences of extra-marital coitus had paralleled the increases which had occurred in pre-marital petting (p. 244) and pre-marital coitus (p. 301) in the post-war generation of American females.

There were, however, no consistent differences in the active median frequencies of extra-marital coitus among the females of the several generations in the sample. It is important to note again, as we already have in connection with the other sexual activities which became more prevalent after the first World War, that the increases in extra-marital coitus lay in the number of females who were involved, and not in any increase in the frequencies with which the average female had had experience.

RELATION TO AGE AT ONSET OF ADOLESCENCE

The incidences of the female's extra-marital coitus in particular periods of her marriage (the active incidences) (Table 118), and the frequencies with which she had had such coitus (the active median frequencies), did not seem to have been affected by the age at which she had turned adolescent.

RELATION TO RELIGIOUS BACKGROUND

The active incidences of extra-marital coitus had been more affected by the religious backgrounds of the females in the sample than by any other factor which we have examined. In every group in which we have sufficient cases for comparing females of different levels of religious devoutness, the lowest incidences of extra-marital coitus had occurred among those who were most devoutly religious, and the highest incidences among those who were least closely connected with any church activity. This was true of all the Protestant, Jewish, and Catholic groups in the sample. The differences in incidences were well enough marked in the younger age groups, but they become even more striking in the older Protestant groups. For example, in the Protestant groups aged twenty-one to twenty-five, some 5 per cent of the religiously active females had had extra-marital coitus, while 13 per cent of the inactive group had had such experience. But during the early thirties the differences lay between 7 per cent for the active Protestants and 28 per cent for the inactive Protestants (Table 121, Figure 80).[20]

While the frequencies of extra-marital experience had varied in the different groups, the variation did not show consistent trends which would warrant the opinion that the differences were related to the degree of religious devotion.

NATURE AND CONDITIONS OF EXTRA-MARITAL COITUS

The times and places and detailed circumstances of extra-marital relationships have been the subject of so much literature throughout written history and fiction, that they need little additional analysis

---

[20] Friedeburg 1950:31 found an inverse relationship between the incidence of extra-marital coitus and regularity of church attendance in a German survey of 579 males and females. The figures ranged from 9 per cent reporting extra-marital experience among those who attended church regularly, to 27 per cent reported by those who never attended church.

in the present context. Because of their previous marital experience, those who engage in extra-marital coitus usually see to it that the contacts are had under conditions which are similar to those usual in marital contacts.[21]

Partners. The extra-marital partners of the females in the sample had, for the most part, been married males somewhat near the females in age; but sometimes they were younger or older males who had not been married or who were widowed or divorced. Not a few of the younger, unmarried males had had their pre-marital coital experience with married females, some of whom were the aggressors in starting the relationships.

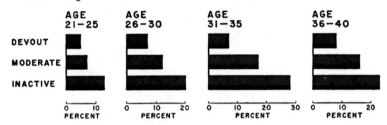

Figure 80. Active incidence: experience in extra-marital coitus, in Protestant groups

Data from Table 121.

Up to the time at which they contributed their histories, some 41 per cent of the females in the active sample had confined their extra-marital coitus to a single partner; another 40 per cent had had contacts with something between two and five partners (Table 122). This means that 19 per cent had had more than five partners and some 3 per cent had had more than twenty partners in their extra-marital relationships.[22] This is a somewhat higher degree of promiscuity than we found in the pre-marital coital histories, primarily because more years had been involved in the extra-marital activity.

Number of Years Involved. Of those females in the sample who had had any experience at all, nearly a third (32 per cent) had had extra-marital coitus ten times or less, up to the date at which they contributed their histories to the present study (Table 123). Some 42 per cent had confined their activity to a single year or less; nearly a quarter (23 per cent) had confined it to two or three years. About a third (35 per cent) had had the coitus for some four years or more, including 10 per cent

[21] Sylvanus Duvall 1952:161 also recognizes that conditions for extra-marital coitus may be better than for pre-marital coitus.

[22] Data on number of partners in extra-marital relations are also to be found in: Hamilton 1929:350–351 (among 24 females, 38 per cent had had only one partner). Gurewitsch and Woroschbit 1931:92 (among 1516 Russian peasant women, 53 per cent had had one partner, 25 per cent had had two, 10 per cent had had three).

who had carried on their extra-marital relationships for more than ten years and in a few instances for more than twenty years.

The length of time over which extra-marital coital activities had been carried on and the number of partners involved were, of course, dependent on the number of years that the females had been married. At the time she contributed her history, the median female in the sample was 34 years old, and had been married 7.1 years (12.5 years among those who had had extra-marital coitus). If the females in the sample had been married longer, they would have had more extensive extra-marital histories. Basing the calculations on the number of years that the females in the active sample had been married, we find that 36 per cent of those who had been married six to ten years had had extra-marital coitus one to ten times; but among those who had been married over twenty years, only 23 per cent had been so confined in their experience (Table 123). Among those in the active sample who had been married six to ten years, 4 per cent had been having extra-marital coitus for six to ten years; but among those who had been married more than twenty years, 19 per cent had been having extra-marital coitus for as long a period of years, and 31 per cent had been having it for more than ten years.

Extra-Marital Petting. There were not a few of the married females and males in the sample who had accepted extra-marital petting even though they had refused to accept vaginal coitus. Such extra-marital petting seems to have increased within recent years,[23] although we do not have sufficient data to establish this point statistically. However, such petting is not confined to younger persons, for it occurs not infrequently in middle-aged histories and even in some of the older married histories. Just as with pre-marital petting, extra-marital petting is accepted because of the satisfactions which are peculiar to it, or in order to avoid a possible pregnancy, or sometimes because the available facilities are inadequate for coitus but sufficient for petting. In fact, at dinner parties, cocktail parties, in automobiles, on picnics, and at dances, a considerable amount of public petting is allowed between married adults when coitus would be unacceptable. Apparently petting is considered less immoral than coitus, even though the petting techniques may be as effective in bringing erotic response and orgasm. Extra-marital petting is not infrequently carried on in social groups which may include the female's husband, and he would usually be less inclined to allow the extra-marital relationships if coitus were involved.

Unfortunately our record on extra-marital petting is incomplete, for we did not realize the extent of such activity when this study first be-

---

[23] Bernard 1938:357 found 49 per cent of females and 59 per cent of males, among 500 college students, tolerant of extra-marital petting.

gan. Information on this point is available, however, on 1090 of the married females in the sample. Of these, some 16 per cent had engaged in extra-marital petting although they had never allowed extra-marital coitus.

The techniques of the extra-marital petting are, of course, identical with those which are used in pre-marital petting (pp. 251 ff.) and in the petting which precedes coitus both in marital (p. 361) and in non-marital (p. 311) relationships. Of those females who had done extra-marital petting without coitus, more than half had accepted breast and genital contacts. In some cases they had accepted mouth-genital contacts. Something short of 15 per cent of the females in the sample had reached orgasm in this extra-marital petting, including 2 per cent who had petted to orgasm although they had never allowed extra-marital coitus. The accumulative incidences of all these activities would undoubtedly have proved to be higher, if our data had been more complete.

Relation to Pre-Marital Coitus. Among the 514 females in the sample who had had extra-marital coitus up to the date at which they contributed their histories, over 68 per cent had also had coitus before marriage (Table 122). Since only 50 per cent of all the married females in the sample had had pre-marital coitus, it would appear that the pre-maritally experienced females were somewhat more inclined to accept coitus with males other than their husbands after marriage.

To put it in another way, 29 per cent of the females with histories of pre-marital coitus had had extra-marital coitus by the time they contributed their histories to this study, but only 13 per cent of those who had not had pre-marital coital experience.

This greater inclination of the females with pre-coital experience to accept extra-marital coitus after marriage, is even more strikingly shown by the accumulative incidences of experienced females, calculated as follows:

By age	% of sample with extra-mar. coitus		Number of cases	
	FEMALES WITHOUT PRE-MAR. COITUS	FEMALES WITH PRE-MAR. COITUS	FEMALES WITHOUT PRE-MAR. COITUS	FEMALES WITH PRE-MAR. COITUS
35	16	33	513	399
40	20	39	364	207
45	20	40	225	87

These correlations between pre-marital and extra-marital experience may have depended in part upon a selective factor: the females who were inclined to accept coitus before marriage may have been the ones who were more inclined to accept non-marital coitus after marriage. A causal relationship may also have been involved, for it is not impossible that non-marital coital experience before marriage had persuaded those females that non-marital coitus might be acceptable after marriage.[24]

However, the females who had had pre-marital coitus seemed to have been no more promiscuous in their extra-marital relationships than the females who had had no pre-marital coitus (Table 122). Among those who had had coitus before they were married, 81 per cent had had extra-marital contacts with one to five partners; and among those who had had no pre-marital experience, 80 per cent had had coitus with that number of partners.

MORAL AND LEGAL STATUS

In nearly all societies and all moral codes, everywhere in the world, extra-marital coitus is restricted more severely than pre-marital coitus.[25] The Jewish, Mohammedan, and more ancient codes attached considerable significance to the virginity of the female at the time of marriage; but the morality of having coitus after marriage with an individual who was not one's spouse was a matter of still more serious concern. In a general way this is true of Christian codes and of the Anglo-American law that grew out of them.

Anglo-American Law. The specific legal restrictions on extra-marital coitus, or *adultery* as it is known in Anglo-American law, largely originate in Jewish and Roman codes, but acquire their force more particularly from the Catholic Church codes which allow fewer exceptions than the original Jewish codes.[26] While there is a general prohibition of extra-marital coitus in Jewish law, there were exceptional circumstances which made such activity acceptable. As we have seen, the code allowed and even required extra-marital activity of the husband

---

[24] A positive relationship between pre-marital and extra-marital coital experience was also reported by: Hamilton 1929:346. Landis et al. 1940:98. Terman 1938:340 reported a similar correlation with *desire* for extra-marital relations.

[25] This greater restriction on extra-marital coitus, in comparison to the restrictions on pre-marital coitus, is surveyed and documented in: Donohue 1931:4. Ohlson 1937:330. Ford and Beach 1951:115.

[26] Extra-marital coitus is penalized in the ancient codes, some of which date back to the 2nd millennium B.C. See: Pritchard 1950:35, Book of the Dead. Pritchard 1950:160, Lipit-Ishtar law code (with some provisions for second wives). Pritchard 1950:171–172, Code of Hammurabi. Pritchard 1950:181–182, Middle Assyrian laws (including the following: "If he has kissed her [the wife of another], they shall draw his lower lip along the edge of the blade of an ax and cut it off"). Pritchard 1950:196, Hittite code. For the Jewish code, see Epstein 1948:201–214. Biblical references are plentiful, ranging from the specific prohibitions in Exodus 20:14, Leviticus 20:10, and Deuteronomy 22:22, to the colorful descriptions in II Samuel 11:4, Proverbs 7:6–23, and Ezekiel 23:1–45. Also see the Koran pt.18, ch. 24:2–20. For the Catholic code, see Davis 1946(2):237–239.

whose wife was sterile, and required the male to take unto himself as an additional wife the wife of his brother who had died.[27]

American law has tended toward the proscription of all extra-marital sexual relationships; but in recognition of the realities of human nature, the penalties for adultery in most states are usually mild and the laws are only infrequently enforced. In 5 of the states the maximum penalty is a fine. There are 3 states which attach no criminal penalty at all to adultery, but civil penalties may be involved in these and in many other states.[28] In every state of the Union proof of adultery is allowed as sufficient grounds for divorce. Adultery is often taken as evidence of desertion or failure to support or a contribution to the delinquency of the children in the home. Both females and males may sometimes be penalized in this indirect fashion. In some states the right of the female to share in the husband's property may be threatened by her adulterous behavior. The broadest definitions of adultery and the heaviest penalties are concentrated in the northeastern section of the United States, all 10 of those states being among the 17 which may impose prison terms for a single act of extra-marital coitus.

In actual practice, such extra-marital coitus is rarely prosecuted because its existence rarely becomes known to any third party. Even when it does become known, the matter is rarely taken to criminal court. Most of the cases which we have seen in penal institutions were prosecuted because of some social disturbance that had grown out of the extra-marital activity, as when a wife had complained, or when the family had been neglected or deserted as a result of the extra-marital relationships, or when arguments, physical fights, or murder was the product of the discovery of the relationships. Not infrequently the prosecutions represented attempts on the part of neighbors or relatives to work off grudges that had developed over other matters. In this, as in many other areas, the law is most often utilized by persons who have ulterior motives for causing difficulties for the non-conformant individuals. Not infrequently the prosecutions represent attempts by sheriffs, prosecutors, or other law enforcement officers to work off personal or political grudges by taking advantage of extra-marital relationships which they may have known about and ignored for some

---

[27] Talmudic students searched assiduously for other special circumstances which might justify exceptions to the rule against extra-marital coitus—as, for instance, when a male, looking over a parapet, accidentally fell onto a passing female, and accidentally effected a genital union as he fell. In such a case the male would be absolved of any blame for having broken the rule against extra-marital coitus. See the Talmud, Yebamoth 54a.

[28] The states in which the maximum penalty for adultery is a fine are: Ky., Md., Tex., Va., and W. Va. The states in which there is no criminal penalty are: La., N. M. (but see §41–702 N.M.Stat. 1941), and Tenn. Also see Bensing 1951:67. In 6 states (Ia., Mich., Minn., N. D., Ore., and Wash.), adultery can be prosecuted only on the complaint of the spouse.

time before they became interested in prosecuting. In Boston, which is the only large city in which there is an active use of the adultery law, the statute appears to serve chiefly as a means of placing heavier penalties on simple prostitution.[29]

In some 14 states, which include a third of the total population of the country, the law specifies that adultery is punishable only when it represents regular contacts or notorious or open cohabitation between two persons who are not lawfully wedded as spouses.[30] Higher courts in such states have consistently pointed out that lone or occasional contacts are not covered by the law. However, in dealing with cases which do not have experienced attorneys to defend them, the lower courts and law enforcement officers in general usually ignore these qualifications in the law.

Social Attitudes. There is a surprising paucity of any open and frank discussion of extra-marital coitus in the serious literature.[31] Public opinion in regard to such coitus represents a mixture of professed disapproval, and spite and malice in which a considerable undercurrent of envy and suppressed desire may become apparent. As we have seen, this envy is more often apparent in the male. The female is more often tolerant of other persons having extra-marital coitus unless it concerns her own spouse. Then her disapproval may represent envy of the female who has distracted her husband's attention, or a general moral disapproval of all such non-marital activity. More often it reflects some fear that the extra-marital relationships will interfere with her own marriage.

As with other types of sexual activity, the most serious objections to extra-marital coitus come from females and males who have never had such experience. Those who have had experience are more often inclined to indicate that they intend to have more. In the sample, when the extra-marital experience had been satisfactory and had not gotten the females into personal or social difficulties, most of the experienced individuals were inclined to continue their activities.

---

[29] This use of the law in Boston is clearly evident from Ploscowe 1951:157, and from Reiman and Schroeter 1951:75.

[30] The states in which the prosecution of adultery is limited to instances where there has been regular, notorious, or open cohabitation are: Ala., Ark., Calif., Colo., Fla., Ill., Ind., Miss., Mo., Mont., Nev., N. C., Ohio, and Wyo. In Arizona and Oklahoma, single acts can be prosecuted only on the complaint of the injured spouse.

[31] The traditional social disapproval is reflected, for instance, in: Inge 1930:366–374. Neumann 1936:109. Fromme 1950:207–225. Sylvanus Duvall 1952:150–171, 292–296. More liberal interpretations may be found in: Bertrand Russell 1929:139–144. Guyon 1948:138. Haire 1948:195. Comfort 1950:103–105. Havelock Ellis 1952:84–90. That extra-marital coitus is less acceptable for females than for males is noted, for instance, in: Talmey 1910:241. Michels 1914:136, 223. Kisch 1918(1):4–6.

Thus, among the married females in the sample who had not had extra-marital experience, some 83 per cent indicated that they did not intend to have it, but in a sample of those who had had extra-marital experience, only 44 per cent indicated that they did not intend to renew their experience (Table 124, Figure 81). Some 5 per cent of those who had not had extra-marital coitus indicated that they wanted to have it, while another 12 per cent indicated that they might at some time consider the possibility of having it. This gave a total of 17 per cent who were not seriously opposed to the idea. In contrast, however, among

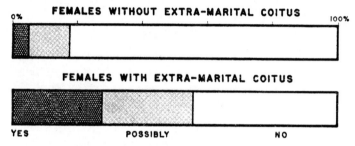

Figure 81. Intent to have extra-marital coitus

Based on females married at time of interview. Data from Table 124.

those who had already had extra-marital coital experience, some 56 per cent indicated that they intended to have more or would consider the possibility of having more.[32]

SOCIAL SIGNIFICANCE

To weigh the social significance of extra-marital sexual activity will require considerable objectivity in order to avoid, on the one hand, the traditional moral interpretations and, on the other hand, biases which are introduced by the human animal's desire for a variety of sexual experience. Certainly any scientific analysis must take into account the fact that there are both advantages and disadvantages to engaging in such activity. We do not yet have sufficient data to undertake an overall appraisal of the problem, but the following aspects of extra-marital activity are pointed up by the experience of the females who have contributed histories to this study:

[32] Specific data on the female's desire to have extra-marital coitus are also in: Davis 1929:409 (16 per cent). Hamilton 1929:368 (35 per cent). Terman 1938:336 (27 per cent). Locke 1951:149 (4 per cent and 12 per cent). For surveys of attitudes, see: Davis 1929:356 (21 per cent of 955 women saw possible justifications for extra-marital relations). Hamilton 1929:364 (73 per cent of 100 married women saw possible justification). Bernard 1938:357–358 (80 per cent of 500 college males and females stated they would pardon one act of infidelity in their spouse). Friedeburg 1950:13 (in German questionnaire survey, found 38 per cent approve some extra-marital relations). Lanval 1950:123 (among 552 chiefly French and Belgian women, 36 per cent considered faithfulness after marriage unimportant).

1. Extra-marital coitus had attracted some of the participants because of the variety of experience it afforded them with new and sometimes superior sexual partners.[33] As in pre-marital coitus, the males in
the extra-marital relationships had usually engaged in more extensive
courting, in more extended sex play, and in more extended coital techniques than the same males had ordinarily employed in their marital
relationships. In consequence many of the females had found the extra-
marital coitus particularly satisfactory. It is true that 24 per cent of the
females in the sample had not reached orgasm in their extra-marital
coitus as often as they had reached it in their marital coitus; but 34 per
cent had reached it with about equal frequency in the two types of
activity; and 42 per cent reported that orgasm had occurred more often
in the extra-marital relations.

2. In many instances the female or male had engaged in the extra-
marital coitus in a conscious or unconscious attempt to acquire social
status through the socio-sexual contacts.

3. In some instances the extra-marital coitus had been accepted
as an accommodation to a respected friend, even though the female
herself was not particularly interested in the relationship.

4. In some instances the female or male had engaged in extra-marital coitus in retaliation for the spouse's involvement in similar activity.
Sometimes the extra-marital activity was in retaliation for some sort of
non-sexual mistreatment, real or imagined, by the other spouse.

5. In some instances, both among the females and males, the extra-
marital coitus had provided a means for the one spouse to assert his
or her independence of the other, or of the social code.[34]

6. For some of the females the extra-marital relationships had provided a new source of emotional satisfactions. Some of them had found
it possible to develop such emotional relationships, while maintaining
good relationships with their husbands. Others, however, had found it
impossible to share such emotional relationships with more than one
partner. In a culture which considers marital fidelity to be the symbol

---

[33] That a low orgasm rate, or other dissatisfactions in marital coitus, may develop a
desire for extra-marital coitus is implied by: Hamilton 1929:389, 395. Strakosch 1934:78. Terman 1938:337, 340, 387. But Landis et al. 1940:172 found
wives with extra-marital experience (14 cases) had made fairly adequate
sexual adjustments with their husbands.

[34] Reasons for extra-marital coitus are also listed in: Hamilton 1929:362. Gurewitsch
and Woroschbit 1931:92. Maslow 1942:279. Sylvanus Duvall 1952:158. These
list such factors as unsatisfactoriness of the marital relationship, absence or
illness of husband, financial advantage, desire for variety, need for emotional
reassurance, emotional attachment to new partner, and attempt to help the
marriage.

and proof of such other things as social conformance, law abidingness, and love, many of the females had found it difficult to engage in non-marital sexual activities without becoming involved in guilt reactions and consequent social difficulties.[35] The females who had accepted their extra-marital activity as another form of pleasure to be shared, did not so often get into difficulties over their extra-marital relationships.

7. Not infrequently the extra-marital activities had led to the development of emotional relationships which had interfered with the relations with the lawfully wedded spouses. This had caused neglect and disagreement which had seriously affected some of the marriages. This is the aspect of extra-marital activity which most societies, throughout the world, have been most anxious to control. We doubt whether such disturbances are inevitable, for there are cases of extra-marital relationships which do not seem to get into difficulties. There are strong-minded and determined individuals who can plan and control their extra-marital relationships in such a way that they avoid possible ill consequences. In such a case, however, the strong-minded spouse has to keep his or her activity from becoming known to the other spouse, unless the other spouse is equally strong-minded and willing to accept the extra-marital activity. Such persons do not constitute a majority in our present-day social organization.

8. Sometimes sexual adjustments with the spouse had improved as a result of the female's extra-marital experience.[36]

9. Extra-marital relationships had least often caused difficulty when the other spouse had not known of them. They had most often caused difficulty at the time that the spouse first discovered them. Some of the extra-marital relationships had been carried on for long periods of years without ill effects on the marital adjustments; but when the other spouse discovered them, difficulties and in some instances divorce proceedings had been immediately begun. In such instances, the extra-marital coitus had not appeared to do as much damage as the knowledge that it had occurred. The difficulties were obviously compounded by the attitudes of our culture toward such non-marital activity.

---

[35] Landis et al. 1940:174 found fewer guilt reactions in extra-marital coitus than in pre-marital coitus.

[36] An actual improvement of the marital relationship following extra-marital experience is also reported in: Landis et al. 1940:98 (in 4 cases out of 12). It is suggested by: Bell 1921:31. Folsom 1937:723. See also Pepys 1668: Nov. 14, who wrote that he had had coitus with his wife "as a husband more times since this falling out [twenty days ago] than in I believe twelve months before. And with more pleasure to her than I think in all the time of our marriage before." Assertions that extra-marital relationships will inevitably do damage to a marriage are typified by the statement in: Kirkendall 1947:29.

The females who had had extra-marital coitus believed that their husbands knew of it, or suspected it, as follows:

	Percent
Husband knew	40
Husband suspected	9
Husband presumably did not know	51
Number of cases	470

The females in the sample reported as follows on the difficulties which developed when their husbands learned of or suspected their wives' extra-marital activities:

	Percent
Serious difficulty	42
Minor difficulty	16
No difficulty	42
Number of cases	221

Adding these cases in which there had been no difficulty to the cases in which the husbands did not know of the extra-marital relationships, it is a total of 71 per cent for whom no difficulty had yet developed.[37]

It is, of course, not impossible that the extra-marital activities may have become sources of marital difficulties at some later date. We have seen instances of what appeared to be a frank and whole-hearted acceptance of the spouse's extra-marital activity when it first began and even for some period after it had begun, with considerable conflict growing out of the activity at some later period. Sometimes extraneous circumstances, such as a new economic situation, the development of some insecurity on the part of the other spouse, or the appearance of a new extra-marital partner, had reopened the issues of extra-marital relationships some five or even ten years after they had begun.

10. Among 16 females we have a record of 18 pregnancies resulting from the extra-marital coitus. The actual incidences of such pregnancies are probably higher than that. In most cases the pregnancies had been terminated by abortions. In some instances the child had been raised in its mother's home, either with or without the husband's knowledge of the child's parentage. In other instances, the pregnancy had led to a divorce.

11. There is a not inconsiderable group of cases in the sample in which the husbands had encouraged their wives to engage in extra-marital activities. This represented a notable break with the centuries-old cultural tradition. In some instances it represented a deliberate

---

[37] Locke 1951:152 reported that 5 per cent of the happily married wives believed their mates knew or suspected that they had had extra-marital relations, while 17 per cent of the divorced wives believed so.

effort to extend the wife's opportunity to find satisfaction in sexual relations. In not a few instances the husband's attitude had originated in his desire to find an excuse for his own extra-marital activity. What is sometimes known as wife swapping usually involves this situation. In another group of cases the husband had encouraged extra-marital relations in order to secure the opportunity for the sort of group activity in which he desired to participate. Sometimes his interest in group participation involved a homosexual element which was satisfied by seeing another male in sexual action, or which he, on occasion, had satisfied by extending his contacts to the male who was having coitus with his wife. Much the same elements were involved when the husband sought the opportunity, through peeping, to surreptitiously observe his wife's extra-marital coitus. In some lower level histories, but infrequently in the better educated or economically better placed groups, the husband had encouraged his wife's extra-marital activity as a form of prostitution which had added to the prestige or the financial income of the family. There were a few instances of husbands who had encouraged their male friends and strangers to have coitus with their wives because of the sadistic satisfaction which they had obtained by forcing their wives into such relationships.

It should, however, be emphasized again that most of the husbands who accepted or encouraged their wives' extra-marital activity had done so in an honest attempt to give them the opportunity for additional sexual satisfaction.

12. Extra-marital coitus had figured as a factor in the divorces of a fair number of the females and males in our histories. We have data on 907 individuals (female and male) who had had extra-marital experience and whose marriages had been terminated by divorce. We have the subjects' judgments of the significance of their extra-marital coitus in 415 cases. In nearly two-thirds (61 per cent) of these cases, the subject did not believe that his or her own extra-marital activity had been any factor in leading to that divorce (Table 125). Some 14 per cent of the females and 18 per cent of the males believed that their extra-marital experience had been prime factors in the disruption of the marriage, and something between 21 and 25 per cent more believed that it had been a contributing factor. It is to be noted, however, that these were the subjects' own estimates of the significance and, as clinicians well know, it is not unlikely that the extra-marital experience had contributed to the divorces in more ways and to a greater extent than the subjects themselves realized.

The subjects' estimates of the importance of the spouses' extra-marital activity were handicapped, both in the case of the females and the males, by the fact that they were ignorant in perhaps half of the cases that the extra-marital activity had occurred (p. 434).

It is particularly notable that the males rated their wives' extra-marital activities as prime factors in their divorces twice as often as the wives made such evaluations of their husbands' activities. Some 51 per cent of the males considered that their wives' non-marital relations had been chief factors in precipitating the divorces, and another 32 per cent considered them factors of some importance. Only 17 per cent considered them minor factors (Table 125). In contrast to this, the females considered that the husbands' extra-marital activities were prime factors in only 27 per cent of the divorces, moderate factors in 49 per cent, and minor factors in a full 24 per cent. It may be a fact that the males' extra-marital activities do not do so much damage to a marriage, or the wives may be more tolerant of their husbands' extra-marital relations, or the wives may not comprehend the extent to which the male activities are actually affecting the stability of their marriages. Contrariwise, like the true mammal that he is, the male shows himself more disturbed and jealous and more ready to take drastic action if he discovers that his wife is having extra-marital relations.

These data once again emphasize the fact that the reconciliation of the married individual's desire for coitus with a variety of sexual partners, and the maintenance of a stable marriage, presents a problem which has not been satisfactorily resolved in our culture. It is not likely to be resolved until man moves more completely away from his mammalian ancestry.

SUMMARY AND COMPARISONS OF FEMALE AND MALE

Extra-Marital Coitus

	IN FEMALES	IN MALES
Mammalian Origins		
Dominant animal acquires several mates	No	Yes
Less dominant animal has difficulty in finding a mate	No	Yes
Seeks coitus with animals other than mate	Sometimes	Yes
More responsive with new partner	Yes	Yes
Prevented from coitus with others by	Mate	Other males
Anthropologic Data		
All societies concerned with maintenance of family		
All societies use marriage to restrain disruptive sexual competition		
In primitive societies, extra-marital coitus		
Permitted rather freely	±10%	
Permitted in special circumstances	±40%	Majority
Completely prohibited	±50%	Minority
Object to extra-marital coitus primarily on social rather than religious grounds	Yes	Yes

	IN FEMALES	IN MALES
Relation to Age		
Accumulative incidence, experience		
By age 20	6%	
By age 30	16%	
By age 40	26%	±50%
Active incidence, experience		
Age 16–20	6%	35%
Age 36–40	17%	28%
Age 51–55	6%	22%
Incidence of orgasm in extra-marital coitus	±85%	±100%
Frequency (active med.), exper., per wk.		
Age 16–20	0.1	0.4
Age 31–35	0.2	0.2
Age 41–45	0.4	0.2
Regularity	Little	Some?
Percentage of total outlet		
Age 21–25	3%	7%
Age 36–40	10%	8%
Age 46–50	13%	9%
Relation to Educational Level		
Accumulative incidence, experience	Little relation	
Active incidence, experience		
Before age 25	No relation	Higher in less educ.
After age 25	Higher in better educ.	Little difference
Frequency	No relation	Higher in less educ.
Relation to Parental Occupational Class	Little or none	
Relation to Decade of Birth		
Accumulative incidence, experience		
By age 25		
Born before 1900	4%	
Born 1900–1909	8%	
Born 1910–1919	10%	
Born 1920–1929	12%	
By age 40		
Born before 1900	22%	
Born 1900–1909	30%	
Active incidence, higher in more recent generat.	Yes	Yes in all but college males
Frequency	No relation	± higher in younger generat.
Relation to Age at Onset of Adolescence	None	None?
Relation to Religious Background		
Active incidence of exper. higher among less devout	Yes	Yes
Frequency	No relation	Higher in less devout
Nature and Conditions of Extra-Marital Coitus		
Partners		
One	41%	
Two to five	40%	

	IN FEMALES	IN MALES
Number of years involved		
Depends on length of marriage	Yes	Yes
One year or less	42%	
Over ten years	10%	
Extra-marital petting		
Petting but no extra-marital coitus	16%	
Petting to orgasm	15%	
Petting to orgasm but never coitus	2%	
Extra-marital vs. pre-marital coitus		
Among those without pre-marital coitus	13%	
Among those with pre-marital coitus	29%	
Number of extra-marital partners	No relation	
Moral and Legal Status		
Most societies restrict extra-mar. more than pre-mar. coitus	Yes	Yes
In Anglo-American law:		
Penalties mild and infrequently enforced	Yes	Yes
Adultery grounds for divorce in all states	Yes	Yes
In 14 states, not punishable unless regular and publicly known	Yes	Yes
Acceptance greater if individual has had exper.	Yes	Yes
Indiv. without exper. intend to or may have	17%	
Indiv. with exper. intend to have more	56%	
Social Significance		
May provide new types of sexual exper.	Yes	Yes
May be done to raise social status	Yes	Yes
May be done as a favor to a friend	Yes	
May be done as retaliation	Yes	Yes
May be done to assert independence	Yes	Yes
May provide emotional satisfaction	Yes	Yes
Emotional involvement may cause diffic.	Yes	Yes
May improve marital adjustment	Yes	Yes
Less often diffic. if spouse unaware	Yes	Yes
Serious difficulty when known	42%	
No difficulty when known	42%	
Pregnancy rarely results	Yes	
Husband may encourage wife's extra-mar. coitus	Yes	Yes
May be a factor in divorce	Yes	Yes
Subject rates own extra-marital coitus a major factor in divorce	14%	18%
Subject rates spouse's extra-marital coitus a major factor in divorce	27%	51%

Table 114. Active Incidence, Frequency, and Percentage of Outlet Extra-Marital Coitus

By Age

AGE DURING ACTIVITY	ACTIVE SAMPLE			TOTAL SAMPLE		CASES IN TOTAL SAMPLE
	Active incid. %	Median freq. per wk.	Mean frequency per wk.	Mean frequency per wk.	% of total outlet	
COITAL EXPERIENCE						
16–20	6	*0.1*	*0.5 ± 0.18*	—		579
21–25	9	0.1	0.5 ± 0.11	—		1654
26–30	14	0.1	0.6 ± 0.14	0.1 ± 0.02		1664
31–35	17	0.2	0.6 ± 0.09	0.1 ± 0.02		1248
36–40	17	0.2	0.6 ± 0.11	0.1 ± 0.02		853
41–45	16	0.4	0.8 ± 0.12	0.1 ± 0.02		500
46–50	11	*0.4*	*0.8 ± 0.20*	0.1 ± 0.03		261
51–55	6			0.1 ± 0.03		119
56–60	4			—		49
COITUS TO ORGASM						
16–20	5	*0.2*	*0.6 ± 0.25*	—	5	577
21–25	7	0.1	0.9 ± 0.25	0.1 ± 0.02	3	1654
26–30	11	0.2	1.2 ± 0.30	0.1 ± 0.03	6	1662
31–35	14	0.2	1.0 ± 0.25	0.1 ± 0.04	6	1244
36–40	14	0.3	1.4 ± 0.37	0.2 ± 0.06	10	851
41–45	14	0.4	1.6 ± 0.45	0.2 ± 0.07	12	497
46–50	9	*0.3*	*2.0 ± 1.18*	0.2 ± 0.11	13	260
51–55	4			0.1 ± 0.03	5	118
56–60	4			—	4	49

Italic figures throughout the series of tables indicate that the calculations are based on less than 50 cases. No calculations are based on less than 11 cases. The dash (—) indicates a percentage or frequency smaller than any quantity which would be shown by a figure in the given number of decimal places.

Table 115. Accumulative Incidence: Extra-Marital Coital Experience By Educational Level

AGE	TOTAL SAMPLE	EDUCATIONAL LEVEL 9–12	13–16	17+	TOTAL SAMPLE	EDUCATIONAL LEVEL 9–12	13–16	17+
	%		*Percent*		*Cases*		*Cases*	
18	8	6	6		190	86	67	
19	7	8	4		375	149	161	
20	6	7	6	6	556	195	260	64
25	9	10	10	7	1338	412	566	294
30	16	16	16	17	1216	343	476	331
35	23	21	26	25	912	238	346	271
40	26	24	31	27	571	140	205	188
45	26	22	29	27	312	85	111	92

For an explanation of the discrepancies between certain of these active incidences and the accumulative incidences shown for the same ages, see p. 44.

Table 116. Active Incidence and Percentage of Outlet: Extra-Marital Coitus By Educational Level

Age during activity	Educ. level	Active incid. exper. %	Active incid. or-gasm %	% of total out-let	Cases in total sample	Age during activity	Educ. level	Active incid. exper. %	Active incid. or-gasm %	% of total out-let	Cases in total sample
16–20	9–12	7	5	2	210	36–40	0–8	6	4	5	52
	13–16	6	5	1	257		9–12	14	11	5	210
	17+	6	5	—	66		13–16	21	18	13	323
21–25	0–8	7	3	6	74		17+	18	15	9	268
	9–12	10	6	3	487	41–45	9–12	16	15	14	117
	13–16	10	9	3	727		13–16	18	15	9	182
	17+	7	6	1	368		17+	14	13	16	163
26–30	0–8	8	6	4	84	46–50	9–12	8	7	4	71
	9–12	14	10	3	489		13–16	15	12	10	91
	13–16	15	12	7	671		17+	11	11	25	76
	17+	15	11	7	421						
31–35	0–8	3	0	0	71						
	9–12	16	12	3	338						
	13–16	19	14	8	480						
	17+	20	17	7	360						

Age during activity	Parental class	Active incid. exper. %	Active incid. orgasm %	% of total outlet	Cases in total sample	Age during activity	Parental class	Active incid. exper. %	Active incid. orgasm %	% of total outlet	Cases in total sample
16–20	2+3	11	9	15	142	31–35	2+3	11	8	1	216
	4	7	5	—	86		4	10	8	4	177
	5	7	5	1	150		5	18	11	3	292
	6+7	6	5	2	225		6+7	20	19	9	572
21–25	2+3	8	7	3	311	36–40	2+3	13	10	2	136
	4	8	6	1	230		4	11	7	1	97
	5	10	7	2	427		5	14	11	5	209
	6+7	9	7	3	732		6+7	20	18	12	420
26–30	2+3	10	8	3	310	41–45	2+3	7	6	8	83
	4	13	10	3	232		4	11	10	10	63
	5	12	9	7	409		5	17	15	8	90
	6+7	16	14	6	738		6+7	18	16	16	262

The occupational classes are as follows: 2+3 = unskilled and semi-skilled labor. 4 = skilled labor. 5 = lower white collar class. 6+7 = upper white collar and professional classes.

Age during activity	Age at adol.	Active incid. exper. %	Active incid. orgasm %	% of total outlet	Cases in total sample	Age during activity	Age at adol.	Active incid. exper. %	Active incid. orgasm %	% of total outlet	Cases in total sample
16–20	8–11	8	6	8	139	31–35	8–11	17	14	8	185
	12	6	5	1	153		12	19	15	5	291
	13	7	6	7	169		13	17	15	7	415
	14	8	7	6	77		14	16	11	3	225
							15+	18	15	12	132
21–25	8–11	10	7	1	333	36–40	8–11	15	14	12	133
	12	10	7	1	458		12	18	14	8	188
	13	8	7	4	488		13	17	16	12	270
	14	7	5	3	254		14	20	14	5	169
	15+	15	10	5	122		15+	19	14	8	93
26–30	8–11	13	11	6	286	41–45	8–11	17	15	25	78
	12	15	11	6	426		12	17	17	11	103
	13	14	12	7	527		13	15	15	10	145
	14	12	8	4	276		14	14	11	7	102
	15+	18	16	5	148		15+	15	13	13	71

Table 119. Accumulative Incidence: Extra-Marital Coital Experience
By Decade of Birth

| AGE | DECADE OF BIRTH | | | | DECADE OF BIRTH | | | |
	Bf. 1900	1900–1909	1910–1919	1920–1929	Bf. 1900	1900–1909	1910–1919	1920–1929
	Percent				Cases			
18			10	9			63	79
19		4	7	10		72	116	153
20	2	3	5	9	63	107	172	214
25	4	8	10	12	197	357	568	216
30	10	16	19		250	453	513	
35	18	26	25		282	453	177	
40	22	30			268	303		
45	21	40			236	76		

For an explanation of the discrepancies between certain of these active incidences and the accumulative incidences shown for the same ages, see p. 44.

Table 120. Active Incidence and Percentage of Outlet: Extra-Marital Coitus
By Decade of Birth

Age	Decade of birth	Active incid. exper. %	Active incid. orgasm %	% of total out-let	Cases in total sample	Age	Decade of birth	Active incid. exper. %	Active incid. orgasm %	% of total out-let	Cases in total sample
16–20	Bf. 1900	2	2	1	62	31–35	Bf. 1900	15	13	8	293
	1900–1909	3	2	3	114		1900–1909	21	16	8	509
	1910–1919	5	4	11	172		1910–1919	16	13	5	448
	1920–1929	9	8	7	230						
21–25	Bf. 1900	4	3	2	208	36–40	Bf. 1900	15	13	10	285
	1900–1909	8	7	6	378		1900–1909	19	16	10	464
	1910–1919	9	7	1	624		1910–1919	16	12	3	105
	1920–1929	11	9	2	447						
26–30	Bf. 1900	9	7	6	275	41–45	Bf. 1900	13	12	11	275
	1900–1909	16	12	8	507		1900–1909	19	17	14	225
	1910–1919	16	13	4	731						
	1920–1929	8	6	2	153						

Table 121. Active Incidence, Frequency, and Percentage of Outlet Extra-Marital Coitus

By Religious Background

AGE DURING ACTIVITY	RELIGIOUS GROUP	COITAL EXPERIENCE		COITUS TO ORGASM			CASES IN TOTAL SAMPLE
		Active incid. %	Active median freq. per wk.	Active incid. %	Active median freq. per wk.	% of total outlet	
16–20	Protestant						
	Devout	2		2		—	93
	Moderate	2		2		—	97
	Inactive	9	0.1	6		1	139
	Jewish						
	Inactive	8		6		3	119
21–25	Protestant						
	Devout	5	0.1	4	0.1	2	319
	Moderate	7	0.1	5	0.1	—	309
	Inactive	13	0.1	9	0.1	3	392
	Catholic						
	Devout	8		5		7	86
	Moderate	15		11		1	54
	Inactive	25	0.1	19	1.0	6	68
	Jewish						
	Moderate	5		3		1	145
	Inactive	12	0.2	9	0.1	3	321
26–30	Protestant						
	Devout	7	0.1	5	0.1	3	332
	Moderate	12	0.1	9	0.1	1	336
	Inactive	20	0.1	15	0.1	7	413
	Catholic						
	Devout	10		7		1	84
	Inactive	16		12		3	58
	Jewish						
	Moderate	11	0.3	9	0.3	10	140
	Inactive	19	0.3	16	0.3	7	277
31–35	Protestant						
	Devout	7	0.1	6	0.1	—	265
	Moderate	17	0.1	13	0.1	3	253
	Inactive	28	0.2	22	0.2	6	315
	Catholic						
	Devout	10		5		—	63
	Jewish						
	Moderate	17	0.3	15	0.4	17	102
	Inactive	21	0.4	19	0.4	10	201
36–40	Protestant						
	Devout	8	0.1	7	0.2	2	201
	Moderate	16	0.2	13	0.2	12	175
	Inactive	23	0.2	19	0.1	5	233
	Jewish						
	Moderate	21	0.3	19	0.5	26	58
	Inactive	23	0.4	20	0.8	16	132
41–45	Protestant						
	Devout	6		6		5	126
	Moderate	18	0.3	17	0.3	14	110
	Inactive	21	0.3	20	0.4	13	129
46–50	Protestant						
	Devout	3		1		—	72
	Moderate	13		12		6	52
	Inactive	16		15		12	62

Table 122. Number of Extra-Marital Partners, Correlated with History of Pre-Marital Coitus

Number of extra-marital partners	Total sample	Females without pre-marital coitus	Females with pre-marital coitus
	%	*Percent*	
1 only	41	46	38
2–5	40	34	43
6–10	11	12	11
11–20	5	4	5
21–30	1	2	1
31+	2	2	2
Number of cases	514	163	351

Table 123. Number of Years Involved: Extra-Marital Coital Experience

YEARS INVOLVED	TOTAL SAMPLE	YEARS MARRIED			
		6–10	11–15	16–20	21–30
	%	*Percent*			
1–10 times	32	36	28	20	23
1 year or less	42	51	37	27	21
2–3 years	23	29	26	16	23
4–5 years	11	16	9	14	6
6–10 years	14	4	22	25	19
11–20 years	8		6	18	21
21+ years	2				10
Number of cases	507	116	128	80	80

"Years married" represents the number of years that the spouses actually lived together.

Table 124. Intent to Have Extra-Marital Coitus

Intend to have or to continue	Total sample	Females without extra-marital coitus	Females with extra-marital coitus
	%	Percent	
Yes	7	5	28
Doubtful	14	12	28
No	79	83	44
Number of cases	1702	1537	165

The table is based on females who were married at the time of interview.

Table 125. Reported Significance of Extra-Marital Coitus in Divorce

SIGNIFICANCE IN DIVORCE	SUBJECT'S ESTIMATE, OWN EXTRA-MARITAL EXPER. REPORTED BY		SUBJECT'S ESTIMATE, SPOUSE'S EXTRA-MARITAL EXPER. REPORTED BY	
	Females	Males	Females	Males
	Percent		Percent	
Major	14	18	27	51
Moderate	15	9'	49	32
Minor	10	12	24	17
None	61	61		
Number of cases	234	181	181	82

Chapter 11

HOMOSEXUAL RESPONSES AND CONTACTS

The classification of sexual behavior as masturbatory, heterosexual, or homosexual is based upon the nature of the stimulus which initiates the behavior. The present chapter, dealing with the homosexual behavior of the females in our sample, records the sexual responses which they had made to other females, and the overt contacts which they had had with other females in the course of their sexual histories.

The term homosexual comes from the Greek prefix *homo*, referring to the sameness of the individuals involved, and not from the Latin word *homo* which means man. It contrasts with the term heterosexual which refers to responses or contacts between individuals of different (*hetero*) sexes.

While the term homosexual is quite regularly applied by clinicians and by the public at large to relations between males, there is a growing tendency to refer to sexual relationships between females as *lesbian* or *sapphic*. Both of these terms reflect the homosexual history of Sappho who lived on the Isle of Lesbos in ancient Greece. While there is some advantage in having a terminology which distinguishes homosexual relations which occur between females from those which occur between males, there is a distinct disadvantage in using a terminology which suggests that there are fundamental differences between the homosexual responses and activities of females and of males.

PHYSIOLOGIC AND PSYCHOLOGIC BASES

It cannot be too frequently emphasized that the behavior of any animal must depend upon the nature of the stimulus which it meets, its anatomic and physiologic capacities, and its background of previous experience. Unless it has been conditioned by previous experience, an animal should respond identically to identical stimuli, whether they emanate from some part of its own body, from another individual of the same sex, or from an individual of the opposite sex.

The classification of sexual behavior as masturbatory, heterosexual, or homosexual is, therefore, unfortunate if it suggests that three different types of responses are involved, or suggests that only different types of persons seek out or accept each kind of sexual activity. There is

446

nothing known in the anatomy or physiology of sexual response and orgasm which distinguishes masturbatory, heterosexual, or homosexual reactions (Chapters 14–15). The terms are of value only because they describe the source of the sexual stimulation, and they should not be taken as descriptions of the individuals who respond to the various stimuli. It would clarify our thinking if the terms could be dropped completely out of our vocabulary, for then socio-sexual behavior could be described as activity between a female and a male, or between two females, or between two males, and this would constitute a more objective record of the fact. For the present, however, we shall have to use the term homosexual in something of its standard meaning, except that we shall use it primarily to describe sexual *relationships*, and shall prefer not to use it to describe the *individuals* who were involved in those relationships.

The inherent physiologic capacity of an animal to respond to any sufficient stimulus seems, then, the basic explanation of the fact that some individuals respond to stimuli originating in other individuals of their own sex—and it appears to indicate that every individual could so respond if the opportunity offered and one were not conditioned against making such responses. There is no need of hypothesizing peculiar hormonal factors that make certain individuals especially liable to engage in homosexual activity, and we know of no data which prove the existence of such hormonal factors (p. 758). There are no sufficient data to show that specific hereditary factors are involved. Theories of childhood attachments to one or the other parent, theories of fixation at some infantile level of sexual development, interpretations of homosexuality as neurotic or psychopathic behavior or moral degeneracy, and other philosophic interpretations are not supported by scientific research, and are contrary to the specific data on our series of female and male histories. The data indicate that the factors leading to homosexual behavior are (1) the basic physiologic capacity of every mammal to respond to any sufficient stimulus; (2) the accident which leads an individual into his or her first sexual experience with a person of the same sex; (3) the conditioning effects of such experience; and (4) the indirect but powerful conditioning which the opinions of other persons and the social codes may have on an individual's decision to accept or reject this type of sexual contact.[1]

---

[1] Various factors which have been supposed to cause or contribute to female homosexual activity are the following: *Fear of pregnancy or venereal disease:* Talmey 1910:149. Krafft-Ebing 1922:397. Norton 1949:62. Cory 1951:88. *Heterosexual trauma or disappointment:* Havelock Ellis 1915(2):323. Stekel 1922:292–305. Krafft-Ebing 1922:397–398. Marañón 1932:200. Caufeynon 1934:31. Hutton 1937:139. Kahn 1939:268. Beauvoir 1952:418. *Sated with males:* Bloch 1908:546–547. Krafft-Ebing 1922:398. Moreck 1929:286. *Society's heterosexual taboos:* Hutton 1937:139–140. Henry 1941(2):1026. English and Pearson 1945:378. Strain 1948:179. *Seeing parents in coitus:* Farnham

MAMMALIAN BACKGROUND

The impression that infra-human mammals more or less confine themselves to heterosexual activities is a distortion of the fact which appears to have originated in a man-máde philosophy, rather than in specific observations of mammalian behavior. Biologists and psychologists who have accepted the doctrine that the only natural function of sex is reproduction, have simply ignored the existence of sexual activity which is not reproductive. They have assumed that heterosexual responses are a part of an animal's innate, "instinctive" equipment, and that all other types of sexual activity represent "perversions" of the "normal instincts." Such interpretations are, however, mystical. They do not originate in our knowledge of the physiology of sexual response (Chapter 15), and can be maintained only if one assumes that sexual function is in some fashion divorced from the physiologic processes which control other functions of the animal body. It may be true that heterosexual contacts outnumber homosexual contacts in most species of mammals, but it would be hard to demonstrate that this depends upon the "normality" of heterosexual responses, and the "abnormality" of homosexual responses.

In actuality, sexual contacts between individuals of the same sex are known to occur in practically every species of mammal which has been extensively studied. In many species, homosexual contacts may

1951:168. *Seduction by older females:* Moll 1912:314. Havelock Ellis 1915 (2):322. Moreck 1929:302. English and Pearson 1945:378. Norton 1949:62. Farnham 1951:167. *Masturbation which leads to homosexuality:* Havelock Ellis 1915(2):277. Krafft-Ebing 1922:286. *This factor is also mentioned for males by:* Taylor 1933:63, and Remplein 1950:246–247. *Endocrine imbalance:* Havelock Ellis 1915 (2): 316. Lipschütz 1924:371. S. Kahn 1937:135. Hyman 1946(3):2491. Negri 1949:197. *Penis envy and castration complex:* Chideckel 1935:14. Brody 1943:56. Deutsch 1944:347. Fenichel 1945:338. Freud 1950 (5):257. *Father-fixation or hatred toward mother:* Blanchard and Manasses 1930:104, 106. Hesnard 1933:208–209. S. Kahn 1937:20. Bergler 1943:48. Fenichel 1945:338–339. *Mother-fixation:* S. Kahn 1937:20. Deutsch 1944: 347–348. Fenichel 1945:338. Farnham 1951:169. *A continuation of a childhood "bisexual" phase, or a fixation at, or a regression to, an early adolescent stage of psychosexual development:* Moll 1912:60–61, 125. Havelock Ellis 1915(2):309–310. Stekel 1922:39. Marañón 1929:172–174. Blanchard and Manasses 1930:104. Hesnard 1933:188. Freud 1933:177–178. Hamilton in Robinson 1936:336, 341. S. Kahn 1937:18–19. Deutsch 1944:330–331. Sadler 1944:91. English and Pearson 1945:379. Negri 1949:203–204. Hutton in Neville-Rolfe 1950:429. London and Caprio 1950:635. Farnham 1951:166, 175. Kallmann 1952:295. Brody 1943:58 (adds that a homosexual would be neurotic even in a society which accepted homosexuality). *A defense against or a flight from incestuous desires:* Hamilton in Robinson 1936:341. Farnham 1951:175. *Constitutional, congenital, or inherited traits or tendencies:* Parke 1906:320. Bloch 1908:489. Carpenter 1908:55. Moll 1912:125, 130; 1931:234. Havelock Ellis 1915(2):308–311, 317. Krafft-Ebing 1922:285, 288. Kelly 1930:132–133, 220. Robinson 1931:230–231. Freud 1933:178. Potter 1933: 151. Caufeynon 1934:34. Hirschfeld in Robinson 1936:326. S. Kahn 1937:89. Henry 1941(2):1023–1026. Sadler 1944:106. Hirschfeld 1944:281. Thornton 1946:94. Negri 1949:163, 187. Benvenuti 1950:168. Kallmann 1952:295 (in a study of twins).

occur with considerable frequency, although never as frequently as heterosexual contacts. Heterosexual contacts occur more frequently because they are facilitated (1) by the greater submissiveness of the female and the greater aggressiveness of the male, and this seems to be a prime factor in determining the roles which the two sexes play in heterosexual relationships; (2) by the more or less similar levels of aggressiveness between individuals of the same sex, which may account for the fact that not all animals will submit to being mounted by individuals of their own sex; (3) by the greater ease of intromission into the female vagina and the greater difficulty of penetrating the male anus; (4) by the lack of intromission when contacts occur between two females, and the consequent lack of those satisfactions which intromission may bring in a heterosexual relationship; (5) by olfactory and other anatomic and physiologic characteristics which differentiate the sexes in certain mammalian species; (6) by the psychologic conditioning which is provided by the more frequently successful heterosexual contacts.

Homosexual contacts in infra-human species of mammals occur among both females and males. Homosexual contacts between females have been observed in such widely separated species as rats, mice, hamsters, guinea pigs, rabbits, porcupines, marten, cattle, antelope, goats, horses, pigs, lions, sheep, monkeys, and chimpanzees.[2] The homosexual contacts between these infra-human females are apparently never completed in the sense that they reach orgasm, but it is not certain how often infra-human females ever reach orgasm in any type of sexual relationship. On the other hand, sexual contacts between males of the lower mammalian species do proceed to the point of orgasm, at least for the male that mounts another male.[3]

In some species the homosexual contacts between females may occur as frequently as the homosexual contacts between males.[4] Every farmer

---

[2] We have observed homosexual behavior in male monkeys, male dogs, bulls, cows, male and female rats, male porcupines, and male and female guinea pigs. Homosexual activities in other animals are noted by: Karsch 1900:128–129 (female antelope, male and female goat, ram, stallion). Féré 1904:78 (male donkey). Havelock Ellis 1910(1):165 (male elephant, male hyena). Hamilton 1914:307 (female monkey). Bingham 1928:126–127 (female chimpanzee). Marshall and Hammond 1944:39 (doe rabbit). Reed 1946:200 (male bat). Beach 1947a:41 (female cat). Beach 1948:36 (male mouse). Beach in Hoch and Zubin 1949:63–64 (female marten, female porcupine, male lion, male rabbit). Gantt in Wolff 1950:1036 (male cat). Ford and Beach 1951:139 (male porpoise), 141 (lioness, mare, sow, ewe, female hamster, female mouse, female dog). Shadle, verbal communication (male porcupine, male raccoon).

[3] Ejaculation resulting from homosexual contact between males of lower mammalian species has been noted in: Karsch 1900:129 (ram and goat). Kempf 1917:134–135 (monkey). Moll 1931:17 (dog). Beach 1948:36 (mouse). Ford and Beach 1951:139 (rat). Brookfield Zoo, verbal communication (baboon). We have observed such ejaculation in the bull.

[4] For the sub-primates, Beach 1947a:40 states that "the occurrence of masculine sexual responses in female animals is more common than is the appearance

who has raised cattle knows, for instance, that cows quite regularly mount cows. He may be less familiar with the fact that bulls mount bulls, but this is because cows are commonly kept together while bulls are not so often kept together in the same pasture.

It is generally believed that females of the infra-human species of mammals are sexually responsive only during the so-called periods of heat, or what is technically referred to as the estrus period. This, however, is not strictly so. The chief effect of estrus seems to be the preparation of the animal to accept the approaches of another animal which tries to mount it. The cows that are mounted in the pasture are those that are in estrus, but the cows that do the mounting are in most instances individuals which are not in estrus (p. 737).[5]

Whether sexual relationships among the infra-human species are heterosexual or homosexual appears to depend on the nature of the immediate circumstances and the availability of a partner of one or the other sex. It depends to a lesser degree upon the animal's previous experience, but no other mammalian species is so affected by its experience as the human animal may be. There is, however, some suggestion, but as yet an insufficient record, that the males among the lower mammalian species are more likely than the females to become conditioned to exclusively homosexual behavior; but even then such exclusive behavior appears to be rare.[6]

The mammalian record thus confirms our statement that any animal which is not too strongly conditioned by some special sort of experience is capable of responding to any adequate stimulus. This is what we find in the more uninhibited segments of our own human species, and this is what we find among young children who are not too rigorously restrained in their early sex play. Exclusive preferences and patterns of behavior, heterosexual or homosexual, come only with experience, or as a result of social pressures which tend to force an individual into an exclusive pattern of one or the other sort. Psychologists and psychiatrists, reflecting the mores of the culture in which they have been

---

of feminine behavior in males." Ford and Beach 1951:143 note, however, that in the class Mammalia taken as a whole, homosexual behavior among males is more frequent than homosexual behavior among females.

[5] Two situations may be involved: (1) the estrual female may be receptive to being mounted and often attempts to elicit such mounting (see Beach in Hoch and Zubin 1949:64; Rice and Andrews 1951:151). (2) If she is not mounted the estrual female may mount another animal of the same or opposite sex. For the latter, see: Beach 1948:66–68 (cow, sow, rabbit, cat, shrew). Ford and Beach 1951:141–142 (rabbit, sow, mare, cow, guinea pig).

[6] Exclusive, although usually temporary male homosexuality is noted in: Hamilton 1914:307–308 (monkey). Beach in Hoch and Zubin 1949:64–65 (lion). Ford and Beach 1951:136, 139 (baboon and porpoise). Shadle, verbal communication (porcupine). No exclusively homosexual patterns have been reported for female mammals.

raised, have spent a good deal of time trying to explain the origins of homosexual activity; but considering the physiology of sexual response and the mammalian backgrounds of human behavior, it is not so difficult to explain why a human animal does a particular thing sexually. It is more difficult to explain why each and every individual is not involved in every type of sexual activity.

ANTHROPOLOGIC BACKGROUND

In the course of human history, distinctions between the acceptability of heterosexual and of homosexual activities have not been confined to our European and American cultures. Most cultures are less acceptant of homosexual, and more acceptant of heterosexual contacts. There are some which are not particularly disturbed over male homosexual activity, and some which expect and openly condone such behavior among young males before marriage and even to some degree after marriage; but there are no cultures in which homosexual activity among males seems to be more acceptable than heterosexual activity.[7] It is probable that in some Moslem, Buddhist, and other areas male homosexual contacts occur more frequently than they do in our European or American cultures, and in certain age groups they may occur more frequently than heterosexual contacts; but heterosexual relationships are, at least overtly, more acceptable even in those cultures.

Records of male homosexual activity are also common enough among more primitive human groups, but there are fewer records of homosexual activity among females in primitive groups. We find some sixty pre-literate societies from which some female homosexual activity has been reported, but the majority of the reports imply that such activity is rare. There appears to be only one pre-literate group, namely the Mohave Indians of our Southwest, for whom there are records of exclusively homosexual patterns among females. That same group is the only one for which there are reports that female homosexual activity is openly sanctioned.[8] For ten or a dozen groups, there are records of female transvestites—i.e., anatomic females who dress and assume the position of the male in their social organization—but transvestism and homosexuality are different phenomena, and our data show that only a portion of the transvestites have homosexual histories (p. 679).

There is some question whether the scant record of female homosexuality among pre-literate groups adequately reflects the fact. It may merely reflect the taboos of the European or American anthropologists who accumulated the data, and the fact that they have been notably

---

[7] Ford and Beach 1951:130 note that 64 per cent of a sample of 76 societies consider homosexuality acceptable for certain persons.
[8] The sexual life of the Mohave was intensively studied by Devereux 1936, 1937.

reticent in inquiring about sexual practices which are not considered "normal" by Judeo-Christian standards. Moreover, the informants in the anthropologic studies have usually been males, and they would be less likely to know the extent of female homosexual activities in their cultures. It is, nonetheless, quite possible that such activities are actually limited among the females of these pre-literate groups, possibly because of the wide acceptance of pre-marital heterosexual relationships, and probably because of the social importance of marriage in most primitive groups.[9]

RELATION TO AGE AND MARITAL STATUS

As in any other type of sexual situation, there are: (1) individuals who have been erotically aroused by other individuals of the same

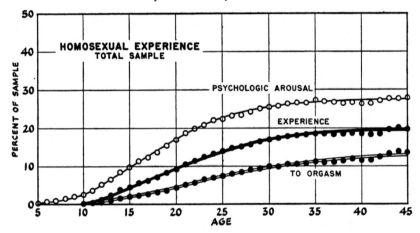

Figure 82. Accumulative incidence: homosexual experience, arousal, and orgasm

Data from Table 131.

sex, whether or no they had physical contact with them; (2) individuals who have had physical contacts of a sexual sort with other individuals of the same sex, whether or no they were erotically aroused in those contacts; and (3) individuals who have been aroused to the point of orgasm by their physical contacts with individuals of the same sex. These three types of situations are carefully distinguished in the statistics given here.

Accumulative Incidence in Total Sample. Some of the females in the sample had been conscious of specifically erotic responses to other females when they were as young as three and four (Table 13). The percentages of those who had been erotically aroused had then steadily

---

[9] Ford and Beach 1951:133, 143, also note that female homosexuality seems less frequent than male homosexuality among pre-literates.

risen, without any abrupt development, to about thirty years of age. By that time, a quarter (25 per cent) of all the females had recognized erotic responses to other females. The accumulative incidence figures had risen only gradually after age thirty. They had finally reached a level at about 28 per cent (Table 131, Figure 82).[10]

The number of females in the sample who had made specifically sexual contacts with other females also rose gradually, again without any abrupt development, from the age of ten to about thirty. By then some 17 per cent of the females had had such experience (Tables 126, 131, Figure 82). By age forty, 19 per cent of the females in the total sample had had some physical contact with other females which was deliberately and consciously, at least on the part of one of the partners, intended to be sexual.[11]

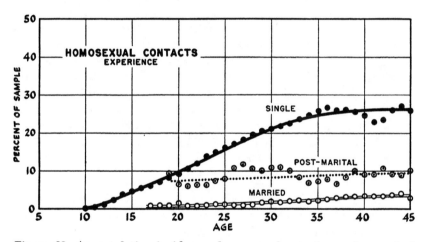

Figure 83. Accumulative incidence: homosexual experience, by marital status

Data from Table 126.

Homosexual activity among the females in the sample had been largely confined to the single females and, to a lesser extent, to previously married females who had been widowed, separated, or divorced. Both the incidences and frequencies were distinctly low among the married females (Table 126, Figure 83). Thus, while the accumu-

---

[10] Our accumulative incidence figures for homosexual responses among females are close to those in two other studies: Davis 1929:247 (26 per cent at age 36). Gilbert Youth Research 1951 (13 per cent, college students).

[11] Our accumulative incidence figures for overt homosexual contacts among females are of the same general order as those from other studies: Davis 1929:247 (20 per cent, unmarried college and graduate females, average age 36). Bromley and Britten 1938:117 (4 per cent, college females). Landis et al. 1940:262, 286 (4 per cent, single females). England acc. Rosenthal 1951:58 (20 per cent, British females). Gilbert Youth Research 1951 (6 per cent, college females).

lative incidences of homosexual contacts had reached 19 per cent in the total sample by age forty, they were 24 per cent for the females who had never been married by that age, 3 per cent for the married females, and 9 per cent for the previously married females. The age at which the females had married seemed to have had no effect on the pre-marital incidences of homosexual activity, even though we found that the pre-marital heterosexual activities (petting and pre-marital coitus) had been stepped up in anticipation of an approaching marriage. The chief effect of marriage had been to stop the homosexual activities, thereby lowering the active incidences and frequencies in the sample of married females.

A half to two-thirds of the females who had had sexual contacts with other females had reached orgasm in at least some of those contacts. By twenty years of age there were only 4 per cent of the total sample who had experienced orgasm in homosexual relations, and by age thirty-five there were still only 11 per cent with such experience (Table 131, Figure 82). The accumulative incidences finally reached 13 per cent in the middle forties. Since there were differences in the incidences among females of the various educational levels (Table 131, Figure 85), and since our sample includes a disproportionate number of the females of the college and graduate groups where the incidences seem to be higher than in the grade school and high school groups, the figures for this sample are probably higher than those which might be expected in the U. S. population as a whole.

Active Incidence to Orgasm. Since there is every gradation between the casual, non-erotic physical contacts which females regularly make and the contacts which bring some erotic response, it has not been possible to secure active incidence or frequency data on homosexual contacts among the females in the sample except where they led to orgasm. However, comparisons of the accumulative incidence data for experience and for orgasm (Table 131, Figure 82) suggest that the active incidences of the homosexual contacts may, at least in the younger groups, have been nearly twice as high as the active incidences of the contacts which led to orgasm.

In the total sample, not more than 2 to 3 per cent had reached orgasm in their homosexual relations during adolescence and their teens (Table 128), although five times that many may have been conscious of homosexual arousal and three times that many may have had physical contacts with other girls which were specifically sexual. After age twenty, the active incidences of the contacts which led to orgasm had gradually increased among the females who were still unmarried, reaching their peak, which was 10 per cent, at age forty. Then they began to drop. Between the ages of forty-six and fifty, about 4 per cent of the still unmarried females were actively involved in homosexual relations that led to orgasm. We do not have complete histories of single females who were reaching orgasm in homosexual relations after fifty years of age, but we do have incomplete information on still older women who were making such contacts with responses to orgasm while they were in their fifties, sixties, and even seventies.

Among the married females, slightly more than 1 per cent had been actively involved in homosexual activities which reached orgasm in each and every age group between sixteen and forty-five (Table 128).

On the other hand, among the females who had been previously married and who were then separated, widowed, or divorced, something around 6 per cent were having homosexual contacts which led to orgasm in each of the groups from ages sixteen to thirty-five (Table 128). After that some 3 to 4 per cent were involved, but by the middle fifties, only 1 per cent of the previously married females were having contacts which were complete enough to effect orgasm.

Frequency to Orgasm. Among the unmarried females in the sample who had ever experienced orgasm from contacts with other females, the average (active median) frequencies of orgasm among the younger adolescent girls who were having contacts had averaged nearly once in five weeks (about 0.2 per week), and they had increased in frequency among the older females who were not yet married (Table 128). In the late twenties they had averaged once in two and a half weeks (0.4 per week), and had stayed on about that level for the next ten years. This means that the active median frequencies of orgasm derived from the homosexual contacts had been higher than the active median frequencies of orgasm derived from nocturnal dreams and from heterosexual petting, and about the same as the active median frequencies of orgasm attained in masturbation.[12]

The active mean frequencies were three to six times higher than the active median frequencies, because of the fact that there were some females in each age group whose frequencies were notably higher than those of the median females (Table 128). The individual variation had depended in part upon the fact that the frequencies of contact had varied, and in part upon the fact that some of these females had regularly experienced multiple orgasms in their homosexual contacts.

In most age groups, three-quarters or more of the single females who were having homosexual experience to the point of orgasm were having it with average frequencies of once or less per week (Figure

---

[12] The limited frequency data previously published were not calculated on any basis comparable to our 5-year calculations.

84). There were individuals, however, in every age group from adolescence to forty-five, who were having homosexual contacts which had led to orgasm on an average of seven or more per week. From ages twenty-one to forty there were a few individuals who had averaged ten or more and in one instance as many as twenty-nine orgasms per week from homosexual sources. In contrast to the record for most other types of sexual activities, the most extreme variation in the homosexual relationships had not occurred in the youngest groups, but in the groups aged thirty-one to forty.

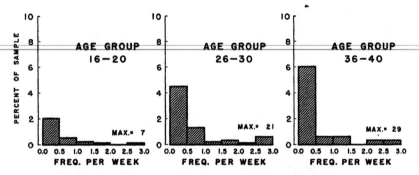

Figure 84. Individual variation: frequency of homosexual contacts to orgasm

For three age groups of single females. Each class interval includes the upper but not the lower frequency. For incidences of females not having homosexual experience or reaching orgasm in such contacts, see Table 128.

As in most other types of sexual activity among females (except coitus in marriage), the homosexual contacts had often occurred sporadically. Several contacts might be made within a matter of a few days, and then there might be no such contacts for a matter of weeks or months. In not a few instances the record was one of intense and frequently repeated contacts over a short period of days or weeks, with a lapse of several years before there were any more. On the other hand, there were a fair number of histories in which the homosexual partners had lived together and maintained regular sexual relationships for many years, and in some instances for as long as ten or fifteen years or even longer, and had had sexual contacts with considerable regularity throughout those years. Such long-time homosexual associations are rare among males. A steady association between two females is much more acceptable to our culture and it is, in consequence, a simpler matter for females to continue relationships for some period of years. The extended female associations are, however, also a product of differences in the basic psychology of females and males (Chapter 16).

Among the married females in the sample, there were a few in each age group—usually not more than one in a hundred or so—who were having homosexual contacts to the point of orgasm (Table 128). Even in those small active samples, however, the range of individual variation was considerable. Most of the married females had never had more than a few such contacts, but in nearly every age group there were married females who were having contacts with regular frequencies of once or twice or more per week. There were a few histories of married females who were completely homosexual and who were not having coitus with their husbands, although they continued to live with them as a matter of social convenience. In some of these cases there were good social adjustments between the spouses even though the sexual lives of each lay outside of the marriage.

Among the females in the sample who had been previously married and who were then widowed, separated, or divorced, the frequencies of homosexual experience were distinctly higher than among the married females (Table 128). In some cases these females, after the dissolution of their marriages, had established homes with other women with whom they had then had their first homosexual contacts and with whom they subsequently maintained regular homosexual relationships. Some of the women had been divorced because of their homosexual interests, although homosexuality in the female is only rarely a factor in divorce. It should be emphasized, however, that a high proportion of the unmarried females who live together never have contacts which are in any sense sexual.

Percentage of Total Outlet. Homosexual contacts are highly effective in bringing the female to orgasm (p. 467). In spite of their relatively low incidence, they had accounted for an appreciable proportion of the total number of orgasms of the entire sample of unmarried females. Before fifteen years of age, the homosexual contacts had been surpassed only by masturbation and heterosexual coitus as sources of outlet, and they were again in that position among the still single females after age thirty (Table 171, Figure 110). Among these single females, orgasms obtained from homosexual contacts had accounted for some 4 per cent of the total outlet of the younger adolescent females, some 7 per cent of the outlet of the unmarried females in their early twenties, and some 19 per cent of the total outlet of the females who were still unmarried in their late thirties (Table 128).

Among the married females in the sample, homosexual contacts had usually accounted for less than one-half of one per cent of all their orgasms (Table 128).

However, among the females who had been previously married, homosexual contacts had become somewhat more important again as a source of outlet. They had accounted for something around 2 per cent of the total outlet of the younger females in the group, and for nearly 10 per cent of the outlet of the females who were in their early thirties (Table 128).

Number of Years Involved. For most of the females in the sample, the homosexual activity had been limited to a relatively short period of time (Table 129). For nearly a third (32 per cent) of those who had had any experience, the experience had not occurred more than ten times, and for many it had occurred only once or twice. For nearly a half (47 per cent, including part of the above 32 per cent), the experience had been confined to a single year or to a part of a single year. For another quarter (25 per cent), the activity had been spread through two or three years. These totals, interesting to note, had not materially differed between females who were in the younger, and females who were in the older age groups at the time they contributed their histories. This means that for most of them, most of the homosexual activity had occurred in the younger years. There were a quarter (28 per cent) whose homosexual experience had extended for more than three years. There were histories of a few females whose activities had extended for as many as thirty or forty years, and more extended samples of older females would undoubtedly show cases which had continued for still longer periods of time.

Number of Partners. In the sample of single females, a high proportion (51 per cent) of those who had had any homosexual experience had had it with only a single partner, up to the time at which they had contributed their histories to the record. Another 20 per cent had had it with two different partners. Only 29 per cent had had three or more partners in their homosexual relations, and only 4 per cent had had more than ten partners (Table 130).[13]

In this respect, the female homosexual record contrasts sharply with that for the male. Of the males in the sample who had had homosexual experience, a high proportion had had it with several different persons, and 22 per cent had had it with more than ten partners (p. 683). Some of them had had experience with scores and in many instances with hundreds of different partners. Apparently, basic psychologic factors account for these differences in the extent of the promiscuity of the female and the male (Chapter 16).

---

[13] Davis 1929:251 gives closely parallel data (63 per cent with one partner, 18 per cent with two partners, 19 per cent with three or more partners). Statistically unsupported impressions of a high degree of promiscuity in female homosexuality may be found in: Bloch 1908:530 (female homosexuals change partners more frequently than male homosexuals). Alibert 1926:22. Kisch 1926:192. Chideckel 1935:122. But the greater durability of relationships among female homosexuals is also noted in: Smitt 1951:102.

RELATION TO EDUCATIONAL LEVEL

The incidences of homosexual activity among the females in the sample had been definitely correlated with their educational backgrounds. This was more true than with any of their other sexual activities.

Accumulative Incidence. Homosexual responses had occurred among a smaller number of the females of the grade school and high school sample, a distinctly larger number of the college sample, and still more

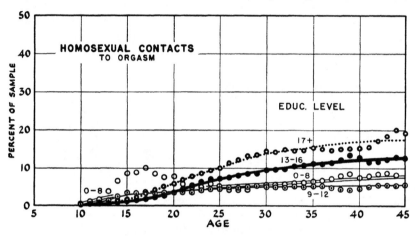

Figure 85. Accumulative incidence: homosexual contacts to orgasm, by educational level

Based on total sample, including single, married, and previously married females. Data from Table 131.

of the females who had gone on into graduate work (Table 131). At thirty years of age, for instance, there were 10 per cent of the grade school sample, 18 per cent of the high school sample, 25 per cent of the college sample, and 33 per cent of the graduate group who had recognized that they had been erotically aroused by other females.[14]

Overt contacts had similarly occurred in a smaller number of the females of the lower educational levels and a larger number of those of the upper educational levels. At thirty years of age, the accumulative incidence figures had reached 9 per cent, 10 per cent, 17 per cent, and 24 per cent in the grade school, high school, college, and graduate groups, respectively.

At thirty years of age, homosexual experience to the point of orgasm had occurred in 6 per cent of the grade school sample, 5 per cent of

[14] Davis 1929:308 also finds a higher incidence of adult homosexual responses among better educated females (38 per cent of college group, 15 per cent of non-college group).

the high school sample, 10 per cent of the college sample, and 14 per cent of the graduate sample (Table 131, Figure 85).

We have only hypotheses to account for the extension of this type of sexual activity in the better educated groups. We are inclined to believe that moral restraint on pre-marital heterosexual activity is the most important single factor contributing to the development of a homosexual history, and such restraint is probably most marked among the younger and teen-age girls of those social levels that send their daughters to college. In college, these girls are further restricted by administrators who are very conscious of parental concern over the heterosexual morality of their offspring. The prolongation of the years

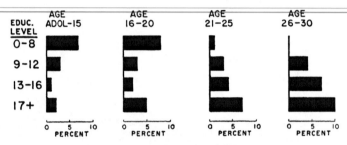

Figure 86. Active incidence: homosexual contacts to orgasm, by educational level

Data based on single females; see Table 127.

of schooling, and the consequent delay in marriage (Figure 46), interfere with any early heterosexual development of these girls. This is particularly true if they go on into graduate work. All of these factors contribute to the development of homosexual histories. There may also be a franker acceptance and a somewhat lesser social concern over homosexuality in the upper educational levels.

Active Incidence to Orgasm. Between adolescence and fifteen years of age, homosexual contacts to orgasm were more common in the sample of high school females and in the limited sample of grade school females (Table 127, Figure 86). However, between the ages of twenty-one and thirty-five, while the active incidences stood at something between 3 and 6 per cent among the high school females, they had risen to something between 7 and 11 per cent in the college and graduate school groups.

Frequency to Orgasm. Between adolescence and fifteen years of age, the active median frequencies of homosexual contacts to orgasm among the females in the sample were higher in the grade school and high school groups, and lower among the sexually more restrained young females of the upper educational levels (Table 127). Subsequently these discrepancies had more or less disappeared, and after age twenty the frequencies had averaged once in two or three weeks for the median females of all the educational levels represented in the sample.

Percentage of Total Outlet. Among the younger teen-age girls, 14 per cent of the orgasms of the grade school group had come from homosexual contacts, while only 1 or 2 per cent of the orgasms of the college and graduate groups had come from such sources (Table 127). Subsequently, these differences were reversed, and between thirty and forty years of age the still unmarried females of the graduate group were deriving 18 to 21 per cent of their total outlet from homosexual sources. If one-fifth of the outlet of this group came from homosexual sources, and only a little more than one-tenth (11 per cent) of the females in the group were having such activity, it is evident that the females who were having homosexual experience were reaching orgasm more frequently than those who were depending on other types of sexual activity for their outlet.

RELATION TO PARENTAL OCCUPATIONAL CLASS

In the available sample there seems to be little or nothing in the accumulative or active incidences, or the frequencies of the homosexual contacts, which suggests that there is any correlation with the occupational classes of the homes in which the females were raised (Tables 132, 133). There is only minor evidence that the accumulative incidences of contacts to the point of orgasm may have involved a slightly higher percentage of the females who came from upper white collar homes, and a smaller percentage of those who came from the homes of laboring groups—at age forty, a matter of 14 per cent in the first instance, and under 10 per cent in the second instance (Table 132).

The active incidences in the younger age groups were higher among the females who had come from the homes of laborers; but after the age of twenty the differences had largely disappeared, and after the age of twenty-five the females who had come from upper white collar homes were the ones most often involved (Table 133).

RELATION TO DECADE OF BIRTH

In the available sample, the accumulative incidences of homosexual contacts to the point of orgasm had been very much the same for the females who were born in the four decades on which we have data. There is no evidence that there are any more females involved in homosexual contacts today than there were in the generation born before

1900 or in any of the intermediate decades (Table 134, Figure 87).[15] Similarly, the number of females having homosexual contacts in particular five-year periods of their lives (the active incidences), the frequencies of such contacts, and the percentages of the total outlet which had been derived from homosexual contacts, do not seem to have varied in any consistent fashion during the four decades covered by the sample (Table 135).

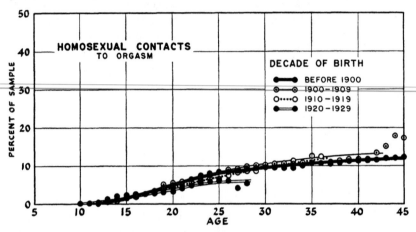

Figure 87. **Accumulative incidence: homosexual contacts to orgasm, by decade of birth**

Based on total sample, including single, married, and previously married females. Data from Table 134.

It is not immediately obvious why this, among all other types of sexual activity, should have been unaffected by the social forces which led to the marked increase in the incidences of masturbation, heterosexual petting, pre-marital coitus, and even nocturnal dreams among American females immediately after the first World War, and which have kept these other activities on the new levels or have continued to keep them rising since then.

RELATION TO AGE AT ONSET OF ADOLESCENCE

There do not seem to be any consistent correlations between either the accumulative incidences, the active incidences, or the frequencies of homosexual contacts, and the ages at which the females in the sample had turned adolescent (Tables 136, 137). Among males we found (1948:320) that those who turned adolescent at earlier ages were more often involved in homosexual contacts as well as in mastur-

---

[15] Statistically unsubstantiated statements that female homosexuality is on the increase may be found, for instance, in: Parke 1906:319. Havelock Ellis 1915 (2):261–262. Potter 1933:6–9, 150. McPartland 1947:143, 150. Norton 1949: 61.

bation and pre-marital heterosexual contacts. The absence of such a correlation among females may be significant (see Chapter 18).

RELATION TO RURAL-URBAN BACKGROUND

The accumulative incidences of homosexual contacts to the point of orgasm were a bit higher among the city-bred females in the sample (Table 138, Figure 88). The active incidences appear to have been a

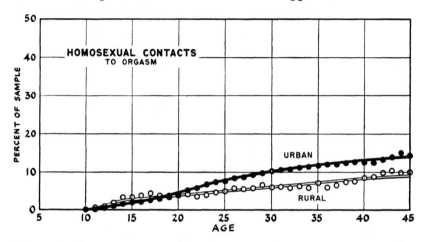

Figure 88. Accumulative incidence: homosexual contacts to orgasm, by rural-urban background

Based on total sample, including single, married, and previously married females. Data from Table 138.

bit higher among the rural females in their teens, but they were higher among urban females after the age of twenty (Table 139). The data, however, are insufficient to warrant final conclusions.

RELATION TO RELIGIOUS BACKGROUND

The educational levels and religious backgrounds of the females in the sample were the social factors which were most markedly correlated with the incidences of their homosexual activity.

Accumulative Incidence. In the Protestant, Catholic, and Jewish groups on which we have samples, fewer of the devout females were involved in homosexual contacts to the point of orgasm, and distinctly more of the females who were least devout religiously (Table 140, Figures 89–91). For instance, by thirty-five years of age among the Protestant females some 7 per cent of the religiously devout had had homosexual relations to orgasm, but 17 per cent of those who were least actively identified with the church had had such relations. The differences were even more marked in the Catholic groups: by thirty-five

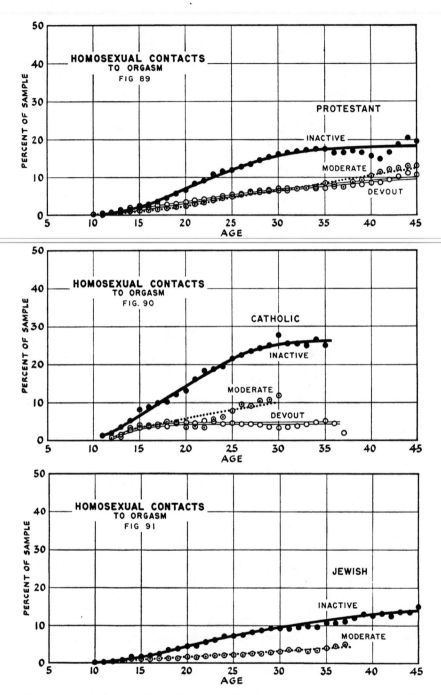

Figures 89–91. Accumulative incidence: homosexual contacts to orgasm, by religious background

Based on total sample, including single, married, and previously married females. Data from Table 140.

464

years of age, only 5 per cent of the devoutly Catholic females had had homosexual relations to the point of orgasm, but some 25 per cent of those who were only nominally connected with the church. The differences between the Jewish groups lay in the same direction.

There is little doubt that moral restraints, particularly among those who were most actively connected with the church, had kept many of the females in the sample from beginning homosexual contacts, just as some were kept from beginning heterosexual activities. On the other

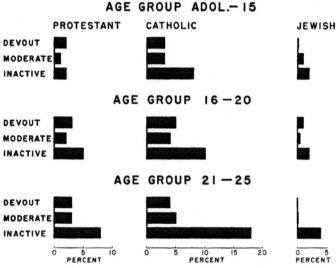

Figure 92. Active incidence: homosexual contacts to orgasm, by religious background

Data based on single females; see Table 141.

hand, as we have already noted, some of the females had become involved in homosexual activities because they were restrained by the religious codes from making pre-marital heterosexual contacts, and such devout individuals had sometimes become so disturbed in their attempt to reconcile their behavior and their moral codes that they had left the church, thereby increasing the incidences of homosexual activity among the religiously inactive groups.

Active Incidence. In eleven out of the twelve groups on which we have data available for comparisons, the active incidences of homosexual contacts to the point of orgasm were lower among the more devout females and higher among those who were religiously least devout (Table 141, Figure 92). For instance, among the younger adolescent groups, there were 3 per cent of the devoutly Catholic females who were having homosexual relations to the point of orgasm, but 8 per cent of the inactive Catholics. Similarly, at ages twenty-six to thirty,

among the still unmarried Protestant groups, 5 per cent of the more
devout females were involved, but 13 per cent of the least devout females (Table 141).

Active Median Frequency to Orgasm. In the sample, there does not
seem to have been any consistent correlation between the active median
frequencies of homosexual activities and the religious backgrounds of
the females in the various groups (Table 141).

Percentage of Total Outlet. The percentage of the total outlet which
had been derived by the various groups of females from their homosexual relations was, in most instances, correlated with the number of
females (the active incidences) who were involved in such activity
(Table 141); but among the religiously more devout females, and
especially in the older age groups, the percentage of the total outlet
derived from homosexual sources was in excess and often in considerable excess of what the incidences might have led one to expect. This
had depended in part upon the fact that an unusually large number of
the religiously devout were not reaching orgasm in any sort of sexual
activity (Table 165), and for those who had accepted homosexual relations and reached orgasm in them, those relations had become a chief
source of all the orgasms experienced by the group. It is also possible
that a selective factor was involved, and that the sexually more responsive females were the ones who had most often accepted homosexual
relations.

TECHNIQUES IN HOMOSEXUAL CONTACTS

The techniques utilized in the homosexual relations among the females in the sample were the techniques that are ordinarily utilized
in heterosexual petting which precedes coitus, or which may serve as
an end in itself. The homosexual techniques had differed primarily in
the fact that they had not included vaginal penetrations with a true
phallus.

The physical contacts between the females in the homosexual relations had often depended on little more than simple lip kissing and
generalized body contacts (Table 130). In some cases the contacts,
even among the females who had long and exclusively homosexual
histories, had not gone beyond this. In many instances the homosexual
partners had not extended their techniques to breast and genital stimulation for some time and in some cases for some period of years after
the relationships had begun. Ultimately, however, among the females
in the sample who had had more extensive homosexual experience,
simple kissing and manual manipulation of the breast and genitalia
had become nearly universal (in 95 to 98 per cent); and deep kissing

(in 77 per cent), more specific oral stimulation of the female breast (in 85 per cent), and oral stimulation of the genitalia (in 78 per cent) had become common techniques. In something more than half of the histories (56 per cent), there had been genital appositions which were designed to provide specific and mutual stimulation (Table 130). But vaginal penetrations with objects which had served as substitutes for the male penis had been quite rare in the histories.[16]

It is not generally understood, either by males or by females who have not had homosexual experience, that the techniques of sexual relations between two females may be as effective as or even more effective than the petting or coital techniques ordinarily utilized in heterosexual contacts. But if it is recalled that the clitoris of the female, the inner surfaces of the labia minora, and the entrance to the vagina are the areas which are chiefly stimulated by the male penetrations in coitus (pp. 574 ff.), it may be understood that similar tactile or oral stimulation of those structures may be sufficient to bring orgasm. However, for females who find satisfaction in having the deeper portions of the vagina penetrated during coitus (pp. 579–584), the lack of this sort of physical stimulation may make the physical satisfactions of homosexual relationships inferior to those which are available in coitus.

Nevertheless, comparisons of the percentages of contacts which had brought orgasm in marital coitus among the females who had been married for five years, and in the homosexual relations of females who had had about the same number of years of homosexual experience, show the following:

% of contacts leading to orgasm	In fifth year of marital coitus	In more extensive homosexual experience
	Percent of females	
0	17	7
1–29	13	7
30–59	15	8
60–89	15	10
90–100	40	68
Number of cases	1448	133

The higher frequency of orgasm in the homosexual contacts may have depended in part upon the considerable psychologic stimulation provided by such relationships, but there is reason for believing that it may also have depended on the fact that two individuals of the same

---

[16] Further data on the nature of female homosexual techniques may be found in: Forberg 1884(2):113–115, 135, 141, 143. Parke 1906:322. Rohleder 1907(2): 466, 484, 494. Bloch 1908:529. Talmey 1910:154–155. Havelock Ellis 1915(2):257–258. Krafft-Ebing 1922:400. Kronfeld 1923:58. Kisch 1926: 195–196. Eberhard 1927:354, 360. Kelly 1930:137. Deutsch 1933:40; 1944: 348. Sadler 1944:96. Hirschfeld 1944:232–233. Bergler 1948:200. See also the classical references in footnote 22.

sex are likely to understand the anatomy and the physiologic responses and psychology of their own sex better than they understand that of the opposite sex. Most males are likely to approach females as they, the males, would like to be approached by a sexual partner. They are likely to begin by providing immediate genital stimulation. They are inclined to utilize a variety of psychologic stimuli which may mean little to most females (Chapter 16). Females in their heterosexual relationships are actually more likely to prefer techniques which are closer to those which are commonly utilized in homosexual relationships. They would prefer a considerable amount of generalized emotional stimulation before there is any specific sexual contact. They usually want physical stimulation of the whole body before there is any specifically genital contact. They may especially want stimulation of the clitoris and the labia minora, and stimulation which, after it has once begun, is followed through to orgasm without the interruptions which males, depending to a greater degree than most females do upon psychologic stimuli, often introduce into their heterosexual relationships (p. 668).

It is, of course, quite possible for males to learn enough about female sexual responses to make their heterosexual contacts as effective as females make most homosexual contacts. With the additional possibilities which a union of male and female genitalia may offer in a heterosexual contact, and with public opinion and the mores encouraging heterosexual contacts and disapproving of homosexual contacts, relationships between females and males will seem, to most persons, to be more satisfactory than homosexual relationships can ever be. Heterosexual relationships could, however, become more satisfactory if they more often utilized the sort of knowledge which most homosexual females have of female sexual anatomy and female psychology.

THE HETEROSEXUAL-HOMOSEXUAL BALANCE

There are some persons whose sexual reactions and socio-sexual activities are directed only toward individuals of their own sex. There are others whose psychosexual reactions and socio-sexual activities are directed, throughout their lives, only toward individuals of the opposite sex. These are the extreme patterns which are labeled homosexuality and heterosexuality. There remain, however, among both females and males, a considerable number of persons who include both homosexual and heterosexual responses and/or activities in their histories. Sometimes their homosexual and heterosexual responses and contacts occur at different periods in their lives; sometimes they occur coincidentally. This group of persons is identified in the literature as bisexual.

That there are individuals who react psychologically to both females and males, and who have overt sexual relations with both females and males in the course of their lives, or in any single period of their lives, is a fact of which many persons are unaware; and many of those who are academically aware of it still fail to comprehend the realities of the situation. It is a characteristic of the human mind that it tries to dichotomize in its classification of phenomena. Things either are so, or they are not so. Sexual behavior is either normal or abnormal, socially acceptable or unacceptable, heterosexual or homosexual; and many persons do not want to believe that there are gradations in these matters from one to the other extreme.[17]

In regard to sexual behavior it has been possible to maintain this dichotomy only by placing all persons who are exclusively heterosexual in a heterosexual category, and all persons who have any amount of experience with their own sex, even including those with the slightest experience, in a homosexual category. The group that is identified in the public mind as heterosexual is the group which, as far as public knowledge goes, has never had any homosexual experience. But the group that is commonly identified as homosexual includes not only those who are known or believed to be exclusively homosexual, but also those who are known to have had any homosexual experience at all. Legal penalties, public disapproval, and ostracism are likely to be leveled against a person who has had limited homosexual experience as quickly as they are leveled against those who have had exclusive experience. It would be as reasonable to rate all individuals heterosexual if they have any heterosexual experience, and irrespective of the amount of homosexual experience which they may be having. The attempt to maintain a simple dichotomy on these matters exposes the traditional biases which are likely to enter whenever the heterosexual or homosexual classification of an individual is involved.

Heterosexual-Homosexual Rating. Only a small proportion of the females in the available sample had had exclusively homosexual histories. An adequate understanding of the data must, therefore, depend upon some balancing of the heterosexual and homosexual elements in each history. This we have attempted to do by rating each individual on a heterosexual-homosexual scale which shows what proportion of

---

[17] Attempts to categorize female homosexuality as congenital, real, genuine, acquired, situational, temporary, latent, partial, complete, total, absolute, regressive, progressive, pseudo-homosexuality, psychosexual hermaphroditism, bisexuality, inversion, perversity, etc., may be found, for instance, in: Féré 1904: 188. Parke 1906:320. Bloch 1908:489. Carpenter 1908:55. Freud 1910:2. Talmey 1910:143, 152. Moll 1912:125–130. Krafft-Ebing 1922:285–289, 336. Kelly 1930:136, 220. Robinson 1931:230–231. Marañón 1932:199. Potter 1933:151. Henry 1941(2):1023–1026. Hirschfeld 1944:281–282. Negri 1949: 163, 187. The concept of a continuum from exclusive heterosexuality to exclusive homosexuality is less often encountered, but is suggested, for instance, in: Freud 1924(2):207–208. Marañón 1929:170. Blanchard and Manasses 1930:109.

her psychologic reactions and/or overt behavior was heterosexual, and what proportion of her psychologic reactions and/or overt behavior was homosexual (Figure 93). We have done this for each year for which there is any record. This heterosexual-homosexual rating scale was explained in our volume on the male (1948:636–659), but before applying it to the data on the female it seems desirable to summarize again the principles involved in the construction and use of the scale.

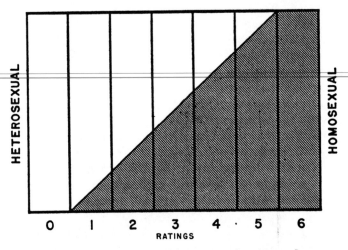

Figure 93. Heterosexual-homosexual rating scale

Definitions of the ratings are as follows: 0 = entirely heterosexual. 1 = largely heterosexual, but with incidental homosexual history. 2 = largely heterosexual, but with a distinct homosexual history. 3 = equally heterosexual and homosexual. 4 = largely homosexual, but with distinct heterosexual history. 5 = largely homosexual, but with incidental heterosexual history. 6 = entirely homosexual.

¶ The ratings represent a balance between the homosexual and heterosexual aspects of an individual's history, rather than the intensity of his or her psychosexual reactions or the absolute amount of his or her overt experience.

¶ Individuals who fall into any particular classification may have had various and diverse amounts of overt experience. An individual who has had little or no experience may receive the same classification as one who has had an abundance of experience, provided that the heterosexual and homosexual elements in each history bear the same relation to each other.

¶ The ratings depend on the psychologic reactions of the individual and on the amount of his or her overt experience. An individual may receive a rating on the scale even if he or she has had no overt heterosexual or homosexual experience.

¶ Since the psychologic and overt aspects of any history often parallel each other, they may be given equal weight in many cases in determining a rating. But in some cases one aspect may seem more significant than the other, and then some evaluation of the relative importance of the two must be made. We find, however, that most persons agree in their ratings of most histories after they have had some experience in the use of the scale. In our own research, where each year of each individual history has been rated independently by two of us, we find that our independent ratings differ in less than one per cent of the year-by-year classifications.

¶ An individual may receive a rating for any particular period of his or her life, whether it be the whole life span or some smaller portion of it. In the present study it has proved important to give ratings to each individual year, for some individuals may materially change their psychosexual orientation in successive years.

¶ While the scale provides seven categories, it should be recognized that the reality includes individuals of every intermediate type, lying in a continuum between the two extremes and between each and every category on the scale.

The categories on the heterosexual-homosexual scale (Figure 93) may be defined as follows:

0. Individuals are rated as 0's if all of their psychologic responses and all of their overt sexual activities are directed toward persons of the opposite sex. Such individuals do not recognize any homosexual responses and do not engage in specifically homosexual activities. While more extensive analyses might show that all persons may on occasion respond to homosexual stimuli, or are capable of such responses, the individuals who are rated 0 are those who are ordinarily considered to be completely heterosexual.

1. Individuals are rated as 1's if their psychosexual responses and/or overt experience are directed almost entirely toward individuals of the opposite sex, although they incidentally make psychosexual responses to their own sex, and/or have incidental sexual contacts with individuals of their own sex. The homosexual reactions and/or experiences are usually infrequent, or may mean little psychologically, or may be initiated quite accidentally. Such persons make few if any deliberate attempts to renew their homosexual contacts. Consequently the homosexual reactions and experience are far surpassed by the heterosexual reactions and/or experience in the history.

2. Individuals are rated as 2's if the preponderance of their psychosexual responses and/or overt experiences are heterosexual, although they respond rather definitely to homosexual stimuli and/or have more than incidental homosexual experience. Some of these individuals may have had only a small amount of homosexual experience, or they may have had a considerable amount of it, but the heterosexual element always predominates. Some of them may turn all of their overt experience in one direction while their psychosexual responses turn largely in the opposite direction; but they are always erotically aroused by anticipating homosexual experience and/or in their physical contacts with individuals of their own sex.

3. Individuals are rated as 3's if they stand midway on the heterosexual-homosexual scale. They are about equally heterosexual and homosexual in their psychologic responses and/or in their overt experience. They accept or equally enjoy
both types of contact and have no strong preferences for the one or the other.

4. Individuals are rated as 4's if their psychologic responses are more often
directed toward other individuals of their own sex and/or if their sexual contacts
are more often had with their own sex. While they prefer contacts with their own
sex, they, nevertheless, definitely respond toward and/or maintain a fair amount
of overt contact with individuals of the opposite sex.

5. Individuals are rated as 5's if they are almost entirely homosexual in their
psychologic responses and/or their overt activities. They respond only incidentally
to individuals of the opposite sex, and/or have only incidental overt experience
with the opposite sex.

6. Individuals are rated as 6's if they are exclusively homosexual in their psychologic responses, and in any overt experience in which they give any evidence
of responding. Some individuals may be rated as 6's because of their psychologic
responses, even though they may never have overt homosexual contacts. None
of these individuals, however, ever respond psychologically toward, or have overt
sexual contacts in which they respond to individuals of the opposite sex.

X. Finally, individuals are rated as X's if they do not respond erotically to either
heterosexual or homosexual stimuli, and do not have overt physical contacts with
individuals of either sex in which there is evidence of any response. After early
adolescence there are very few males in this classification (see our 1948:658), but
a goodly number of females belong in this category in every age group (Table 142,
Figure 95). It is not impossible that further analyses of these individuals might
show that they do sometimes respond to socio-sexual stimuli, but they are unresponsive and inexperienced as far as it is possible to determine by any ordinary
means.

Percentage With Each Rating. It should again be pointed out, as we
did in our volume on the male (1948:650), that it is impossible to determine the number of persons who are "homosexual" or "heterosexual."
It is only possible to determine how many persons belong, at any particular time, to each of the classifications on a heterosexual-homosexual
scale. The distribution of the available female sample on the heterosexual-homosexual scale is shown in Table 142 and Figure 94. These
incidence figures differ from the incidence figures presented in the
earlier part of this chapter, because the heterosexual-homosexual ratings are based on psychologic responses and overt experience, while
the accumulative and active incidences previously shown are (with the
exception of Table 131 and Figure 82) based solely on overt contacts.

The following generalizations may be made concerning the experience of the females in the sample, up to the time at which they contributed their histories to the present study.

Something between 11 and 20 per cent of the unmarried females and
8 to 10 per cent of the married females in the sample were making
at least incidental homosexual responses, or making incidental or more
specific homosexual contacts—*i.e.*, **rated 1 to 6**—in each of the years

between twenty and thirty-five years of age. Among the previously married females, 14 to 17 per cent were in that category (Table 142).

Something between 6 and 14 per cent of the unmarried females, and 2 to 3 per cent of the married females, were making more than incidental responses, and/or making more than incidental homosexual contacts —*i.e.*, **rated 2 to 6**—in each of the years between twenty and thirty-five years of age. Among the previously married females, 8 to 10 per cent were in that category (Table 142).

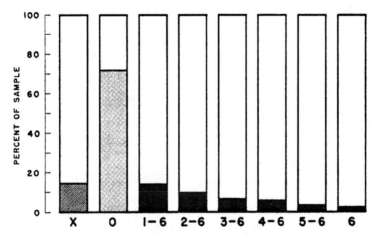

Figure 94. Active incidence: heterosexual-homosexual ratings, single females, age twenty-five

For definitions of the ratings, see p. 471. Data from Table 142.

Between 4 and 11 per cent of the unmarried females in the sample, and 1 to 2 per cent of the married females, had made homosexual responses, and/or had homosexual experience, at least as frequently as they had made heterosexual responses and/or had heterosexual experience—*i.e.*, **rated 3 to 6**—in each of the years between twenty and thirty-five years of age. Among the previously married females, 5 to 7 per cent were in that category (Table 142).

Between 3 and 8 per cent of the unmarried females in the sample, and something under 1 per cent of the married females, had made homosexual responses and/or had homosexual experience more often than they had responded heterosexually and/or had heterosexual experience—*i.e.*, **rated 4 to 6**—in each of the years between twenty and thirty-five years of age. Among the previously married females, 4 to 7 per cent were in that category (Table 142).

Between 2 and 6 per cent of the unmarried females in the sample, but less than 1 per cent of the married females, had been more or less exclusively homosexual in their responses and/or overt experience —*i.e.*, **rated 5 or 6**—in each of the years between twenty and thirty-five years of age. Among the previously married females, 1 to 6 per cent were in that category (Table 142).[18]

Between 1 and 3 per cent of the unmarried females in the sample, but less than three in a thousand of the married females, had been exclusively homosexual in their psychologic responses and/or overt

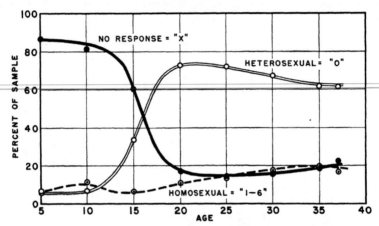

Figure 95. Active incidence: heterosexual-homosexual ratings, single females
For definitions of the ratings, X, 0, and 1–6, see p. 471. Data from Table 142.

experience—*i.e.*, **rated 6**—in each of the years between twenty and thirty-five years of age. Among the previously married females, 1 to 3 per cent were in that category (Table 142).

Between 14 and 19 per cent of the unmarried females in the sample, and 1 to 3 per cent of the married females, had not made any socio-sexual responses (either heterosexual or homosexual)—*i.e.*, **rated X**—in each of the years between twenty and thirty-five years of age. Among the previously married females, 5 to 8 per cent were in that category (Table 142).

Extent of Female vs. Male Homosexuality. The incidences and frequencies of homosexual responses and contacts, and consequently the incidences of the homosexual ratings, were much lower among the females in our sample than they were among the males on whom we have previously reported (see our 1948:650–651). Among the females, the accumulative incidences of homosexual responses had ultimately reached 28 per cent; they had reached 50 per cent in the males. The

---

[18] That fewer females than males are exclusively homosexual is also noted in: Havelock Ellis 1915(2):195. Potter 1933:151. Hesnard 1933:189. Cory 1951:88.

accumulative incidences of overt contacts to the point of orgasm among the females had reached 13 per cent (Table 131, Figure 82); among the males they had reached 37 per cent. This means that homosexual responses had occurred in about half as many females as males, and contacts which had proceeded to orgasm had occurred in about a third as many females as males. Moreover, compared with the males, there were only about a half to a third as many of the females who were, in any age period, primarily or exclusively homosexual.

A much smaller proportion of the females had continued their homosexual activities for as many years as most of the males in the sample.

A much larger proportion (71 per cent) of the females who had had any homosexual contact had restricted their homosexual activities to a single partner or two; only 51 per cent of the males who had had homosexual experience had so restricted their contacts. Many of the males had been highly promiscuous, sometimes finding scores or hundreds of sexual partners.

There is a widespread opinion which is held both by clinicians and the public at large, that homosexual responses and completed contacts occur among more females than males.[19] This opinion is not borne out by our data, and it is not supported by previous studies which have been based on specific data.[20] This opinion may have originated in the fact that females are more openly affectionate than males in our culture. Women may hold hands in public, put arms about each other, publicly fondle and kiss each other, and openly express their admiration and affection for other females without being accused of homosexual interests, as men would be if they made such an open display of their interests in other men. Males, interpreting what they observe in terms of male psychology, are inclined to believe that the female behavior reflects emotional interests that must develop sooner or later into overt sexual relationships. Nevertheless, our data indicate that a high proportion of this show of affection on the part of the female does not reflect any psychosexual interest, and rarely leads to overt homosexual activity.

Not a few heterosexual males are erotically aroused in contemplating the possibilities of two females in a homosexual relation; and the opin-

[19] For instance, Clark 1937:70, and Bergler 1951:317, feel that the incidences of homosexuality among females exceed those among males. Others differentiate various types of homosexuality, and feel that incidental or temporary homosexuality is commoner in the female, as in: Bloch 1908:525, and Hirschfeld 1944:281. Others who estimate that homosexuality is equally common in both sexes include: Havelock Ellis 1915(2):195. Krafft-Ebing 1922:397. Freud 1924(2):202. Kelly 1930:143. Sadler 1944:92.

[20] All specific studies have arrived at incidence figures for the male which exceed those for the female: Hamilton 1929: 492–493 (57 per cent male, 37 per cent

ion that females are involved in such relationships more frequently than males may represent wishful thinking on the part of such heterosexual males. Psychoanalysts may also see in it an attempt among males to justify or deny their own homosexual interests.

The considerable amount of discussion and bantering which goes on among males in regard to their own sexual activities, the interest which many males show in their own genitalia and in the genitalia of other males, the amount of exhibitionistic display which so many males put on in locker rooms, in shower rooms, at swimming pools, and at informal swimming holes, the male's interest in photographs and drawings of genitalia and sexual action, in erotic fiction which describes male as well as female sexual prowess, and in toilet wall inscriptions portraying male genitalia and male genital functions, may reflect homosexual interests which are only infrequently found in female histories. The institutions which have developed around male homosexual interests include cafes, taverns, night clubs, public baths, gymnasia, swimming pools, physical culture and more specifically homosexual magazines, and organized homosexual discussion groups; they rarely have any counterpart among females. Many of these male institutions, such as the homosexually oriented baths and gymnasia, are of ancient historic origin, but there do not seem to have been such institutions for females at any time in history. The street and institutionalized homosexual prostitution which is everywhere available for males, in all parts of the world, is rarely available for females, anywhere in the world.[21] All of these differences between female and male homosexuality depend on basic psychosexual differences between the two sexes.

SOCIAL SIGNIFICANCE OF HOMOSEXUALITY

Society may properly be concerned with the behavior of its individual members when that behavior affects the persons or property of other members of the social oganization, or the security of the whole group. For these reasons, practically all societies everywhere in the world attempt to control sexual relations which are secured through the use of force or undue intimidation, sexual relations which lead to unwanted pregnancies, and sexual activities which may disrupt or prevent marriages or otherwise threaten the existence of the social organization itself. In various societies, however, and particularly in our own Judeo-

---

female). Bromley and Britten 1938:117, 210 (13 per cent male, 4 per cent female). Gilbert Youth Research 1951 (12 per cent male, 6 per cent female).
[21] In addition to our own data, female homosexual clubs and bars are recorded in: Bloch 1908:530. Caufeynon 1934:22. Hirschfeld 1944:285. McPartland 1947:149–150. Cory 1951:122 (more rare than male homosexual clubs and bars). Female homosexual prostitution is also noted, for example, in: Martineau 1886:31. Parke 1906:313. Rohleder 1907(2):493; 1925:338–339. Bloch 1908:530. Hirschfeld 1944:282.

Christian culture, still other types of sexual activity are condemned by religious codes, public opinion, and the law because they are contrary to the custom of the particular culture or because they are considered intrinsically sinful or wrong, and not because they do damage to other persons, their property, or the security of the total group.

The social condemnation and legal penalties for any departure from the custom are often more severe than the penalties for material damage done to persons or to the social organization. In our American culture there are no types of sexual activity which are as frequently condemned because they depart from the mores and the publicly pretended custom, as mouth-genital contacts and homosexual activities. There are practically no European groups, unless it be in England, and few if any other cultures elsewhere in the world which have become as disturbed over male homosexuality as we have here in the United States. Interestingly enough, there is much less public concern over homosexual activities among females, and this is true in the United States and in Europe and in still other parts of the world.[22]

In an attempt to secure a specific measure of attitudes toward homosexual activity, all persons contributing histories to the present study were asked whether they would accept such contacts for themselves, and whether they approved or disapproved of other females or males engaging in such activity. As might have been expected, the replies to these questions were affected by the individual's own background of experience or lack of experience in homosexual activity, and the following analyses are broken down on that basis.

Acceptance for Oneself. Of the 142 females in the sample who had had the most extensive homosexual experience, some regretted their experience and some had few or no regrets. The record is as follows:

Regret	Percent
None	71
Slight	6
More or less	3
Yes	20
Number of cases	142

[22] For ancient Greece, Rome, and India, female homosexuality is recorded in: Ovid [1st cent. B.C., Roman]: Heroides, XV, 15–20, 201 (1921:183, 195) (Sappho recounts her past loves). Plutarch [1st cent. A.D., Greek]: Lycurgus, 18.4 (1914:(1)265). Martial [1st cent. A.D., Roman]: I,90(1919(1):85–87; 1921:33); VII, 67 (1919(1):469–471; 1921:193–194); VII, 70 (1919(1):471; 1921:194). Juvenal [1st–2nd cent. A.D., Roman]: Satires, VI, 308–325 (1789:272–275; 1817:239–240). Lucian [2nd cent. A.D., Greek]: Amores (1895:190); Dialogues of Courtesans, V (1895:100–105). Kama Sutra of Vatsyayana [1st–6th cent. A.D., Sanskrit] 1883–1925:62, 124. For additional accounts of Sappho of Lesbos, see: Wharton 1885, 1895. Miller and Robinson 1925. Weigall 1932.

Among the females who had never had homosexual experience, there were only 1 per cent who indicated that they intended to have it, and 4 per cent more who indicated that they might accept it if the opportunity were offered (Table 144).

But among the females who had already had some homosexual experience, 18 per cent indicated that they expected to have more. Another 20 per cent were uncertain what they would do, and some 62 per cent asserted that they did not intend to continue their activity. Some of the 18 per cent who indicated that they would continue were making a conscious and deliberate choice based upon their experience and their decision that the homosexual activity was more satisfactory than any other type of sexual contact which was available to them. Some of the others were simply following the path of least resistance, or accepting a pattern which was more or less forced upon them.

The group which had had homosexual experience and who expected to continue with it represented every social and economic level, from the best placed to the lowest in the social organization. The list included store clerks, factory workers, nurses, secretaries, social workers, and prostitutes. Among the older women, it included many assured individuals who were happy and successful in their homosexual adjustments, economically and socially well established in their communities and, in many instances, persons of considerable significance in the social organization. Not a few of them were professionally trained women who had been preoccupied with their education or other matters in the day when social relations with males and marriage might have been available, and who in subsequent years had found homosexual contacts more readily available than heterosexual contacts. The group included women who were in business, sometimes in high positions as business executives, in teaching positions in schools and colleges, in scientific research for large and important corporations, women physicians, psychiatrists, psychologists, women in the auxiliary branches of the Armed Forces, writers, artists, actresses, musicians, and women in every other sort of important and less important position in the social organization.[23] For many of these women, heterosexual relations or marriage would have been difficult while they maintained their professional careers. For many of the older women no sort of socio-sexual contacts would have been available if they had not worked out sexual adjustments with the companions with whom they had lived, in some instances for many years. Considerable affection or strong emotional attachments were involved in many of these relationships.

---

[23] As examples of the statistically unsupported opinion that homosexuality is more common among females in aesthetic professions, see: Eberhard 1924:548. Rohleder 1925:381–382. Moreck 1929:312. Hesnard 1933:189. Chesser 1947:257. Martinez 1947:103. McPartland 1947:154. Beauvoir 1952:411.

On the other hand, some of the females in the sample who had had homosexual experience had become much disturbed over that experience. Often there was a feeling of guilt in having engaged in an activity which is socially, legally, and religiously disapproved, and such individuals were usually sincere in their intention not to continue their activities. Some of them, however, were dissatisfied with their homosexual relations merely because they had had conflicts with some particular sexual partner, or because they had gotten into social difficulties as a result of their homosexual activities.

Some 27 per cent of those who had had more extensive homosexual experience had gotten into difficulty because of it (Table 145). Some of these females were disturbed because they had found it physically or socially impossible to continue relationships with the partner in whom they were most interested, and refused to contemplate the possibility of establishing new relationships with another partner. In a full half of these cases, the difficulties had originated in the refusal of parents or other members of their families to accept them after they had learned of their homosexual histories.

On the other hand, among those who had had homosexual experience, as well as among those who had not had experience, there were some who denied that they intended to have or to continue such activity, because it seemed to be the socially expected thing to disavow any such intention. Some of these females would actually accept such contacts if the opportunity came and circumstances were propitious. It is very difficult to know what an individual will do when confronted with an opportunity for sexual contact.

Approval for Others. As a further measure of female reactions to homosexual activity, each subject was asked whether she approved or disapproved or was neutral in regard to other persons, of her own or of the opposite sex, having homosexual activity. Each of the female subjects was also asked to indicate whether she would keep friends, female or male, after she had discovered that they had had homosexual experience. Since the question applied to persons whom they had previously accepted as friends, it provided a significant test of current attitudes toward homosexual behavior. From these data the following generalizations may be drawn:

1. The approval of homosexual activity for other females was much higher among the females in the sample who had had homosexual experience of their own. Some 23 per cent of those females recorded definite approval, and only 15 per cent definitely disapproved of other females having homosexual activity (Table 144, Figure 96).

2. Females who had had experience of their own approved of homosexual activity for males less often than they approved of it for females. Only 18 per cent completely approved of the male activity, and 22 per cent definitely disapproved (Table 144, Figure 96).

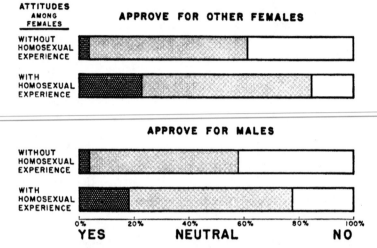

Figure 96. Attitudes of females toward homosexual contacts for others
Data from Table 144.

3. The females who had never had homosexual experience were less often inclined to approve of it for other persons. Some 4 per cent expressed approval of homosexual activity for males, but approximately 42 per cent definitely disapproved (Table 144). Some 4 per cent approved of activities for females, and 39 per cent disapproved.

4. Among the females who had had homosexual experience, some 88 per cent indicated that they would keep female friends after they had discovered their homosexual histories; 4 per cent said they would not (Table 144, Figure 97). Some of these latter responses reflected the subject's dissatisfaction with her own homosexual experience, but some represented the subject's determination to avoid persons who might tempt her into renewing her own activities.

5. Among the females who had had homosexual experience, 74 per cent indicated that they would continue to keep male friends after they had discovered that they had homosexual histories, and 10 per cent said they would not (Table 144, Figure 97). The disapproval of males with homosexual histories often depends upon the opinion that such males have undesirable characteristics, but this objection could not have been a factor in the present statistics because the question had concerned males whom the subject had previously accepted as friends.

6. Females who had never had homosexual experience were less often willing to accept homosexual female friends. Only 55 per cent said they would keep such friends, and 22 per cent were certain that they would not keep them (Table 144, Figure 97). This is a measure of the intolerance with which our Judeo-Christian culture views any type of sexual activity which departs from the custom.

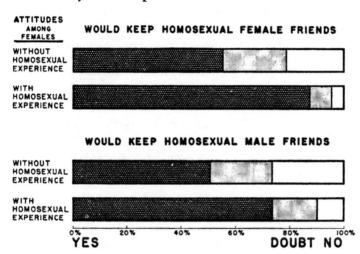

Figure 97. Attitudes of females toward keeping homosexual friends
Data from Table 144.

7. Some 51 per cent of the females who had never had homosexual experience said that they would keep homosexual males as friends, 26 per cent said they would not, and 23 per cent were doubtful (Table 144, Figure 97). As we have noted before (1948:663–664), this sort of ostracism by females often becomes a factor of considerable moment in forcing the male who has had some homosexual experience into exclusively homosexual patterns of behavior.

Moral Interpretations. The general condemnation of homosexuality in our particular culture apparently traces to a series of historical circumstances which had little to do with the protection of the individual or the preservation of the social organization of the day. In Hittite, Chaldean, and early Jewish codes there were no over-all condemnations of such activity, although there were penalties for homosexual activities between persons of particular social status or blood relationships, or homosexual relationships under other particular circumstances, especially when force was involved.[24] The more general condemnation of

---

[24] For the rather broad acceptance of homosexuality in many parts of the ancient Near East, see: Pritchard 1950:73–74, 98–99, for the Gilgamesh Epic (2nd millennium B.C. or earlier) which contains passages suggesting homosexual relations between the heroes Gilgamesh and Enkidu. Homosexuality *is not*

all homosexual relationships originated in Jewish history in about the seventh century B.C., upon the return from the Babylonian exile. Both mouth-genital contacts and homosexual activities had previously been associated with the Jewish religious service, as they had been with the religious services of most of the other peoples of that part of Asia, and just as they have been in many other cultures elsewhere in the world.[25] In the wave of nationalism which was then developing among the Jewish people, there was an attempt to dis-identify themselves with their neighbors by breaking with many of the customs which they had previously shared with them. Many of the Talmudic condemnations were based on the fact that such activities represented the way of the Canaanite, the way of the Chaldean, the way of the pagan, and they were originally condemned as a form of idolatry rather than a sexual crime. Throughout the middle ages homosexuality was associated with heresy.[26] The reform in the custom (the mores) soon, however, became a matter of morals, and finally a question for action under criminal law.

Jewish sex codes were brought over into Christian codes by the early adherents of the Church, including St. Paul, who had been raised in the Jewish tradition on matters of sex.[27] The Catholic sex code is an almost precise continuation of the more ancient Jewish code.[28] For centuries in Medieval Europe, the ecclesiastic law dominated on all questions of morals and subsequently became the basis for the English common law, the statute laws of England, and the laws of the various states of the United States. This accounts for the considerable conformity between

---

mentioned in the codes of Lipit-Ishtar or Hammurabi, and the injunction in the Hittite code (Pritchard 1950:196) is aimed only at men who have contact with their sons. The Middle Assyrian laws (12th century B.C. or earlier) likewise mention male homosexuality which was punishable by castration (see: Barton 1925:Chapter 15, item 19), but a more modern translation suggests that this punishment was preceded by homosexual contact between the convicted man and his punishers (Pritchard 1950:181). Epstein 1948:135–136 assumes a general taboo on male homosexuality among the ancient Hebrews, but admits that this taboo is not to be found in the Covenant Code or in Deuteronomy, but only in the somewhat later Leviticus 18:22 and 20:13. See also Genesis 19:1–25, and Judges 19:17–25, for the protection of a male guest from forced homosexual relations. Deuteronomy 23:17–18 simply prohibits men of the Israelites from becoming temple prostitutes, but goes no further.

[25] Male homosexual temple prostitutes, "kadesh," were at one time a part of Jewish religion, as may be gathered from II Kings 23:7, and from the warning in Deuteronomy 23:17–18. This is discussed by Westermarck 1917(2):488, and by Epstein 1948:135–136. The subsequent condemnation of homosexuality occurs repeatedly, as in: I Kings 14:24; 15:12; 22:46. Leviticus 18:22; 20:13. See also the Talmud, Sanhedrin 54a, 78a, 82a. Yebamoth 25a, 54b. Sotah 26b, etc.

[26] The condemnation of homosexuality as idolatry is noted by Westermarck 1917(2):487–488, and by Epstein 1948:136.

[27] For St. Paul's condemnation of homosexuality, see: Romans 1:26–27. I Corinthians 6:9. I Timothy 1:10.

[28] The Catholic codes explicitly condemn male and female homosexuality. See such accepted Catholic sources as: Arregui 1927:153. Davis 1946(2):246.

the Talmudic and Catholic codes and the present-day statute law on sex, including the laws on homosexual activity.[29]

Condemnations of homosexual as well as some other types of sexual activity are based on the argument that they do not serve the prime function of sex, which is interpreted to be procreation, and in that sense represent a perversion of what is taken to be "normal" sexual behavior. It is contended that the general spread of homosexuality would threaten the existence of the human species, and that the integrity of the home and of the social organization could not be maintained if homosexual activity were not condemned by moral codes and public opinion and made punishable under the statute law. The argument ignores the fact that the existent mammalian species have managed to survive in spite of their widespread homosexual activity, and that sexual relations between males seem to be widespread in certain cultures (for instance, Moslem and Buddhist cultures) which are more seriously concerned with problems of overpopulation than they are with any threat of underpopulation. Interestingly enough these are also cultures in which the institution of the family is very strong.

Legal Attitudes. While it is, of course, impossible for laws to prohibit homosexual interests or reactions, they penalize, in every state of the Union, some or all of the types of contact which are ordinarily employed in homosexual relations. The laws are variously identified as statues against sodomy, buggery, perverse or unnatural acts, crimes against nature, public and in some instances private indecencies, grossly indecent behavior, and unnatural or lewd and lascivious behavior. The penalties in most of the states are severe, and in many states as severe as the penalties against the most serious crimes of violence.[30] The penalties are particularly severe when the homosexual relationships involve an adult with a young minor.[31] There is only one state, New York, which, by an indirection in the wording of its statute, appears to attach no penalty to homosexual relations which are carried on between adults in private and with the consent of both of the participating parties; and this sort of exemption also appears in Scandinavia and in many other European countries. There appears to be no other major culture in the world in which public opinion and the statute law so severely penalize homosexual relationships as they do in the United States today.

It might be expected that the moral and legal condemnations of homosexual activity would apply with equal force to both females and

---

[29] For the relationship between Jewish and Catholic codes, and the statute law, see also: Westermarck 1917 (2): 480–489. May 1931: ch. 2, 3.

[30] For a convenient and almost complete summary of the statutes concerning homosexuality in the forty-eight states, see: Cory 1951: appendix B.

[31] For the problem involved in the relationships of adults and minors, see: Guttmacher and Weihofen 1952: 156.

males. The ancient Hittite code, however, condemned only male homosexual activity and then only when it occurred under certain circumstances, and made no mention of homosexual activity among females.[24] Similarly the references to homosexual activity in the Bible and in the Talmud apply primarily to the male. The condemnations were severe and usually called for the death of the transgressing male, but they rarely mentioned female activity, and when they did, no severe penalties were proposed.[32] In medieval European history there are abundant records of death imposed upon males for sexual activities with other males, but very few recorded cases of similar action against females.[33] In modern English and other European law, the statutes continue to apply only to males [34]; but in American law, the phrasing of the statutes would usually make them applicable to both female and male homosexual contacts.[35] The penalties are usually invoked against "all persons," "any person," "whoever," "one who," or "any human being" without distinction of sex. Actually there are only five states [36] in the United States where the statutes do not cover female homosexual relationships, and it is probable that the courts would interpret the statutes in nearly all of the other states to apply to females as well as to males.

These American statutes appear, however, to have gone beyond public opinion in their condemnation of homosexual relations between females, for practically no females seem to have been prosecuted or convicted anywhere in the United States under these laws. In our total sample of several hundred females who had had homosexual experience, only three had had minor difficulties and only one had had more serious difficulty with the police (Table 145), and none of the cases had been brought to court. We have cases of females who were

---

[32] The stringent penalty for homosexuality given in Leviticus 18:22 and 20:13 applies only to the male. Reference to female homosexuality does not appear until much later: Romans 1:26, where it is considered a "vile affection." The Talmud is relatively lenient regarding females, stating that female homosexual activity is a "mere obscenity" disqualifying a woman from marrying a priest. See Yebamoth 76a. Maimonides, according to Epstein 1948:138, felt that a female guilty of homosexuality should be flogged and excluded from the company of decent women, which is a penalty far less severe than the death penalty required for the male.

[33] Such medieval penalties for homosexuality are mentioned, for instance, in: Havelock Ellis 1915(2):346–347. Westermarck 1917(2):481–482. For a case of capital punishment levied on a female, see: Wharton 1932(1):1036–1037, footnote 18.

[34] There are specific statutes against female homosexuality only in Austria, Greece, Finland, and Switzerland.

[35] The applicability of the laws to both females and males are also noted in: Sherwin 1951:13. Ploscowe 1951:204. Pilpel and Zavin 1952:220.

[36] The states in which the statutes apparently do not apply to female homosexuality are: Conn., Ga., Ky., S. C., and Wis. Heterosexual cunnilingus has been held not "the crime against nature" in Illinois, Mississippi, and Ohio, and the decisions would supposedly apply to homosexual cunnilingus. In Arkansas, Colorado, Iowa, and Nebraska there is also some doubt as to the status of female homosexuality.

disciplined or more severely penalized for their homosexual activities in penal or other institutions, or while they were members of the Armed Forces of the United States, and we have cases in which social reactions constituted a severe penalty, but no cases of action in the courts.

Our search through the several hundred sodomy opinions which have been reported in this country between 1696 and 1952 has failed to reveal a single case sustaining the conviction of a female for homosexual activity. Our examination of the records of all the females admitted to the Indiana Women's Prison between 1874 and 1944 indicates that only one was sentenced for homosexual activity, and that was for activity which had taken place within the walls of another institution. Even in such a large city as New York, the records covering the years 1930 to 1939 show only one case of a woman convicted of homosexual sodomy, while there were over 700 convictions of males on homosexual charges, and several thousand cases of males prosecuted for public indecency, or for solicitation, or for other activity which was homosexual.[37] In our own more recent study of the enforcement of sex law in New York City we find three arrests of females on homosexual charges in the last ten years, but all of those cases were dismissed, although there were some tens of thousands of arrests and convictions of males charged with homosexual activity in that same period of time.

It is not altogether clear why there are such differences in the social and legal attitudes toward sexual activities between females and sexual activities between males. They may depend upon some of the following, and probably upon still other factors:

1. In Hittite, Jewish and other ancient cultures, women were socially less important than males, and their private activities were more or less ignored.

2. Both the incidences and frequencies of homosexual activity among females are in actuality much lower than among males. Nevertheless, the number of male cases which are brought to court are, even proportionately, tremendously higher than the number of female cases that reach court.

3. Male homosexual activity more often comes to public attention in street solicitation, public prostitution, and still other ways.

4. Male homosexual activity is condemned not only because it is homosexual, but because it may involve mouth-genital or anal contacts. It is not so widely understood that female homosexual techniques may also involve mouth-genital contacts.

---

[37] New York City data are to be found in the report of the Mayor's Committee on Sex Offenses 1944:75.

5. Homosexual activities more often interfere with the male's, less often interfere with the female's marrying or maintaining a marriage.

6. The Catholic Code emphasizes the sin involved in the wastage of semen in all male activities that are non-coital; it admits that female non-coital activities do not involve the same species of sin.

7. There is public objection to the effeminacy and some of the other personality traits of certain males who have homosexual histories; there is less often objection to the personalities of females who have homosexual histories.[38]

8. The public at large has some sympathy for females, especially older females, who are not married and who would have difficulty in finding sexual contacts if they did not engage in homosexual relations.

9. Many heterosexual males are erotically aroused when they consider the possibilities of two females in sexual activities. In not a few instances they may even encourage sexual contacts between females. There are fewer cases in our records of females being aroused by the contemplation of activities between males.

10. There are probably more males and fewer females who fear their own capacities to respond homosexually. For this reason, many males condemn homosexual activities in their own sex more severely than they condemn them among females.

11. Our social organization is presently much concerned over sexual relationships between adults and young children. This is the basis for a considerable portion of the action which is taken against male homosexual contacts; but relationships between older women and very young girls do not so often occur.

Basic Social Interests. When a female's homosexual experience interferes with her becoming married or maintaining a marriage into which she has entered, social interests may be involved. On the other hand, our social organization has never indicated that it is ready to penalize, by law, all persons who fail to become married.

When sexual relationships between adult females do not involve force

---

[38] The statistically unsupported opinion that females with homosexual histories frequently or usually exhibit masculine physical characters, behavior, or tastes appears, however, in such authors as the following: Féré 1904:189. Parke 1906:266, 300–301, 321. Bloch 1908:526. Carpenter 1908:30–31. Talmey 1910:158–161. Freud 1910:11. Havelock Ellis 1915(2):251–254. Krafft-Ebing 1922:336, 398–399. Kisch 1926:192. Kelly 1930:138. Moll 1931:226 ff. Potter 1933:158. Hesnard 1933:186. Caufeynon 1934:132. S. Kahn 1937:69, 134. Hutton 1937:126, 129. Henry 1941(2):1062, 1075, 1081. Deutsch 1944:325. Negri 1949:187. Keiser and Schaffer 1949:287, 289. Bergler 1951:318. Higher "masculinity" ratings on masculinity-femininity tests are reported by: Terman and Miles 1936:577–578. Henry 1941(2):1033–1034.

or undue coercion, and do not interfere with marital adjustments that might have been made, many persons, both in Europe and in our American culture, appear to be fairly tolerant of female homosexual activities. At any rate, many of those who feel that a question of morality may be involved, fail to believe that the basic social interests are sufficient to warrant any rigorous legal action against females who find a physiologic outlet and satisfy their emotional needs in sexual contacts with other females.

SUMMARY AND COMPARISONS OF FEMALE AND MALE

Homosexual Responses and Contacts

	IN FEMALES	IN MALES
Physiologic and Psychologic Bases		
Inherent capacity to respond to any sufficient stimulus	Yes	Yes
Preference developed by psychologic conditioning	Yes	Yes
Among mammals homosexual behavior widespread	Yes	Yes
Anthropologic Background		
Data on homosexual behavior	Very few	Some
Heterosexual more acceptable in most cultures	Yes	Yes
Homosexual behavior sometimes permitted	Yes	Yes
Social concern over homosexual behavior	Less	More
Relation to Age and Marital Status		
Accumulative incidence		
Homosexual response, by age 45	28%	±50%
Homosexual experience, by age 45	20%	
Single	26%	±50%
Married	3%	±10%
Previously married	10%	
Homo. exper. to orgasm, by age 45	13%	±37%
Active incidence, to orgasm		
Single		
Age 16–20	3%	22%
Age 36–40	10%	40%
Age 46–50	4%	36%
Married	1–2%	2–8%
Previously married, age 16–50	3–7%	5–28%
Frequency to orgasm, per week		
Single		
Age Adol.–15	0.2	0.1
Age 21–30	0.3–0.4	0.4–0.7
Age 31–40	0.3–0.4	0.7–1.0
Married freq. lower than in single	Somewhat	Markedly
Percent of total outlet, before age 40		
Single, gradual increase	4–19%	5–22%
Married	Under 1%	Under 1%
Previously married, gradual increase	2–10%	9–26%
Number of years involved		
1 year or less	47%	
2 to 3 years	25%	

	IN FEMALES	IN MALES
Number of partners		
1–2	71%	51%
Over 10	4%	22%
Relation to Educational Level *		
Accumul. incid. to orgasm, by age 30		
Grade school	6%	27%
High school	5%	39%
College	10%⎫	34%
Graduate	14%⎭	
Act. incid. and % of outlet, higher		
Before age 20	In less educ.	In less educ.
After age 20	In better educ.	In less educ.
Frequency to orgasm higher	In less educ.	In less educ.
Relation to Parental Occupational Class	Little	Little or none
Relation to Decade of Birth	None	Little or none
Relation to Age at Onset of Adolescence	None	Higher incid. and freq. in early-adol.
Relation to Rural-Urban Background	Little	Incid. and freq. higher in urban
Relation to Religious Background		
Accum. and act. incid. higher among less dev.	Yes	Yes
Frequency to orgasm (active median)	No relation	Little relation
Percentage of total outlet	Higher among devout	No relation
Techniques in Homosexual Contacts		
Essentially same as in hetero. petting	Yes	Yes
Kissing and general body contacts	Extensive	
Genital techniques utilized	Later or never	Early and ± always
More effective than marital coitus	Yes	No
Hetero.-Homo. Ratings, *e.g.*, ages 20–35		
X : no socio-sexual response		
Single	14–19%	3–4%
Married	1–3%	0%
Previously married	5–8%	1–2%
0: entirely heterosexual experience		
Single	61–72%	53–78%
Married	89–90%	90–92%
Previously married	75–80%	
1–6: at least some homosexual	11–20%	18–42%
2–6: more than incidental homosexual	6–14%	13–38%
3–6: homo. as much or more than hetero.	4–11%	9–32%
4–6: mostly homosexual	3–8%	7–26%
5–6: ± exclusively homosexual	2–6%	5–22%
6: exclusively homosexual	1–3%	3–16%
Social Significance of Homosexuality		
Social concern in Anglo-Amer. culture	Little	Great
Most exper. indiv. regret least	Yes	Yes
Intent to have. highest among those with exper.	Yes	

* Beginning at this point, the data apply to single females and males only, unless otherwise indicated.

Summary and Comparisons (*Continued*)

	IN FEMALES	IN MALES
Approval for others, most often:		
By those with experience	Yes	Yes
For own sex	Yes	No
Moral and Legal Aspects of Homosexuality		
Injunction against, in:		
Ancient Near Eastern codes	No	Sometimes
Old Testament	No	Yes
Talmud	Yes	Yes
St. Paul and Christian codes	Yes	Yes
Formerly considered heresy	Yes	Yes
Death in ancient and medieval hist.	Rarely	Yes
Legally punishable in	43 states	48 states
Laws enforced	Almost never	Frequently

Table 126. Accumulative Incidence: Homosexual Contacts
By Marital Status

Age	Total sample	While single	While married	Post-marital	Total sample	While single	While married	Post-marital
	%		*Percent*				*Cases*	
12	1	1			5733	5732		
15	5	5			5685	5681		
20	9	9	1	7	4318	3941	556	77
25	14	16	1	8	2779	1464	1338	174
30	17	21	2	11	2045	670	1216	221
35	19	26	2	7	1470	381	912	205
40	19	24	3	9	951	207	571	179
45	20	26	3	10	572	128	312	130

Table based on total sample, including single, married, and previously married females.

Table 127. Active Incidence, Frequency, and Percentage of Outlet
Homosexual Contacts to Orgasm

Single Females, by Educational Level

Age during activity	Educ. level	Active incid. %	Active median freq. per wk.	% of total outlet	Cases in total sample	Age during activity	Educ. level	Active incid. %	Active median freq. per wk.	% of total outlet	Cases in total sample
Adol. −15	0–8	7	*0.7*	14	162	26–30	9–12	4		12	181
	9–12	3	*0.3*	6	983		13–16	7	0.5	10	313
	13–16	1	*0.1*	1	3271		17+	10	0.3	13	531
	17+	2	*0.2*	2	1128	31–35	9–12	6		6	65
16–20	0–8	8	*0.7*	9	143		13–16	7		11	139
	9–12	3	*0.3*	4	976		17+	11	*0.3*	18	309
	13–16	2	0.1	3	3299	36–40	13–16	12		17	68
	17+	5	0.3	6	1149		17+	11	*0.3*	21	205
21–25	9–12	3	*0.5*	7	537	41–45	17+	7		7	122
	13–16	4	*0.4*	8	1204	46–50	17+	5		5	80
	17+	7	0.3	7	1002						

Italic figures throughout the series of tables indicate that the calculations are based on less than 50 cases. No calculations are based on less than 11 cases. The dash (—) indicates a percentage or frequency smaller than any quantity which would be shown by a figure in the given number of decimal places.

Table 128. Active Incidence, Frequency, and Percentage of Outlet Homosexual Contacts to Orgasm

By Age and Marital Status

AGE DURING ACTIVITY	ACTIVE SAMPLE			TOTAL SAMPLE		CASES IN TOTAL SAMPLE
	Active incid. %	Median freq. per wk.	Mean frequency per wk.	Mean freq. per wk.	% of total outlet	
SINGLE FEMALES						
Adol.–15	2	0.2	0.6 ± 0.09	—	4	5677
16–20	3	0.2	0.6 ± 0.08	—	4	5613
21–25	5	0.3	1.0 ± 0.16	—	7	2810
26–30	8	0.4	1.3 ± 0.30	0.1 ± 0.03	11	1064
31–35	9	0.3	1.6 ± 0.63	0.1 ± 0.06	14	539
36–40	10	0.4	2.5 ± 0.96	0.3 ± 0.10	19	315
41–45	6			0.1 ± 0.04	6	179
46–50	4			—	4	109
51–55	0			0.0	0	58
56–60	0			0.0	0	27
MARRIED FEMALES						
16–20	1			—	—	578
21–25	1	0.3	0.9 ± 0.29	—	—	1654
26–30	1	0.2	1.2 ± 0.64	—	1	1662
31–35	1	0.3	1.0 ± 0.51	—	—	1246
36–40	2	0.1	0.4 ± 0.19	—	—	851
41–45	1			—	—	497
46–50	—			—	—	260
51–55	1			—	1	118
56–60	2			—	3	49
PREVIOUSLY MARRIED FEMALES						
16–20	6			—	2	72
21–25	6	0.7	1.1 ± 0.38	0.1 ± 0.03	4	239
26–30	6	2.0	2.3 ± 0.59	0.1 ± 0.05	9	328
31–35	7	1.1	2.4 ± 0.80	0.2 ± 0.06	10	304
36–40	3			—	2	245
41–45	4			0.1 ± 0.05	6	195
46–50	4			—	3	126
51–55	1			—	—	82
56–60	0			0.0	0	53

491

Table 129. Number of Years Involved in Homosexual Contacts

Including activity with and without orgasm

NUMBER OF TIMES OR YEARS	TOTAL SAMPLE	AGE AT REPORTING			
		Adol.–20	21–30	31–40	41–50
		Percent			
1–10 times	32	26	35	32	33
1 year or less	47	48	51	44	44
2–3 years	25	26	27	21	25
4–5 years	10	15	7	9	9
6–10 years	9	11	11	8	7
11–20 years	7		4	14	8
21+ years	2			4	7
Number of cases	709	137	202	202	122

Table 130. Partners and Techniques in Homosexual Contacts

PARTNERS		TECHNIQUES		
Number	%	Technique utilized	By females with limited exper.	By females with extensive exper.
			%	%
1 only	51	Kissing: simple		95
2	20	Kissing: deep		77
3	9	Breast: manual stimul.	27	97
4	5	Breast: oral stimul.	7	85
5	4	Genital: manual stimul.	67	98
6–10	7	Genital: oral stimul.	16	78
11–20	3	Genital apposition	24	56
21+	1			
Cases with exper.	591		499	145

Data on kissing unavailable on females with limited experience.

492

Table 131. Accumulative Incidence: Homosexual Arousal, Experience, and Orgasm

By Educational Level

AGE	TOTAL SAMPLE	EDUCATIONAL LEVEL				TOTAL SAMPLE	EDUCATIONAL LEVEL			
		0–8	9–12	13–16	17+		0–8	9–12	13–16	17+
				HOMOSEXUAL AROUSAL						
	%		Percent			Cases		Cases		
8	2	1	1	1	2	5720	179	999	3226	1124
10	3	2	3	2	4	5699	179	999	3226	1124
12	5	3	6	4	6	5674	179	999	3226	1124
15	10	11	10	9	11	5614	173	998	3226	1124
20	17	13	14	17	21	4267	127	849	2168	1123
25	23	9	16	24	28	2743	117	678	1020	928
30	25	10	18	25	33	2017	109	494	697	717
35	27	10	21	28	33	1447	91	317	487	552
40	27	10	18	28	32	937	67	189	301	380
45	28		17	28	36	565		124	165	231
			HOMOSEXUAL CONTACTS: EXPERIENCE							
	%		Percent			Cases		Cases		
12	1	2	1	1	1	5733	179	1007	3267	1142
15	5	9	5	4	4	5685	175	1006	3267	1142
20	9	11	7	9	12	4318	129	854	2194	1141
25	14	8	8	14	19	2779	119	683	1035	942
30	17	9	10	17	24	2045	111	500	707	727
35	19	9	12	17	25	1470	93	322	493	562
40	19	10	10	19	24	951	68	193	304	386
45	20		8	20	27	572		127	166	234
			HOMOSEXUAL CONTACTS TO ORGASM							
	%		Percent			Cases		Cases		
12	1	2	1	—	—	5779	178	1012	3301	1152
15	2	9	3	1	2	5733	174	1011	3301	1152
20	4	8	4	3	6	4359	128	860	2220	1151
25	7	5	5	7	10	2803	118	687	1046	952
30	10	6	5	10	14	2058	110	502	713	733
35	11	7	5	11	15	1480	92	323	498	567
40	12	7	5	13	15	956	68	194	305	389
45	13		6	13	19	574		127	166	236

Table based on total sample, including single, married, and previously married females.

Table 132. Accumulative Incidence: Homosexual Contacts to Orgasm By Parental Occupational Class

AGE	PARENTAL CLASS				PARENTAL CLASS			
	2+3	4	5	6+7	2+3	4	5	6+7
	Percent				*Cases*			
12	1	—	—	1	973	812	1526	2684
15	5	2	1	2	943	810	1523	2671
20	5	3	4	4	711	631	1139	2018
25	6	6	7	9	511	421	722	1225
30	9	6	9	12	379	313	506	907
35	10	6	9	13	268	197	345	692
40	9	4	10	14	174	119	206	470
45	7	4	15	17	108	71	117	282

Table based on total sample, including single, married, and previously married females.

The occupational classes are as follows: 2+3 = unskilled and semi-skilled labor. 4 = skilled labor. 5 = lower white collar class. 6+7 = upper white collar and professional classes.

Table 133. Active Incidence and Percentage of Outlet Homosexual Contacts to Orgasm
Single Females, by Parental Occupational Class

Age	Parental class	Active incid. %	% of total outlet	Cases in total sample	Age	Parental class	Active incid. %	% of total outlet	Cases in total sample
Adol.					26–30	2+3	5	11	195
-15	2+3	5	8	947		4	6	8	181
	4	2	3	796		5	7	6	275
	5	2	2	1506		6+7	10	14	447
	6+7	1	2	2654					
16–20	2+3	5	5	881	31–35	2+3	4	14	96
	4	3	4	796		4	3	1	91
	5	2	3	1512		5	9	11	148
	6+7	3	4	2649		6+7	13	17	224
21–25	2+3	4	3	461	36–40	2+3	5	25	62
	4	4	6	422		5	9	6	85
	5	4	4	735		6+7	12	23	141
	6+7	5	9	1283					

Table 134. Accumulative Incidence: Homosexual Contacts to Orgasm
By Decade of Birth

AGE	DECADE OF BIRTH				Bf. 1900	DECADE OF BIRTH		
	Bf. 1900	1900–1909	1910–1919	1920–1929		1900–1909	1910–1919	1920–1929
	Percent					*Cases*		
12	—	—	—	1	456	783	1341	3072
15	2	2	2	3	456	783	1341	3058
20	5	6	4	3	456	783	1340	1780
25	8	9	7	6	456	783	1191	373
30	9	10	10		456	783	819	
35	11	11	13		456	754	270	
40	11	12			456	499		
45	12	17			435	139		

Table based on total sample, including single, married, and previously married females.

Table 135. Active Incidence and Percentage of Outlet
Homosexual Contacts to Orgasm

Single Females, by Decade of Birth

Age	Decade of birth	Active incid. %	% of total outlet	Cases in total sample	Age	Decade of birth	Active incid. %	% of total outlet	Cases in total sample
Adol. −15	Bf. 1900	2	1	436	26–30	Bf. 1900	6	18	218
	1900–1909	1	3	760		1900–1909	8	10	344
	1910–1919	2	3	1319		1910–1919	9	10	448
	1920–1929	2	6	3049		1920–1929	7	9	54
16–20	Bf. 1900	4	3	451	31–35	Bf. 1900	8	19	151
	1900–1909	5	6	772		1900–1909	8	14	228
	1910–1919	3	5	1328		1910–1919	11	8	160
	1920–1929	2	4	2999	36–40	Bf. 1900	8	19	123
21–25	Bf. 1900	5	6	366		1900–1909	9	17	165
	1900–1909	6	7	617					
	1910–1919	6	7	987	41–45	Bf. 1900	4	6	116
	1920–1929	3	7	843		1900–1909	8	3	63

Table 136. Accumulative Incidence: Homosexual Contacts to Orgasm
By Age at Onset of Adolescence

AGE	ADOLESCENT					ADOLESCENT				
	By 11	At 12	At 13	At 14	At 15+	By 11	At 12	At 13	At 14	At 15+
	Percent					*Cases*				
12	2	1				1203	1681			
15	3	2	2	2	2	1185	1666	1738	792	348
20	5	3	5	5	4	876	1227	1321	649	283
25	8	6	8	8	8	511	741	848	465	235
30	11	9	9	11	10	355	488	650	368	196
35	12	11	11	11	11	250	322	470	281	156
40	14	11	12	12	10	151	199	297	191	117
45	13	14	14	13	11	93	111	174	119	76

Table based on total sample, including single, married, and previously married females.

Table 137. Active Incidence and Percentage of Outlet
Homosexual Contacts to Orgasm

Single Females, by Age at Onset of Adolescence

Age during activity	Age at adol.	Active incid. %	% of total outlet	Cases in total sample	Age during activity	Age at adol.	Active incid. %	% of total outlet	Cases in total sample
Adol.–15	8–11	3	6	1203	26–30	8–11	10	12	196
	12	2	3	1684		12	9	24	266
	13	2	2	1747		13	5	5	323
	14	2	6	796		14	10	11	192
	15+	2	3	262		15+	7	6	87
16–20	8–11	3	4	1166	31–35	8–11	10	11	99
	12	2	4	1638		12	7	28	122
	13	3	4	1700		13	8	2	162
	14	5	6	777		14	13	16	109
	15+	3	4	345					
21–25	8–11	5	11	526	36–40	8–11	11	9	65
	12	3	5	770		12	14	45	66
	13	5	6	851		13	6	2	96
	14	7	7	460		14	13	25	62
	15+	6	5	203					

Table 138. Accumulative Incidence: Homosexual Contacts to Orgasm
By Rural-Urban Background

Age	Rural	Urban	Rural	Urban	Age	Rural	Urban	Rural	Urban
	Percent		*Cases*			*Percent*		*Cases*	
12	1	1	399	5200	30	6	10	168	1796
15	3	2	397	5161	35	7	12	124	1285
20	4	4	335	3873	40	8	13	84	825
25	5	8	223	2465	45	10	14	60	488

Table based on total sample, including single, married, and previously married females.

Table 139. Active Incidence and Percentage of Outlet
Homosexual Contacts to Orgasm
Single Females, by Rural-Urban Background

Age during activity	Back-grnd.	Active incid. %	% of total outlet	Cases in total sample	Age during activity	Back-grnd.	Active incid. %	% of total outlet	Cases in total sample
Adol.–15	Rural	3	7	388	26–30	Rural	2	2	104
	Urban	2	4	5132		Urban	9	12	915
16–20	Rural	4	5	386	31–35	Rural	2	—	64
	Urban	3	4	5080		Urban	10	15	453
21–25	Rural	3	3	229					
	Urban	5	7	2484					

Table 140. Accumulative Incidence: Homosexual Contacts to Orgasm
By Religious Background

AGE	PROTESTANT			CATHOLIC			JEWISH	
	Dev.	Moder.	Inact.	Dev.	Moder.	Inact.	Moder.	Inact.
	Percent							
12	1	1	1	1	1	2	—	—
15	2	1	2	3	4	8	1	2
20	3	2	7	5	4	13	2	5
25	5	5	12	5	8	22	2	7
30	7	6	16	3	12	28	3	9
35	7	9	17	5		25	4	11
40	8	10	16					13
45	11	13	20					15
	Cases							
12	1234	1166	1090	390	153	171	575	985
15	1217	1156	1083	385	151	171	574	981
20	922	923	976	284	114	146	334	692
25	566	587	739	177	76	107	187	413
30	424	435	552	124	51	79	136	294
35	337	315	419	78		52	77	205
40	237	212	274					127
45	150	132	153					74

Table based on total sample, including single, married, and previously married females.

Table 141. Active Incidence, Frequency, and Percentage of Outlet

Homosexual Contacts to Orgasm

Single Females, by Religious Background

Age during activity	Religious group	Active incid. %	Active median freq. per wk.	% of total outlet	Cases in total sample
Adol.–15	Protestant				
	Devout	2	0.3	6	1218
	Moderate	1	0.4	4	1147
	Inactive	2	0.2	2	1063
	Catholic				
	Devout	3	0.3	13	382
	Moderate	3		6	150
	Inactive	8	0.4	6	169
	Jewish				
	Devout	0		0	107
	Moderate	1		—	571
	Inactive	2	0.1	2	978
16–20	Protestant				
	Devout	3	0.2	6	1197
	Moderate	2	0.2	2	1133
	Inactive	5	0.2	5	1065
	Catholic				
	Devout	5	0.6	15	372
	Moderate	4		6	139
	Inactive	10	0.7	8	160
	Jewish				
	Devout	1		—	107
	Moderate	—		—	571
	Inactive	2	0.2	2	972
21–25	Protestant				
	Devout	3	0.3	10	604
	Moderate	3	0.4	4	615
	Inactive	8	0.4	8	676
	Catholic				
	Devout	4		15	196
	Moderate	5		12	57
	Inactive	18	0.5	13	91
	Jewish				
	Moderate	0		0	192
	Inactive	4	0.3	6	396
26–30	Protestant				
	Devout	5	0.4	17	221
	Moderate	4	0.4	4	249
	Inactive	13	0.5	13	309
	Catholic				
	Devout	5		29	79
31–35	Protestant				
	Devout	7		34	121
	Moderate	6		2	127
	Inactive	15	0.4	11	159
36–40	Protestant				
	Devout	7		37	76
	Moderate	6		2	77
	Inactive	16	0.5	18	102

Table 142. Active Incidence: Heterosexual-Homosexual Ratings By Marital Status

AGE	0	1–6	1	2	RATING 3	4	5	6	X	Cases

SINGLE FEMALES

AGE	Percent 0	1–6	1	2	Percent 3	4	5	6	%	Cases
5	7	6	—	—	2	—	—	3	87	5914
10	7	11	—	1	2	—	—	8	82	5820
15	34	6	2	1	1	—	—	2	60	5714
20	72	11	5	2	1	1	1	1	17	3746
25	72	14	4	3	1	3	1	2	14	1315
30	67	18	5	4	2	3	2	2	15	622
35	61	20	6	3	3	2	3	3	19	370
37	61	17	5	3	2	2	2	3	22	290

MARRIED FEMALES

AGE	Percent 0	1–6	1	2	Percent 3	4	5	6	%	Cases
17	80	11	9	1	0	0	0	1	9	89
20	89	8	5	1	1	—	—	—	3	545
25	90	8	6	1	—	—	—	—	2	1331
30	90	9	6	2	—	—	—	—	1	1215
35	89	10	7	2	—	—	—	—	1	908
40	89	9	6	2	—	0	—	—	2	569
45	89	9	6	2	—	0	—	1	2	311
50	88	8	4	3	0	0	1	0	4	154

PREVIOUSLY MARRIED FEMALES

AGE	Percent 0	1–6	1	2	Percent 3	4	5	6	%	Cases
20	80	14	6	1	3	3	0	1	6	81
25	75	17	7	3	0	1	3	3	8	178
30	78	14	6	2	1	2	1	2	8	224
35	78	17	9	3	0	1	2	2	5	204
40	76	14	8	2	2	0	1	1	10	177

Definitions of the ratings are as follows: 0 = entirely heterosexual. 1–6 = with homosexual history, of any sort. 1 = largely heterosexual, but with incidental homosexual history. 2 = largely heterosexual, but with a distinct homosexual history. 3 = equally heterosexual and homosexual. 4 = largely homosexual, but with distinct heterosexual history. 5 = largely homosexual, but with incidental heterosexual history. 6 = entirely homosexual. X = without either. See p. 471.

Table 143. Active Incidence: Females with Some Homosexual Rating
By Educational Level and Marital Status

AGE	TOTAL SAMPLE	EDUCATIONAL LEVEL			TOTAL SAMPLE	EDUCATIONAL LEVEL		
		9–12	13–16	17+		9–12	13–16	17+
				SINGLE FEMALES				
	%		*Percent*		*Cases*		*Cases*	
5	6	5	6	8	5914	1014	3302	1152
10	11	12	11	12	5820	1014	3301	1151
15	6	6	5	8	5714	1005	3299	1150
20	11	7	10	13	3746	641	1943	1079
25	14	8	14	17	1315	229	423	621
30	18	8	17	21	622	85	167	343
35	20		28	20	370		82	237
				MARRIED FEMALES				
	%		*Percent*		*Cases*		*Cases*	
20	8	7	9	11	545	194	253	63
25	8	5	10	11	1331	408	563	296
30	9	6	10	11	1215	342	475	331
35	10	6	11	13	908	235	346	271
40	9	4	12	10	569	139	204	188
45	9	4	10	13	311	83	111	93
				PREVIOUSLY MARRIED FEMALES				
	%		*Percent*		*Cases*		*Cases*	
25	17	11	18		178	64	65	
30	14	14	12	16	224	76	73	61
35	17	16	20	17	204	55	72	60
40	14		13	18	177		61	62

The table includes all females with heterosexual-homosexual ratings of 1 to 6. These females had had some homosexual history, either psychologic or overt, in the particular year shown. Those with a rating of 1 had minimum homosexual histories; those with a rating of 6 had the maximum and therefore exclusive homosexual histories. See Table 142.

Table 144. Attitudes of Females Toward Homosexual Activity

Correlation of attitudes with subject's own homosexual experience

ATTITUDES	ACCEPT FOR SELF		APPROVE HOMOSEXUAL ACTIVITY				WOULD KEEP FRIENDS WHO HAD HOMOSEXUAL EXPERIENCE			
			For other females		For males		Female friends		Male friends	
	Subject		Subject		Subject		Subject		Subject	
	With exp.	No exp.	With exp.	No exp.	With exp.	No exp.	With exp.	No exp.	With exp.	No exp.
	%	%	%	%	%	%	%	%	%	%
Yes	18	1	23	4	18	4	88	55	74	51
Uncertain	20	4	62	57	60	54	8	23	16	23
No	62	95	15	39	22	42	4	22	10	26
Number of cases	683	4500	653	4758	616	4718	251	941	122	935

Table 145. Social Difficulties Resulting from Homosexual Experience

Source of difficulty	Total cases	Cases of major difficulty	Cases of minor difficulty
Home	21	13	8
School	8	4	4
Business	6	2	4
Police	4	1	3
Institutional	3	1	2
No. of cases of difficulties	42	21	21
No. of females with diffic.	38		
No. of females without diffic.	104		
% of females with diffic.	27		

There were 710 females in the sample with homosexual experience, but only the 142 with the most extensive experience are included in these calculations. None of those with more minor experience had run into such social difficulties.

Chapter 12

ANIMAL CONTACTS

Universally, human males have shown a considerable interest in unusual, rare, and sometimes fantastically impossible types of sexual activity. In consequence there is a great deal more discussion and a more
extensive literature about such things as incest, transvestism, necro-
philia, extreme forms of fetishism, sado-masochism, and animal contacts
than the actual occurrence of any of these phenomena would justify.

From the earliest recorded history, and from the still more ancient
archives of folklore and mythology, there are man-made tales of sexual
relations between the human female and no end of other species of
animals. The mythology of primitive, pre-literate peoples in every part
of the world has included such tales.[1] Classic Greek and Roman mythology had accounts of lovers appearing as asses, Zeus appearing as a
swan, females having sexual relations with bears, apes, bulls, goats,
horses, ponies, wolves, snakes, crocodiles, and still lower vertebrates.
The literary and artistic efforts of more recent centuries have never
abandoned these themes; erotic literature and drawings, including some
of the world's great art, have repeatedly come back to the same idea.[2]

Much of this interest in rare or non-existent forms of sexual performance may represent the male's wishful thinking, a projection of his own
desire to engage in a variety of sexual activities, or his erotic response
to the idea that other persons, especially females, may be involved in
such activities. This stems from the male's capacity to be aroused
erotically by a variety of psychosexual stimuli (Chapter 16). Females,
because of their lesser dependence on psychologic stimulation, are less
inclined to be interested in activities which lie beyond the immediately
available techniques, and rarely, either in their conversation, in their
written literature, or in their art, deal with fantastic or impossible sorts
of sexual activity. Human males, and not the females themselves, are
the ones who imagine that females are frequently involved in sexual
contacts with animals of other species. In fact, human males may be

---

[1] Folk tales and myths of human females in contact with male animals are summarized in: Dubois-Desaulle 1933:31–47. Ford 1945:31. Leach and Fried
1949(1):61.

[2] As examples of the persistence of this theme into more modern life, note the hundreds of representations of "Leda and the Swan," the magazine cartoons showing a female abducted by an ape, the still-current "Prince Charming" nursery
tale, and motion pictures of gorillas interested in human females.

responsible for initiating some of the animal contacts and especially the exhibitionistic contacts in which some females (particularly prostitutes) engage.[3]

Considerable confusion has been introduced into our thinking by this failure to distinguish between sexual activities that are frequent and a fundamental part of the pattern of behavior, and sexual activities which are rare and of significance only to a limited number of persons. Psychologic and psychiatric texts are as likely to give as much space to overt sado-masochistic or necrophilic activity as they give to homosexual and mouth-genital activities, but the last two are widespread and significant parts of the lives of many females and males, while many of the other types of behavior are in actuality rare.

BASES OF INTER-SPECIFIC SEXUAL CONTACTS

As we have already seen, males may be more often involved than females in a variety of non-coital sexual activities. Sexual contacts between the human male and animals of other species are not rare in the rural segments of our American population, and probably not infrequent in other parts of the world. Some 17 per cent of the farm boys in our sample had had some sexual contact with farm animals to the point of orgasm, while half or more of the boys from certain rural areas of the United States had had such experience (see our 1948:671–673). It will be profitable to try to analyze the factors which account for the lesser frequencies of animal contacts among the females in the present sample.

In discussing this matter in our volume on the male (1948:667–668), we pointed out that there is no sufficient explanation, either in biologic or psychologic science, for the confinement of sexual activity to contacts between females and males of the same species. We have no sufficient knowledge to explain why an insect of one species should not mate or attempt to mate with many other species, why different species of birds do not indiscriminately interbreed, or why any species of mammal should confine its sexual activity as often as it does. There are obvious anatomic problems which prevent the indiscriminate, inter-specific mating of some forms, but no known anatomic or psychologic factors which would prevent most of the more closely related species from trying to make inter-specific matings.[4] We have also pointed out

---

[3] The observation of exhibitions of coitus between prostitutes and animals is frequently recorded in our male histories; they are also mentioned in: Kisch 1907:201. Bloch 1908:644, 646. Krafft-Ebing 1922:562. Rohleder 1925:370. Kelly 1930:184–185. Robinson 1936:46. Negri 1949:217. London and Caprio 1950:21–22.

[4] But records of sexual activity attempted between animals of gross morphologic disparity are in: Karsch 1900:129 (female eland with ostrich). Féré 1904:79–80 (male dog with chicken). Hamilton 1914:308 (male monkey with snake). Zell 1921(1):238 (stallion with human). Bingham 1928:71–72 (female chimpanzee with cat). Williams 1943:445 (cow with human).

that evidence is beginning to accumulate that individuals of quite un-
related species do make inter-specific contacts more often than biologists have heretofore allowed. The intensive study of the movements of
individual birds, which bird-banding techniques have made possible,
has shown that there is a great deal more inter-specific mating among
birds than we have previously realized. More intensive taxonomic and
genetic work in the field has shown the existence of a large number of
inter-specific hybrids, and these provide evidence that inter-specific
mating occurred at some time or other and, more than that, that such
matings were viable and gave rise to fertile offspring. The successful
matings must represent only a small proportion of the inter-specific
contacts which are actually attempted or made.

It is not a problem of explaining why individuals of different species
should be attracted to each other sexually. The real problem lies in
explaining why individuals do not regularly make contacts with species
other than their own. In actuality, it is probable that the human animal
makes inter-specific sexual contacts less often than some of the other
species of mammals, primarily because he has no close relative among
the other mammals, and secondarily because of the considerable significance which psychologic stimuli have in limiting his sexual activity.

We have previously pointed out (1948:675–676) that the farm boy
may begin his sexual contacts with animals because he responds sympathetically upon observing their sexual activities. With the mating
animals he can, to a considerable degree, identify his own anatomic
and physiologic capacities. Moreover, the boy may come into contact
with freer discussions of sex at an earlier age than most boys who are
not raised on farms, and in many instances he has an example set for
him by other boys whom he discovers having sexual contacts with the
farm animals. Not infrequently he hears adults in the community discuss such matters. The comments are usually bantering and not too
severely condemnatory.

But none of these factors are of equal significance to the female. At
earlier ages, girls do not discuss sexual activities as freely or as frequently as boys do (p. 675), and they less often observe sexual activity
among other girls or even among the farm animals. The specific record
(p. 663) shows that some 32 per cent of the adult males in the sample
had been erotically aroused when they saw animals in coitus, while
only 16 per cent of the females had been so aroused. The histories indicate that many of the farm-bred females had been oblivious to the
coital activities which went on about them. Quite frequently they had
been kept away from breeding animals by their parents, and we find
that a good many of the rural females in the sample had not learned

that coitus was possible in any animal, let alone the human, until they were adolescent or still older. As a result, the animal contacts which the females had made were usually the consequence of their own discovery of such possibilities, whether the first experiences were had in pre-adolescence or in more adult years. Most of the farm boys had acquired that much information some years before adolescence.

It is not surprising then, to find that the incidences and frequencies of the animal contacts made by the females in the sample were much lower than the incidences and frequencies which we found among the males in the sample.[5]

INCIDENCES AND FREQUENCIES

In Pre-Adolescence. A few of the females—1.5 per cent of the total sample—had had some sort of sexual relation with other animals in pre-adolescence, usually as a result of some accidental physical contact with the household pet, a cat or a dog, or as a result of curiosity which had led to the exploration of the animal's anatomy, or through some deliberate approach on the part of the animal itself. Among the pre-adolescent females who had had any such experience, the contacts were incidental in 38 per cent of the cases. In most of the pre-adolescent cases (92 per cent), however, the girl had had contacts which had aroused her erotically, and in 20 per cent of these pre-adolescent cases she had reached orgasm. Among the 659 females in the total sample who had reached orgasm prior to adolescence, 1.7 per cent had experienced their first orgasm in contact with other species of animals.

Among the 89 females who had had pre-adolescent animal experience, general body contacts and masturbation of the animal had been involved in most cases. But out of the 5940 females in the sample, 23 had had dogs put their mouths on their genitalia, 6 had had cats similarly perform, and 2 had had coitus with dogs.

Among Adult Females. Some 3.6 per cent of the females in the sample had had sexual contacts of some sort with animals of other species after they had become adolescent. Some 3.0 per cent of the females in the total sample had been erotically aroused by their animal contacts. In only 1.2 per cent of the total sample had there been repeated genital contacts which aroused the female erotically, or mouth-genital relations or actual coitus. This means that only one female in each eighty in the sample had had such specific sexual contacts with animals after the onset of adolescence. In addition to the overt experience with animals,

---

[5] A number of other authors have also recognized that females make animal contacts less often than males. See, for instance: Casañ ca.1900(4):67. Bloch 1908:642; 1909:704. Kelly 1930:183. Dubois-Desaulle 1933:143. Chideckel 1935:312.

1 per cent of the females had fantasied such contacts while they masturbated (Table 38), and 1 per cent had dreamt of having animal contacts (Table 55).

Half of the females had had their contacts with animals after the onset of adolescence and before the age of twenty-one, but there were 95 older females who had had such contacts, in some instances even in their late forties. Most of the contacts had occurred among single females, although there were 44 cases among married or previously married females.[6] Contacts had occurred among females of every educational level, although most of them (81 per cent) had occurred in the better educated segments of the sample, largely because that group is better represented in the sample.

Nearly all of the contacts had occurred with dogs or cats which were household pets. Nearly three-quarters (74 per cent) of the females had had contacts with dogs.[7] Over half of the relationships had involved only general body contacts with the animal. In some instances, the females had only touched the animal's genitalia; in other instances, there had been more specific masturbation of the animal. For some 21 per cent of the females, the animal had manipulated the human genitalia with its mouth, but in only one of the adult cases had there been actual coitus with the animal. There were, however, additional cases of coitus in other segments of the female sample which were not utilized in the calculations for the present volume (p. 22).

In 25 out of the 5793 adult histories on which we have data concerning animal contacts, the human female had been brought to orgasm by her sexual contacts with the animal, chiefly as a result of the animal's manipulation of her genitalia with its mouth.

The frequencies of the animal contacts had been low, amounting to only a single experience in about half of the cases. In 47 per cent of the 91 cases which had involved the most specific sort of contact there had been two or more experiences, and in 23 per cent of the cases there had been six or more contacts. There were only 13 females in the sample of 5793 who had reached orgasm by contact with an animal of another species more than three times up to the age at which they contributed their records to the study. There were 6 females, each of whom

---

[6] Talmey 1910:162 notes that animal contacts are commoner among unmarried females. Bloch 1933:185 feels that the married and unmarried are equal in this respect.

[7] That the majority of female contacts with animals are had with dogs is also noted in: Mantegazza 1885:128–131. Moraglia 1897:6. Féré 1904:184. Havelock Ellis 1906(5):83. Hoyer 1929:252. Kelly 1930:184. Chideckel 1935:315. Haire 1937:484. Hirschfeld 1940:138.

had reached orgasm more than 125 times in her animal contacts, and there was one female who had reached orgasm perhaps 900 times in such contacts.

SIGNIFICANCE OF ANIMAL CONTACTS

The incidences and frequencies and significance of animal contacts as a source of outlet for the human female are obviously a minute fraction of what most human males have guessed them to be. The present data consequently illuminate some of the basic differences between the sexual psychology of the human female and male, and show something of the effects that such differences in psychology may have on the overt behavior of the two sexes.

In ancient codes and laws, there were frequent references to human males having animal contacts, and judgments and penalties were prescribed for such activity. The more ancient codes, however, appear to have ignored the possibilities of females having sexual contacts with animals,[8] and there are apparently only two references, both of them in Leviticus (parts of which represent a later development in Biblical law), concerning females who have sexual contacts with animals.[9] The Biblical references involve the prohibition of such acts, and demand death as the penalty for both the female and the animal. The Talmud, however, makes more frequent reference to such female activity, repeating the Biblical injunctions against it and imposing the same penalties. Finally the Talmud goes so far as to prohibit a female being alone with an animal because of the possible suspicion that she might have sexual contact with it, and this is unusual because it gives the matter more attention than is ordinarily given it in any of the other codes.[10]

The Catholic code on animal contacts logically follows the general concept that sexual function is justified only as a means of procreation, primarily in marriage, and all contacts between the human female or male and an animal of another species are consequently contrary to nature, a perversion of the primary function of sex, and sinful in deed or desire. The judgment would appear to apply to female as well as to

---

[8] See: Pritchard 1950:196–197. The code of Lipit-Ishtar, the code of Hammurabi, the Middle Assyrian Laws, and the Neo-Babylonian Laws contain no references to animal contacts, but the Hittite Laws have five references to male contacts with animals. Since the Hittite Laws are in essential agreement with other Near-Eastern codes, it may well be that the makers of these codes held the same opinion as the Hittites in regard to animal contacts. It must be recalled, however, that some of these codes are not known *in toto*.

[9] The injunction against and penalty for contact between a human female and an animal is in Leviticus 18:23 and 20:16. Also Exodus 22:19 might be considered as applying to both sexes.

[10] The Talmudic references to female contacts with animals are in: Kethuboth 65a, Yebamoth 59b, Sanhedrin 2a, 15a, 53a, and 55a, Abodah Zarah 22b–23a.

male contacts.[11] Touching the genitalia of an animal even out of curiosity may be a sin, and touching it with lust may be a grave sin.[12] The opinion is expressed that experience in animal contacts might be sufficient grounds for a separation.[13]

The legal codes of essentially all of the states prohibit sexual relations between the human animal and animals of other species, usually rating them as bestiality or sodomy, and usually attaching the same penalties that are attached to homosexual relations. In a few instances the penalties are lower than those for homosexual relations; in some instances they are very severe.[14] When there is no specific statute covering the matter, the common law ruling against bestiality would sometimes apply.[15] In some instances the statutes specifically indicate that they are applicable to both females and males.[16] In many instances they do not specifically designate the sex to which they apply, but in most such cases they would be interpreted to cover both sexes. It is probable, however, that the lawmakers in most instances had male activity in mind when they framed their statutes; and the question is quite academic, for cases of females who have been prosecuted for animal contacts are practically unknown in the legal record.

There are in the older literature a few records of females receiving the death penalty for such contacts, particularly in medieval history.[17]

---

[11] Catholic interpretations of animal contacts are in: Arregui 1927:153–154, and Davis 1946(2):247, who specifically include the female. The "penitentials," partially secular and partially religious codes dating before the 13th century, occasionally refer to contacts between human females and animals. According to Havelock Ellis 1906(5):87–88, Burchard's penitential stipulates a seven-year penance for a female who has had sexual contact with a horse.

[12] Touching an animal's genitalia may vary from a light to a grave sin depending on the motivation, according to Davis 1946(2):249. Arregui 1927:156 adds that when such touching is necessary, as in animal breeding, it is best that it be done by older or married persons.

[13] Using the term "divorce" to mean permanent separation with "the conjugal bond remaining," Noldin 1904(3):sec. 665 states: According to the probable opinion of learned men any alien sexual intercourse, even that which happens through sodomy or bestiality, suffices for instituting a divorce.

[14] Forty-four states specifically forbid sex relations with animals, and cases in three of the remaining (Ark., Del., and Vt.) indicate it is a crime, leaving only New Hampshire where such activity is not a felony or its equivalent. In Georgia up until 1949 the minimum penalty for the crime against nature when committed with another human was "imprisonment at labor in the penitentiary for and during the natural life of the person convicted," whereas the penalty for "bestiality" was five to twenty years imprisonment. In eight states (Calif., Colo., Ida., Mo., Mont., Nev., N. M., and S. C.) the possible maximum sentence is life imprisonment.

[15] For the application of the common law ruling, see: State v. LaForrest 1899:45 Atl. (Vt.) 225.

[16] For the application of the statutes to both female and male, see, for example: Georgia Code 1933:Title 26 §5903. Maryland Code 1951:Article 27 §627.

[17] The death penalty for females making contacts with animals in medieval times is noted in: Mantegazza 1885:128–131. Havelock Ellis 1906(5):88. Hernandez [Fleuret and Perceau] 1920:83–94. Dubois-Desaulle 1933:58, 81–89. Robinson 1936:42–44.

We do not have any instance of legal action against any of the cases in our sample, and we find only one case in the published court records here in the United States.[18]

Considering the rarity of sexual contacts between females and animals of other species, it is interesting to find specific recognition of such contacts in the moral and legal codes.

SUMMARY AND COMPARISONS OF FEMALE AND MALE
ANIMAL CONTACTS

	IN FEMALES	IN MALES
Among mammals, inter-specific contacts common	Yes	Yes
Anthropologic Background		
Animal contacts in other cultures	Rare in all	Occasional in some
Animal contacts in myth and folklore	Commonly	Less often
Incidence and Frequency		
Accumulative incid. in pre-adol.	1.5%	3%
Erotic response	1.4%	
Orgasm	0.3%	Rare
As source of first orgasm	Rare	Rare
Accumulative incid. in adult	3.6%	8% to orgasm
Erotic response	3.0%	
Orgasm	0.4%	8%
Primarily before age 21	50%	Yes
Frequency, active sample, bf. age 21	Usually 1–2 times	0.1 per wk.
Animals chiefly involved	Dog, cat	Farm animals, pets
Techniques		
General body contact	Common	Some
Masturbation of animal	Some	Common
Anim. mouth on human genit.	Some	Some
Coitus	Very rare	Common
Social Significance		
As a source of outlet	Insignif.	1%
Legal and religious injunctions		
In Hammurabi's code	No	No
In Hittite code	No	Yes
In Old Testament	No	Yes
In Talmud	Yes	Yes
In American statute law	Yes	Yes

[18] The only published case in the U.S. of a female convicted because she had had contact with an animal is in: State v. Tarrant 1949:80 N.E.2d (Ohio) 509.

Chapter 13

TOTAL SEXUAL OUTLET

In the present study, we have tried to secure data on (1) the incidences and frequencies of sexual activities among the females in the available sample; (2) the incidences and frequencies of their responses to socio-sexual contacts and to psychosexual stimuli; and (3) the incidences and frequencies of the responses which led to orgasm.

From most of the subjects it has been possible to secure incidence data on the overt, physical contacts which were recognizably sexual because they were genital or because they brought specific erotic response. From most of the subjects it has also been possible to secure frequency data on most of those contacts, but this has not always been possible because there are situations in which the genital anatomy is not involved, and then it is sometimes difficult to determine whether the contacts or emotional responses are sexual in any real sense of the term (Chapter 15). It has been difficult, for instance, to secure exact data on the incidences and frequencies of self-stimulation which was non-genital, on the frequencies of sexual dreams which did not lead to orgasm, and on the incidences and frequencies of the non-genital socio-sexual contacts. As we have already pointed out (p. 235), there is every gradation between a simple good night kiss or a friendly embrace, and a kiss or an embrace which is definitely sexual in its intent and consequences.

But whenever physical contacts or psychologic stimuli had led to orgasm, there was rarely any doubt of the sexual nature of the situation, and it has in consequence been possible to secure incidence and frequency data which were as reliable as the interview technique would allow. For these reasons, the statistical data in the present volume, just as in our volume on the male, have been largely concerned with the incidences and frequencies of sexual activities that led to orgasm. The procedure may have overemphasized the importance of orgasm, but it would have been impossible in any large-scale survey to have secured as precise records on some of the other, less certainly identifiable aspects of sexual behavior.

For these same reasons, we have defined the total sexual outlet of an individual as the sum of the orgasms derived from the various types

510

of sexual activity in which that individual had engaged. Since all sexual responses, whether they are the product of psychologic stimulation or of some physical contact, may involve some sort of physiologic change (Chapter 15) and therefore some expenditure of energy, all sexual responses might be considered a part of the individual's total sexual outlet; but the term *total outlet* as we have used it has covered only those contacts and/or responses which had led to orgasm.

There is, moreover, a reality involved in any such summation of orgasms, for all orgasms appear to be physiologically similar quantities, whether they are derived from masturbatory, heterosexual, homosexual, or other sorts of activity (Chapters 14–15). For most females and males, there appear to be basic physiologic needs which are satisfied by sexual orgasm, whatever the source, and the sum total of such orgasms may constitute a significant entity in the life of an individual.[1]

It is, of course, true that sexual experience may have a significance which lies beyond the physiologic release that it provides, and each type of sexual experience may have its own peculiar significance. For instance, many persons will consider the psychologic significance of an orgasm attained in masturbation very different from the psychologic significance of an orgasm derived from a socio-sexual source. The social significance of orgasms attained in non-marital coitus may be different from those attained in marital coitus. In many ways, it may be more significant to know the frequencies and incidences of the particular types of sexual activity (Chapters 5–12) than it is to total them as we do in the present chapter. But social interests are still involved when an individual finds satisfaction for a physiologic and psychologic need; and the data on total sexual outlets may, therefore, deserve as much consideration as any of the data on particular types of sexual activity.

DEVELOPMENT OF SEXUAL RESPONSIVENESS

As we have previously indicated, specifically erotic responses and sometimes orgasm may be observed in very young infants, both female and male (Chapter 4, and our 1948:175). Among infant females, the incidences of response and completed orgasm were about as high as they were among infant males; but the number of males who had responded sexually had gradually and steadily increased through the early pre-adolescent years, and then had risen abruptly in the later pre-adolescent years. On the other hand, the number of females who had been aroused erotically appears to have increased somewhat more

---

[1] The concept of a total sexual outlet appears not to have been employed in other studies. The nearest approximation seems to be in Davis 1929:233, who classifies her sample on the basis of the number of sources of outlet which they were utilizing.

gradually through the pre-adolescent and adolescent years (Table 146, Figure 98).

First Erotic Response in the Female. The average female in the available sample had begun to turn adolescent by twelve years and four months of age (p. 123). By that age *about* 30 per cent had been aroused erotically by some sort of psychologic stimulation or physical contact, and under conditions which they subsequently recalled as definitely sexual (Table 146, Figure 98). First menstruation for the average fe-

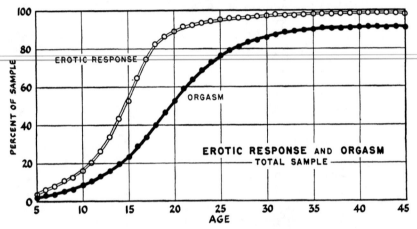

Figure 98. Accumulative incidence: erotic response and orgasm from any source

Data from Tables 146 and 147.

male in the sample had come just as she was turning thirteen (p. 123); and by that age about a third (34 per cent) of the sample had recognized some sort of erotic arousal. By fifteen years of age, half (53 per cent) of the females had been aroused erotically in some form or fashion. By twenty years of age, 89 per cent had been so aroused, but there were some who had not recognized their first arousal until some later age.

Ultimately 98 per cent of the females in the sample had had at least one experience in which they recognized arousal, but even in the late forties there were still some 2 per cent who had never recognized any sexual arousal, under any sort of condition (Table 146, Figure 98). These females reported that they had never been aroused by self-stimulation of their own genitalia or of any other part of their bodies, they had not been aroused by thinking of sexual situations or by dreaming of them at night, and they had not been aroused by any other sort of psychologic stimulation. Neither had they been aroused by physical contacts with other persons, and in most instances they had never had any contacts which could be identified as sexual. It is, of course, not impossible that some of them had reacted erotically without being aware of the nature of their emotional responses; and it is possible that some other method of gathering data, or specific physiologic measurements, might have shown that some of these females had, on occasion, responded to erotic stimuli. On the other hand, their responses must have been so mild, infrequent, or non-specific that it would have been difficult to have identified them as sexual.[2]

Sources of First Arousal. A third of the females (30 to 34 per cent) had first been aroused erotically in heterosexual petting (Table 148). Nearly as many (27 to 30 per cent) had first become aware of the meaning of erotic arousal as a result of their own self-masturbation. Another third (30 to 32 per cent) had first been aroused through psychologic stimulation, chiefly in connection with their social contacts with male friends. Nocturnal dreams, pre-marital coitus, marital coitus, and homosexual and animal contacts had been only minor sources of first arousal.

The sources of first arousal had remained remarkably constant through the four decades represented in the sample (Table 148). They had, however, been more affected by the age at which the female had married: 37 per cent of the females who married between the ages of sixteen and twenty had been aroused first in petting, in contrast to 28 per cent of those who married after age thirty. Those who married at later ages had more often found their first arousal in masturbation or psychologic stimulation. There may have been some causal relationship between the occurrence of the first arousal in heterosexual contacts and the earlier age of marriage among those females.

Age at First Orgasm in the Female. The percentages of females who had experienced orgasm (the accumulative incidences) had risen steadily during pre-adolescence, but they were still relatively low (about 14 per cent) at the onset of adolescence (Table 147, Figure 98). There was some increase after the onset of adolescence, but less than a quarter (23 per cent) of the sample had had such experience by fifteen years of age. A little more than a half (53 per cent) had had orgasm by twenty, three-quarters (77 per cent) by twenty-five, and about 90 per cent by thirty-five years of age. The accumulative incidence curves had leveled off at that point, and there appear to be some 9 per cent who would probably not reach orgasm in the course of their lives. Some 64 per cent of the females in the sample had experienced their first orgasm prior to marriage (Table 150). About 33

---

[2] Discussions of females who had never been erotically aroused are also in: Dickinson and Beam 1931:128 (found no such cases among 100 frigid wives). Landis and Bolles 1942:96 (cases among physically handicapped females).

per cent of the females who had married had experienced their first orgasm after they were married.

The females who had never reached orgasm had, in many instances, been aroused erotically, but none of them had recognized the high levels of tension which ordinarily precede orgasm, the sudden release which is orgasm, and the physical relaxation which follows orgasm (Chapter 15). Most of them had had socio-sexual contacts, and some of them were married and had had coitus regularly for some period of years—but none of them seem ever to have experienced orgasm. For a further discussion of this so-called frigidity, see pages 373 ff.

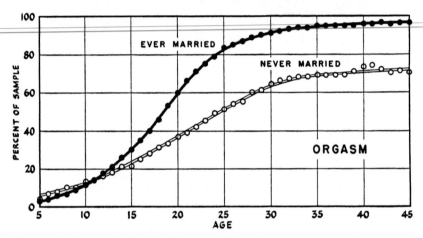

Figure 99. Accumulative incidence: orgasm from any source, by marital status

The second curve is based on those who had never married and were age forty or over at time of interview. Data from Table 150.

Factors Affecting Age at First Orgasm. The age of the female at marriage had been the prime factor affecting the age at which she first experienced orgasm, primarily because marital coitus had provided the first orgasmic experience for most of those who had not had such experience prior to marriage. Before marriage, and until the age of fifteen, there were few differences in the several groups. At fifteen, only 31 per cent of the females who were to marry in the next five years had experienced orgasm, but by the age of twenty, 82 per cent of that group had responded to orgasm, chiefly after marriage (Table 151, Figure 101). Similarly, among those who married at later ages, there were rather definite rises in the curves just before or after the age of marriage.

There seem to have been no significant differences in the accumulative incidence curves for the first experience in orgasm, between the females of the high school, college, and graduate school groups; but the curve for the females who had never gone beyond grade school was definitely lower at nearly every point (Table 147). In the younger age groups, more of the females who were raised in parental homes of the lower occupational classes had experienced orgasm, fewer of the girls who were raised in upper white collar homes (Table 152).

Before the age of twenty, there were no significant differences in the percentages reaching orgasm among the females who were born in the several decades covered by the sample (Table 157, Figure 103). After the age of twenty, the generation which was born before 1900 seems to have included a larger number of females who were slower in reaching their first orgasm and a larger number who had never responded to orgasm, even by the age of forty-five. Some 5 to 10 per cent more of the females who were born after 1900 had ultimately experienced orgasm, and at an earlier age. The females of the generation born after 1920 seem to have been somewhat slower in having experience prior to the age of eighteen, but after that a larger proportion of the group had experienced orgasm, and experienced it at an earlier age than the females born in the previous decades (p. 522). The present generation will probably complete its history with a much smaller percentage of females who remain totally inexperienced in orgasm throughout their lives.

The accumulative incidence data do not indicate that there was any correlation between the ages of the females when they first responded to orgasm, and the ages at which they had turned adolescent (Table 159).

The accumulative incidence curves for first orgasm were generally lower for the rural sample and higher for the urban females (Table 162); but the differences were not great, and the rural sample is not large enough to allow any final conclusions.

The religious factors were of considerable importance in determining the ages at which the females in the sample had first responded to orgasm. For instance, the females who were devoutly Catholic had, on an average, not reached their first orgasm until some six or seven years after the females who were only nominally Catholic (Table 164, Figure 107). There were 21 per cent of the devoutly Catholic females who had not reached orgasm by thirty-five years of age, even though most of them were then married and regularly having coitus in their marriages. It was not more than 2 per cent of the nominal, non-religious Catholics who had not reached orgasm by that age.

The differences were not so extreme between the religiously active and inactive Protestants, but they lay in the same direction (Table 164,

Figure 106). For instance, at forty-five years of age there were still 15 per cent of the devout Protestant females who had never experienced orgasm in their lives, but only 5 per cent of the inactive Protestants who belonged in that category. As far as the more limited record goes, the differences between the religiously devout and inactive Jewish groups appear to have been of about the same order (Table 164). There seems no doubt that the moral restraints which lead a female to avoid sexual contacts before marriage, and to inhibit her responses when she does make contacts, may also affect her capacity to respond erotically later in her life. We shall not solve the problem of female "frigidity" until we realize that it is a man-made situation, and not the product of innate physiologic incapacities in those females.

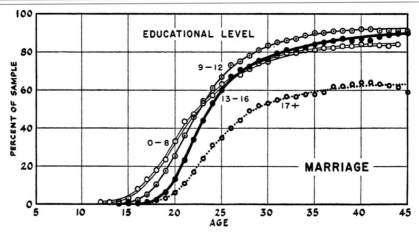

Figure 100. Accumulative incidence: age at first marriage, by educational level

Data from Table 166.

Sources of First Orgasm. If we base our calculations on the females in the sample who had been married, and who therefore had had a maximum opportunity to experience orgasm, the sources of such experience were as follows (Table 148): Some 37 per cent had reached their first orgasm in masturbation. Some 18 per cent had reached their first orgasm in pre-marital petting, and 30 per cent in coitus after marriage. Only smaller percentages had reached their first orgasm in nocturnal dreams, pre-marital coitus, homosexual relations, animal contacts, or psychologic stimulation.

The source of first orgasm had depended somewhat on the age of the female at the time she married (Table 148). Masturbation had been the source of first orgasm for some 32 per cent of the females who married between the ages of sixteen and twenty, but for 41 per cent of those who had married after the age of twenty-five. The first orgasm had been reached in marital coitus among 33 per cent of those who were married between the ages of sixteen and twenty, but among only 21 per cent of those who had been married between thirty-one and thirty-five. There were essentially no differences between the educational levels when the comparisons were made for females who had married at the same age.

The sources of first orgasm had differed somewhat among the females who were born in the different decades represented in the sample (Table 148). There had been no essential differences between the generations when the first orgasms were derived from masturbation;

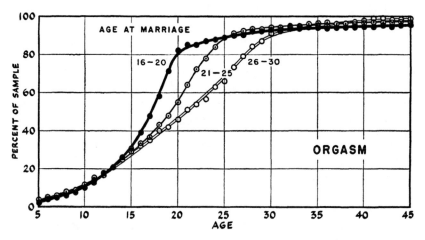

Figure 101. **Accumulative incidence: orgasm from any source, by age at first marriage**

Data from Table 150.

but when the first orgasm had come from pre-marital petting, 10 per cent of those born before 1900, but 25 per cent of those born between 1920 and 1929, had been involved. From pre-marital coitus, some 3 per cent of the older generation, but 10 per cent of the generation born between 1920 and 1929, had derived their first orgasm. On the other hand, the first orgasm had come from coitus after marriage among 37 per cent of the older generation, but among only 26 per cent of the group born between 1920 and 1929. These figures are an interesting reflection of changes in the patterns of female behavior during the four decades. In the younger generations, there had been a rise in the importance of pre-marital petting and pre-marital coitus as sources of first orgasm, and some drop in the importance of masturbation. There had been a marked decrease in the number of females who had waited until after marriage to secure their first experience in orgasm.

SINGLE FEMALES: TOTAL OUTLET

Accumulative Incidence. Among the unmarried females in the sample, the accumulative incidence curve for those who had ever responded to orgasm had included 14 per cent by twelve years and four months of age, which was the average age at the onset of adolescence. It had risen to 23 per cent by age fifteen, and to 49 per cent by age twenty (Table 149). It had approached its ultimate level of about 75 per cent among the still unmarried females near thirty-five years of age.

Active Incidence. Among the single females in the sample, the active incidences had rather definitely risen during pre-adolescence and through adolescence into the mid-twenties (Table 154, Figure 102). Some 22 per cent were responding to the point of orgasm between the ages of adolescence and fifteen. Some 47 per cent were responding in the later teens, and 60 per cent between the ages of twenty-one and twenty-five. Between the ages of twenty-one and fifty, the active incidences among the single females had stood between 60 and 71 per cent. The peak had come between the ages of thirty-one and forty, where 70 to 71 per cent of the still unmarried females were reaching orgasm.

There was some decline in the active incidences after the age of forty, and a steady decline through the fifties (Table 154), but no marked decline until after the age of sixty. In the white, non-prison sample which has been used in the making of the present volume, we have no instance of a female experiencing orgasm after the age of seventy-five; but among our prison and non-white histories we have cases of females responding in their seventies and eighties and, in one instance, responding to the point of orgasm with frequencies which averaged between once a month and once a week at the age of ninety.

Relation of Frequency to Age. The frequencies of orgasm among the single females in the sample who were having any experience at all, had stayed more or less on a level from the youngest to the oldest age group. The median females in the active samples had averaged one orgasm in two weeks (0.5 per week) between adolescence and fifteen years of age (Table 154, Figures 102, 143). The average frequencies then lay between 0.3 and 0.5 per week in every subsequent group up to the age of sixty. We have previously found that the frequencies of female masturbatory activities, nocturnal dreams, petting to the point of orgasm, and still other activities similarly remain near a level throughout this same period of years. This is in marked contrast to the steady decline in the frequencies of total outlet among the males, from their late teens into old age (Figure 143). Hormonal factors probably contribute to these differences between the sexes (Chapter 18).

Female vs. Male Frequency Prior to Marriage. These differences in frequencies of orgasm between unmarried females and unmarried males are of considerable social significance. To summarize again: at the age of marriage there were some 36 per cent of the females in the sample who had not yet responded to orgasm (Table 150), while all of the

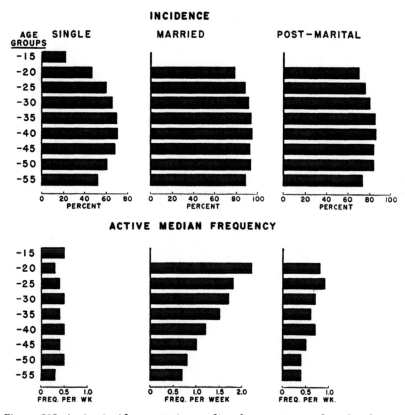

Figure 102. Active incidence, active median frequency: total outlet, by age and marital status

Data from Table 154.

males (essentially 100 per cent) at marriage had not only long since had their first experience in orgasm, but had already passed the peak of their sexual capacity. Among those born after 1910, the average male had experienced orgasm over 1500 times (the mean frequencies) before marriage; the average female appears not to have experienced orgasm more than about 223 times before her marriage. Practically all of the males born since 1910 had had a regular sexual outlet before marriage, with mean frequencies of about 2.9 orgasms per week for ten and a half years. There were not more than 10 to 20 per cent of the females who were having any outlet from any source which averaged as much as once per week for as long as five years before marriage. Between the ages of sixteen and twenty, the average (median) male was having experience with more than three (3.4) different types of sexual activity (our 1948:228), while the average female in that age period was having experience with less than half as many (1.4) types of activity (Table 169). **At the time of marriage,** the mean number of orgasms which the average female and male (born after 1910) had ever had, amounted to the following:

Activity to orgasm	Accumulative incidence % to orgasm		Mean number of orgasms	
	FEMALE	MALE	FEMALE	MALE
Total outlet	64	100	223	1523
Masturbation	41	94	130	872
Nocturnal dreams	12	82	6	175
Petting	37	26	37	64
Coitus	27	80	39	330
Homosexual	5	30	11	75
Animal contacts	—	8	—	7

Although there is, of course, considerable individual variation in these matters, and although there are factors which often lead the more responsive females to marry earlier, many marriages involve even greater differences than those which we have just shown between the average female and male. Many males are disappointed after marriage to find that their wives are not responding regularly and are not as interested in having as frequent sexual contact as they, the males, would like to have; and a great many of the married females may be disappointed and seriously disturbed when they find that they are not responding in their coitus, and not enjoying sexual relations as they had anticipated they would. Not a few of the divorces which occur within the first year or two of marriage are the product of these discrepancies between the sexual backgrounds of the average female and the average male. However, in view of the diverse pre-marital backgrounds of the spouses in the average marriage, it is not surprising that they sometimes find it difficult to adjust sexually. It is more surprising that so many married couples are ever able to work out a satisfactory sexual arrangement (p. 375).

Other Factors Affecting Incidence and Frequency. The educational backgrounds of the unmarried high school and college-bred females in the sample seem to have had no consistent effect upon the active incidences or active median frequencies of their total outlet (Table 155), even though they had had some definite effect upon the incidences of masturbation and homosexual contacts. The grade school sample is too small for certain interpretation.

The occupational classes of the parents in whose homes the females were raised had not been particularly correlated with the active incidences or active median frequencies of their total outlet before marriage (Table 156). At younger ages, the incidences had been a bit higher among those females who were raised in homes of laboring groups (Parental Classes 2 and 3). Apparently, girls from these lower level homes get started earlier in their sexual activity, but by the twenties their total outlets were not particularly different from the total outlets of the females in the other groups.

There were more definite differences among the females born in the successive decades represented in the sample (Table 160, Figure 105). Except in the youngest age group, the active incidences were lower among the unmarried females who were born before 1900. For instance, between the ages of twenty-one and twenty-five, 45 per cent of the older generation were experiencing some orgasm, as against something between 61 and 63 per cent of the later generations. After age thirty the differences were less extreme, even though they lay in the same direction. There were no significant differences in the active median frequencies of the females born in the four decades covered by the sample.

The active incidences of total outlet were correlated with the ages at which the females in the sample had turned adolescent, only for the group which had turned adolescent by eleven—where the active incidences were higher—, and for the group which turned adolescent at fifteen or later—where the active incidences were lower (Table 161). The age at onset of adolescence did not seem to have any effect on the frequencies of total outlet among the unmarried females in the sample (Table 161).

Because of the small size of the rural sample we cannot be certain of our comparisons of the total outlets of the single females of the rural and urban groups. What data we do have, however, indicate that the active incidences were somewhat higher among the unmarried urban females after age fifteen, and the active frequencies of orgasm were a bit higher among the unmarried urban groups after age twenty (Table 163).

Active Incidence and Frequency, and Religious Background. In contrast to the above, the religious backgrounds of the single females in the sample had consistently affected the active incidences and active median frequencies of their total sexual outlet. In every one of the eleven groups which are available for comparisons, definitely smaller percentages of the religiously devout females and higher percentages of the religiously less devout or inactive females were experiencing orgasm from any source prior to marriage (Table 165, Figure 108). This was more or less equally true of the Protestant, Catholic, and Jewish groups in the sample. It was true in every age group, from the youngest to the oldest single females on whom we have sufficient data.

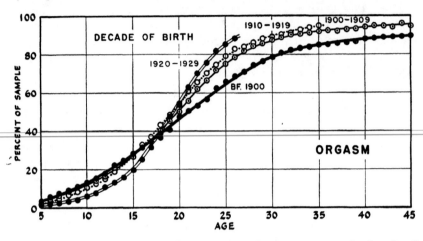

Figure 103. Accumulative incidence: orgasm from any source, by decade of birth

Data from Table 157.

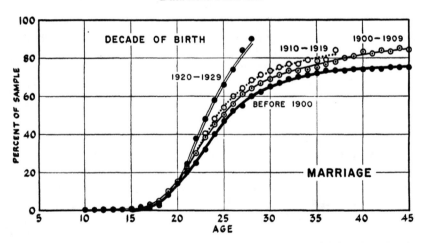

Figure 104. Accumulative incidence: age at first marriage, by decade of birth

Data from Table 167.

For instance, between the ages of sixteen and twenty, 41 per cent of the more devout Protestant females were experiencing some orgasm, in contrast to 53 per cent of the inactive Protestants. In the same age period, 34 per cent of the devout Catholics, but 70 per cent of the inactive Catholics were experiencing some orgasm. Also in that age period, 46 per cent of the devout Jewish females but 56 per cent of the inactive Jewish females were having some experience in orgasm. In older age groups, among the still unmarried females, the differences were not obliterated and, in general, were as great as those just cited.

Similarly, the active median frequencies for the unmarried females in the devout groups were always lower than those in the religiously

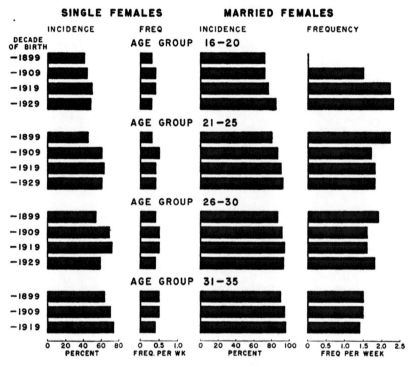

Figure 105. Active incidence, active median frequency: total outlet, by marital status and decade of birth

Data from Table 160.

inactive groups, and in some instances the differences were considerable. For instance, between the ages of twenty-one and twenty-five, the median frequency of orgasm for the devout Protestant group was 0.3 per week against 0.5 per week for the inactive Protestants (Table 165). For the Catholic group in the same age period it was 0.4 for the devout, in contrast to 1.0 per week for the inactive Catholics.

We have seen that the age of the female had affected her total outlet only in the youngest and the very oldest groups; we have seen that the total outlet of the unmarried females in the sample had not been particularly affected by their educational level, the occupational level of the home in which they had been raised, the decade in which they had been born, the age at which they had turned adolescent, or their rural-urban background. Among all of the cultural and biologic factors which might affect their sexual activity, and which in actuality had consid-

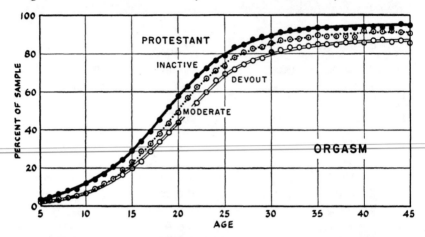

Figure 106. Accumulative incidence: orgasm from any source, in Protestant groups

Data from Table 164.

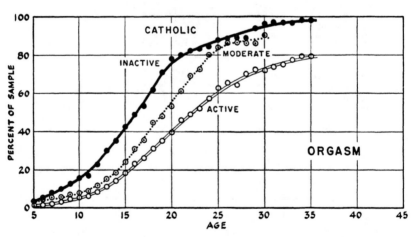

Figure 107. Accumulative incidence: orgasm from any source, in Catholic groups

Data from Table 164.

erably affected the sexual activities of the males in the sample, only the religious backgrounds of the unmarried females had had any material relation to their acceptance of either solitary or socio-sexual contacts.

Sources of Total Outlet for Single Females. The sources from which the single females in the sample had derived their total outlet had varied considerably in the different age groups (Table 171, Figure 110). As examples, the following tabulation will show the three most important outlets in four of the age groups:

AGE: ADOL.–15	%	AGE: 21–25	%
Masturbation	84	Masturbation	46
Pre-marital coitus	6	Pre-marital coitus	26
Homosexual contacts	4	Petting to orgasm	18
AGE: 16–20		AGE: 31–35	
Masturbation	60	Masturbation	42
Petting to orgasm	18	Pre-marital coitus	33
Pre-marital coitus	15	Homosexual contacts	14

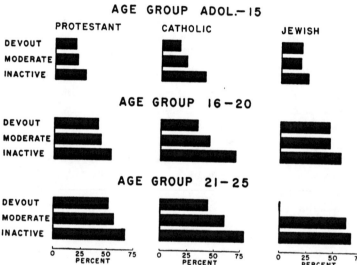

Figure 108. Active incidence: total outlet, by religious background, among single females

Data from Table 165.

At all ages, masturbation had been the most important source of total outlet for the unmarried females in the sample (Table 171, Figure 110). Coitus had been the second chief source of the pre-marital outlet in all of the groups after age twenty, and for those females who were still unmarried in their late thirties and forties, it was nearly as important a source of outlet as masturbation. Homosexual contacts had provided a rather important portion of the total outlet for the unmarried females in the sample between the ages of twenty-six and forty. Petting to orgasm had been an important source of pre-marital outlet for the females between the ages of sixteen and thirty, but it had become much less important after that age. Nocturnal dreams to orgasm had never accounted for more than 2 to 4 per cent of the total outlet of the unmarried females in any of the age groups.

Single Females Without Orgasm. A considerable portion of the unmarried females in the sample, in each and every age group, was not
experiencing orgasm from any source. Thus, between adolescence and
fifteen years of age there were 78 per cent, and among the older teen-
age girls there were 53 per cent who were not reaching orgasm in any
type of sexual activity (Table 171). Although the incidences of unresponding individuals had dropped steadily from that point, there were
29 per cent of the still unmarried females who were not reaching orgasm from any source between the ages of thirty-six and forty, and
there were higher percentages again in the still older groups. Some 39
per cent of the unmarried females were without such experience between the ages of forty-six and fifty. There were more than a quarter
(28 per cent) of the older unmarried females who had never experienced orgasm at any time in their lives (Table 150, Figure 99).

This existence of such a large group of females who are not having
any sexual outlet poses a problem of some social importance. Some of
these females in the sample had been frustrated in their attempts to
make social adjustments, and resented the fact that they had not been
able to marry. Many of them were sexually responsive enough, but they
were inhibited, chiefly by their moral training, and had not allowed
themselves to respond to the point of orgasm. Many of them had been
psychologically disturbed as a result of this blockage of their sexual
responses.[3] But others, including many of the 28 per cent who had
never reached orgasm, were sexually unresponsive individuals who had
not felt the lack of a sexual outlet. All of these females, however, were
limited in their understanding of the nature of sexual responses and
orgasm, and many of them seemed unable to comprehend what sexual
activity could mean to other persons. They disapproved of the sexual
activities of females who had high rates of outlet, and they were particularly incapable of understanding the rates of response which we
have reported (1948) for the males in the population.[4]

When such frustrated or sexually unresponsive, unmarried females
attempt to direct the behavior of other persons, they may do considerable damage. There were grade school, high school, and college teachers among these unresponsive or unresponding females. Some of them
had been directors of organizations for youth, some of them had been

---

[3] Discussions of the sexual problems of the mature single female may be found in:
Hellmann acc. Weissenberg 1924b:213 (notes diverse types of cases). Parshley 1933:300. Van der Hoog 1934:47–58. Hutton 1935:168. Smith 1951:48
(continence and health not workable for majority). Sylvanus Duvall 1952:186–
188.

[4] An attitude found among some females was expressed by one woman who in-
dignantly wrote, after the publication of our volume on the male, that the study
was a waste of effort for it merely confirmed her previous opinion "that the
male population is a herd of prancing, leering goats."

directors of institutions for girls or older women, many of them had
been active in women's clubs and service organizations, and not a few
of them had had a part in establishing public policies. Some of them
had been responsible for some of the more extreme sex laws which
state legislatures had passed. Not a few of them were active in religious

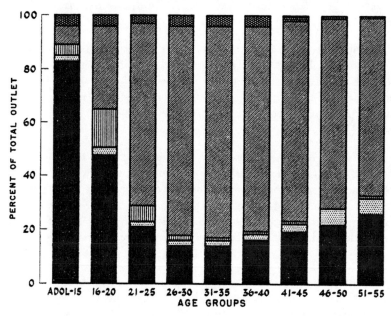

Figure 109. Percentage of total outlet: sources of orgasm, in total active
sample

Data from Table 170.

work, directing the sexual education and trying to direct the sexual
behavior of other persons. Some of them were medically trained, but
as physicians they were still shocked to learn of the sexual activities
of even their average patients. If it were realized that something between a third and a half of the unmarried females over twenty years of
age have never had a completed sexual experience (Table 149), parents
and particularly the males in the population might debate the wisdom
of making such women responsible for the guidance of youth. There
were, however, another half to two-thirds of the unmarried females
who did understand the significance of sex, and were not living the
blank or sexually frustrated lives which our culture, paradoxically, had
expected them to live.

MARRIED FEMALES: TOTAL OUTLET

The median female in our white, non-prison sample had married at the age of twenty-three. The median age given by the U. S. Census for the *total* female population (white and non-white) for the last forty years was about twenty-one. In this respect our sample may, therefore, be taken as fairly typical of white American females in general.

Among the females in the sample, about 97 per cent had experienced erotic arousal before marriage, but 3 per cent had never been so aroused before marriage. Some 64 per cent had experienced orgasm at least once before marriage, but 36 per cent married without understanding, through actual experience, the meaning of sexual orgasm (Table 150).

Relation to Age. After marriage the frequencies of total outlet for the females in the sample had increased considerably over the frequencies which single females of the same age would have had. This depended, of course, primarily upon the fact that marital coitus had begun to provide such a regular and frequent source of sexual activity and outlet as few females had found in any type of activity before marriage.

The number of females reaching orgasm from any source after marriage (the active incidences) had begun at 78 per cent between the ages of sixteen and twenty (Table 154, Figure 102). They had then increased steadily to 95 per cent at ages thirty-six to forty, after which they had begun to drop, reaching 89 per cent by age fifty-five and, to judge by our small sample, 82 per cent by age sixty.

The median frequencies of total outlet for the married females who were ever reaching orgasm show marked "aging effects." The active median frequencies between sixteen and twenty had amounted to 2.2 orgasms per week, from which point they had steadily declined, reaching 1.0 per week between the ages of forty-one and forty-five, and 0.5 per week by age sixty (Table 154, Figure 102).

The mean frequencies were much higher than the corresponding median frequencies because they included the activities of a few highly responsive and unusually active females. This had been true in all of the age groups, from the youngest to the oldest. These "aging effects," however, must be largely dependent, as we have noted elsewhere (p. 353), upon the aging processes which occur in the male, and not upon any physiologic or psychologic aging in the female between the ages of twenty and fifty-five.

Relation to Decade of Birth. In the chapter on marital coitus, we found that the percentages of married females who were responding to the point of orgasm in that coitus (the active incidences) had risen more or less steadily in the four decades represented in the sample. This accounted for the fact that the active incidences of the total outlet of the married females in the sample had steadily risen in that period of time. For instance, among the females in the sample who were between the ages of twenty-one and twenty-five, some 80 per cent of those born before 1900 had reached orgasm; but of those who were born in the successive decades, 86, 90, and 92 per cent had so responded (Table 160, Figure 105). Something of the same differences had been maintained in the subsequent age groups.

We have also pointed out (p. 358) that the frequencies of marital coitus had been reduced during the four decades represented in the sample. Consequently the active median frequencies of total outlet were, for the most part, reduced in that period of time (Table 160, Figure 105). There were some inconsistencies in the data and the reductions were not great. For instance, between the ages of twenty-one and twenty-five, the active median frequencies had been 2.2 in the generation born before 1900, but they had dropped to 1.8 in the generation born after 1920.

Relation to Other Factors. The incidences of total outlet for the married females in the sample had generally been a bit higher for the better educated groups, and lower for the grade school and high school groups (Table 155). There seem to have been no consistent correlations between the active median frequencies of the total outlet among the married females in the sample and their educational backgrounds, except that the graduate school group had slightly higher frequencies up to the age of thirty-five (Table 155).

There seem to have been no correlations at all between the occupational classes of the parental homes in which the females in the sample had been raised and the incidences and frequencies of their total outlet (Table 156).

On the other hand, the religious backgrounds of the females in the sample had definitely and consistently affected their total outlet after marriage. In nearly every age group, and in nearly all the samples that we have from Protestant, Catholic, and Jewish females, smaller percentages of the more devout and larger percentages of the inactive groups had responded to orgasm after marriage (Table 165). Similarly, the median frequencies of orgasm for those who were responding at all were, in most instances, lower for those who were devout and higher for those who were religiously inactive. In many groups the differences

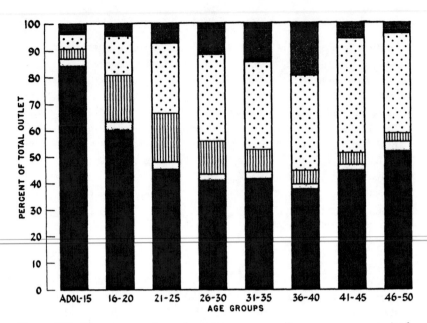

Figure 110. Percentage of total outlet: sources of orgasm, among single females

Data from Table 171.

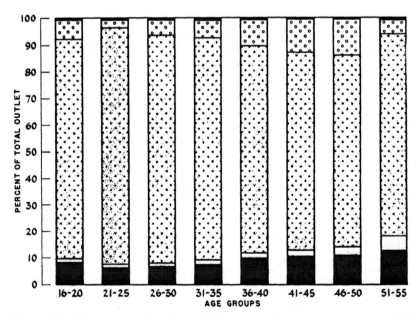

Figure 111. Percentage of total outlet: sources of orgasm, among married females

Data from Table 171.

530

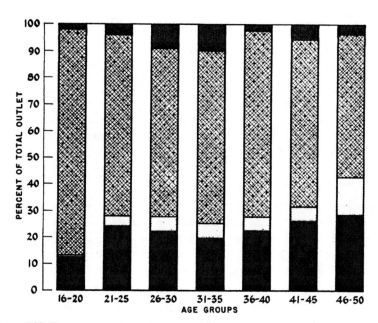

Figure 112. Percentage of total outlet: sources of orgasm, among previously married females

Data from Table 171.

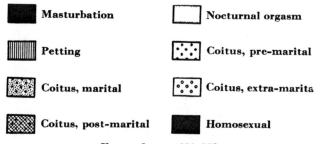

	Masturbation		Nocturnal orgasm
	Petting		Coitus, pre-marital
	Coitus, marital		Coitus, extra-marita
	Coitus, post-marital		Homosexual

Key to figures 110–112

had not been great; but in some instances, as among the Catholic females who were married between the ages of twenty-one and twenty-five, the differences were of some magnitude: an active median frequency of 1.1 orgasms per week for the devoutly Catholic females, and 2.4 for the inactive Catholic females.

Sources of Total Outlet. Coitus in marriage had accounted for something between 84 and 89 per cent of the total outlet of the married females in the sample who were between the ages of sixteen and thirty-five (Table 171, Figure 111). After the middle thirties, the importance of marital coitus had decreased. In the age group forty-six to fifty, only

531

73 per cent of the total number of orgasms were coming from that source.

Masturbation was the second most important source of sexual outlet for the married females in the sample, providing something between 7 and 10 per cent of the total number of orgasms for each of the age groups between sixteen and forty (Table 171, Figure 111). Although 11 per cent of the total outlet had come from this source in the next ten years, the increase in importance of extra-marital coitus had reduced masturbation to a third place in the list.

Extra-marital coitus and orgasms derived in extra-marital petting had accounted, in various age groups, for something between 3 and 13 per cent of the total outlet of the married females in the sample (Table 171, Figure 111). This had become the second most important source of outlet after age forty, providing 12 to 13 per cent of the total orgasms in that period.

Nocturnal dreams had provided between 1 and 3 per cent of the total outlet of the married females in each of the age groups in the sample. Homosexual contacts had never provided more than a fraction of 1 per cent of the orgasms experienced by the married females in the sample (Table 171, Figure 111).

Married Females Without Orgasm. There had been an appreciable percentage of the married females who were not reaching orgasm either in their marital coitus or in any other type of sexual activity while they were married. The percentage had been highest in the younger age groups where 22 per cent of the married females between the ages of sixteen and twenty, and 12 per cent of the married females between the ages of twenty-one and twenty-five, had never experienced any orgasm from any source (Table 171). The number of unresponsive individuals had dropped steadily in the successive age groups, reaching 5 per cent in the late thirties; but it had risen again to 6 and 7 per cent in the forties.

PREVIOUSLY MARRIED FEMALES: TOTAL OUTLET

According to the 1950 census, there are some 15 per cent of the females in the United States who have been previously married and who are no longer living with their husbands because they are widowed, separated, or divorced. Many persons, both clinicians and the public at large, have realized that women who have previously had coitus with some frequency in marriage may be faced with a problem of readjustment when they are left without any legalized or socially approved source of sexual outlet.[5] It is generally presumed that the problem is

---

[5] That previously married females, whether widowed or divorced, often face distinctive sexual problems, is also recognized in: Bienville 1771:13. Tissot

more extreme for a female who has been previously married than for one who has never been married. Our sample of previously married females is not large, but it may be sufficient to warrant some generalizations concerning the group.

There was, of course, considerable individual variation in the problems which these previously married females had faced. Some of them who were not particularly responsive seem not to have missed the regular outlet which coitus had provided, and they had been quite satisfied with the outlet which masturbation or homosexual contacts had afforded. Some of them had even been satisfied to live without any sort of sexual outlet. Many of them had, however, faced difficult problems of adjustment. Even when they had not been physiologically disturbed, they had felt a lack of socio-sexual contacts. On the other hand, a fair number of these previously married females had had socio-sexual contacts as frequently, in some instances, as they had had them in their marriages. Some of the females had welcomed the opportunity that divorce had provided to secure a wider variety of sexual experience than was possible while they were married.

The most notable aspect of the histories of these previously married females was the fact that their frequencies of activity had not dropped to the levels which they had known as single females, before they had ever married. It will be recalled that the median frequencies of orgasm for the single females in the sample lay between 0.3 and 0.5 per week in every age group between adolescence and age fifty-five (Table 154, Figure 102). The median frequencies of total outlet for the females who had been previously married had ranged between 0.4 and 0.9 per week. In every age group, however, they were well below those of the married females in the sample.

Relation to Age. Among the females who were still in their teens or early twenties when they were widowed, separated, or divorced, 70 to 76 per cent had still found some source of sexual outlet (Table 154, Figure 102). Just as in the single and married females, the highest incidences had not developed until after age twenty-five, by which time they leveled off at something between 80 and 86 per cent between ages twenty-six and fifty.

The active median frequencies of orgasm among these previously married females had declined gradually from the youngest to the oldest age groups in the sample. Between the ages of sixteen and twenty-five, the median frequencies had been 0.8 or 0.9 per week. They had

1773(1):52. Forel 1922:96. Dickinson and Beam 1931:270–287. Baber 1939: 487. Goode 1949:396 (450 divorced mothers, no consistent trauma). Shultz 1949:243–265. Waller 1951:553–559 (trauma following divorce or death of spouse).

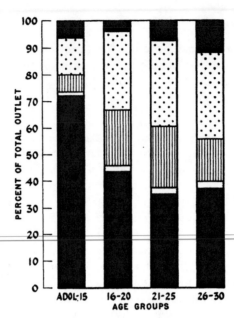

Figure 113. Percentage of total outlet: sources of orgasm, among single females of high school level

Data from Table 172.

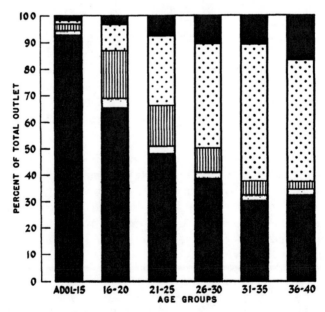

Figure 114. Percentage of total outlet: sources of orgasm, among single females of college level

Data from Table 172.

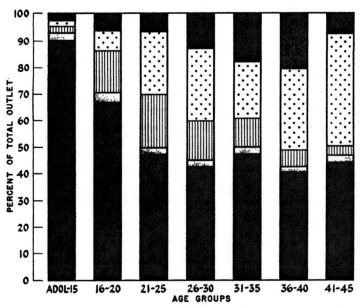

Figure 115. Percentage of total outlet: sources of orgasm, among single
females of graduate level

Data from Table 172.

■ **Masturbation**	▫ **Nocturnal orgasms**	
▨ **Coitus**	▥ **Petting**	■ **Homosexual**

Key to figures 113–115

dropped in the middle age groups, reaching 0.4 by fifty, and 0.3 by
sixty years of age (Table 154, Figure 102). In all of these groups, however, there were some females who were engaging in sexual activities
which had brought orgasm with much higher frequencies, and the
active mean frequencies and the total mean frequencies were in consequence definitely higher than the corresponding median frequencies.

Relation to Educational Level. There seem to have been no important correlations between the educational levels of the previously married females in the sample and the incidences and median frequencies
of their total outlet (Table 155). During the twenties the active median
frequencies of orgasm had been higher for the previously married
females of the high school sample, and lower for those of the college
and graduate samples; but the differences had not been maintained in
all of the age groups.

We do not have sufficient material for analyzing the effects of the
other background factors on the total outlet of the previously married
females in the sample.

Sources of Total Outlet. Among these previously married females, the active median frequencies of the post-marital coital contacts had ranged from once in two and a half weeks to nearly twice in three weeks (0.4 to 0.6 per week) in most of the age groups (Table 168). Consequently heterosexual coitus and/or petting to orgasm had, in many of the groups, provided three-quarters as much of the total outlet as it had for the females who were married (Table 171, Figures 111, 112). For the previously married females between the ages of sixteen and twenty, about 85 per cent of the outlet had been derived from the heterosexual contacts. The percentage of the outlet which had come from those sources had, however, then dropped, reaching 54 per cent by fifty years of age.

We have previously found (1948:294–296) that the post-marital patterns for the male had approached those of the married males, just as the post-marital patterns for the females are rather close to those of the married females. The patterns clearly reflect the marital experience of the previously married females and males, and they suggest that some of the females had been more interested in coitus than one would have concluded from an examination of their histories prior to divorce. In some instances, however, these females may have been more interested in the social aspects of the post-marital coital contacts than in the sexual experience itself.

In every age group, masturbation had been the second most important source of outlet and in actuality a material source of outlet for most of the previously married females (Table 171, Figure 112). For those who were still in their teens, 13 per cent of the outlet had come from masturbation, but the importance of this source of orgasm had increased in later age groups until it had accounted for 29 per cent of the total outlet of the previously married females between the ages of forty-six and fifty.

The third most important source of outlet for the previously married females had been either nocturnal dreams to orgasm or homosexual contacts. In some of the older groups, nocturnal dreams had accounted for two to three times as much of the outlet as such dreams had provided for the single or married females. This had been due to the slightly higher active median frequencies and the much higher incidences of dreams to orgasm in the previously married group. The previously married females may have dreamt of sex more often because of their desire, conscious or unconscious, for a more active sex life.

The homosexual activities of the previously married females seem to have been most important between the ages of twenty-six and thirty-five, at which time they had provided between 9 and 10 per cent of the total outlet for each of the age groups. The importance of the homosexual contacts had then declined, but the samples for the older age groups are so small that we cannot be sure that this conclusion is generally applicable.

In the earlier age groups, as much as 87 per cent of the outlet of these previously married females had been derived from socio-sexual contacts, either heterosexual or homosexual; but by age fifty only 57 per cent of the contacts were socio-sexual. This provides a measure of the considerable problem which many an older individual has in making a sexual adjustment after the termination of a marriage. Although sexual relationships may be significant because they satisfy a physiologic need, they are more significant as factors in the development and the maintenance of an individual's personality and, consequently, may contribute to her value in the total social organization.

Previously Married Females Without Orgasm. In the teen-age group in the sample, some 30 per cent of the previously married females had found no outlet in any type of sexual activity. This was not markedly higher than the percentage among the married females of that age, but it was only half as high as the percentage of unresponsive individuals among the single females in the teen-aged sample. In the older age groups, however, the number of previously married females who had lived without orgasm had decreased, reaching 20 per cent by age thirty and 16 per cent by age fifty. Some of these previously married females seemed almost as unresponsive, and as incapable of understanding responsive females, as though they had never been married. This, however, had been less often true than it had been for the single females in the sample. Consequently, some of these previously married women had shown a considerable understanding of the problems involved when they had become responsible as teachers, counselors, or institutional administrators for directing the behavior of other persons.

INDIVIDUAL VARIATION

It is obvious, but it needs to be pointed out again, that no population of variant individuals can ever have their behavior characterized by any simple description. It is more nearly possible to write anatomic descriptions which cover a whole group of individuals, but physiologic functions are usually more variable than anatomic structures, and behavioral characters are still more variable than physiologic functions.

In our previous volume (1948) we have shown the extent of the sexual variation that occurs among males, and in the present volume we have recorded the variation that occurs in the sexual behavior of the female. We have noted that the range of variation in the female far exceeds the range of variation in the male. Consequently, it would be unfortunate if any of the comparisons made in the present volume should be taken to indicate that there are sexual qualities which are found only in one or the other sex. The record will have misled the reader if he fails to note this emphasis on the range of variation, and fails to realize that he or she probably does not fit any calculated median or mean, and may in actuality depart to a considerable degree from all of the averages which have been presented. The difficulty lies in the fact that one has to deal with averages in order to compare the most characteristic aspects of two different groups, but such averages do not adequately emphasize the individual variation which is the most persistent reality in human sexual behavior (Figures 116, 117).

Since sexual matters are less frequently discussed among females than other aspects of their lives (p. 675), most females have little knowledge of the sexual habits of any large number of other females. Only clinicians who have seen a considerable variety of sexual histories, and are able to think in terms of statistical averages, can have foreseen the sort of record which we have presented here.

Interestingly enough, we may predict that the persons who will be most often incapable of accepting our description of American females will be some of the promiscuous males who have had the largest amount of sexual contact with females. Most of these males do not realize that it is only a select group of females, and usually the more responsive females, who will accept pre-marital or extra-marital relationships. Some of these males will find it difficult to believe that the incidences of extra-marital coitus are as low as our data indicate; some of them will not easily be persuaded that there is such a percentage of females who fail to reach orgasm in their coitus; some of them will find it difficult to believe that there is such a large proportion of the female population which is not aroused in anticipation of a sexual relationship, and which is not dependent upon having a regular sexual outlet. They will fail to take into account the large number of females who never make socio-sexual contacts, and never become involved in the sort of non-marital relationships from which these males have acquired most of their information about females.

Because there is such wide variation in the sexual responsiveness and frequencies of overt activity among females, many females are incapable of understanding other females. There are fewer males who are incapable of understanding other males.[6] Even the sexually least re-

---

[6] Cases of females who fail to comprehend the sexuality of other females are frequently noted in psychiatric and sociologic literature, as in: Burgess and Cottrell 1939:230 (female at twenty-one shocked to find there are women who

sponsive of the males can comprehend something of the meaning of the frequent and continuous arousal which some other males experience. But the female who goes through life or for any long period of years with little or no experience in orgasm, finds it very difficult to comprehend the female who is capable of several orgasms every time she has sexual contact, and who may, on occasion, have a score or more orgasms in an hour. To the third or more of the females who have rarely been aroused by psychologic stimuli, it may seem fantastic to believe that there are females who come to orgasm as the result of sexual fantasy, without any physical stimulation of their genitalia or of any other part of their body. Sensing something of this variation in capacities and experience, many females—although not all—hesitate to discuss their sexual histories with other females, and may prefer to carry their sexual problems to male clinicians. Because they fail to comprehend this variation in female sexual capacities, some of the females in positions of authority in schools and penal and other institutions may be more harsh than males are in their judgments of other females. There are many social problems which cannot be understood unless one comprehends the tremendous range of variation which is to be found in sexual behavior among both females and males, but particularly among females.

As a conclusion to this section of the present volume, it seems appropriate, therefore, to summarize the record of individual variation which has thus far been presented.

1. **Incidences and Frequencies of Erotic Response.** We have noted that there were 2 per cent of the females in the sample who had never been aroused erotically (Table 146, Figure 98). There were some who had been aroused only once or twice or a very few times in their lives. At the other extreme there were females who had been aroused almost daily, and sometimes many times per day, for long periods of years. There was, of course, every gradation between those extremes.

2. **Intensity of Erotic Response.** The same degree of tactile or psychologic stimulation had brought very different responses from different females. There were individuals who had responded mildly, with only mild physiologic reactions and without reaching orgasm; but there

desire coitus). Kroger and Freed 1951:382 (female of forty-three asks if women "actually enjoyed sex or were just talking"). Two contrasting statements may be found in: Gray 1951:193, 195–196 (a female author who states "There is no such thing, in a women at least, as 'sex starvation.'. . . There is no hunger for sex in the sense that there is hunger for food. . . . lack of a sex life will not hurt you no matter how young you are"). English and Pearson 1945:364 (male authors who state that when coitus occurs only once or twice a month, "if the wife is sexually normal, she is bound to find incompatibility in such a marriage").

were other females who had responded instantaneously to a wide
variety of stimuli, with intense physiologic reactions which had quickly
led to orgasm.

3. **Physical vs. Psychologic Sources of Arousal.** Among those who
had ever been aroused, there was no female who had been totally unre-

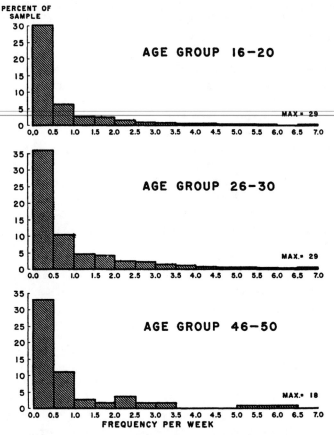

Figure 116. **Individual variation: frequency of total outlet, among single
females**

For three age groups. Each class interval includes the upper but not the lower
frequency. For incidences of females not reaching orgasm from any source, see Table
154.

sponsive to tactile stimulation, but there were females in the sample
who appear to have never been aroused by any sort of psychosexual
stimuli (Chapter 16). At the other extreme there were females who
had been aroused by a great variety of psychologic stimuli, and some
who had responded to the point of orgasm from psychologic stimulation alone. There were females in the sample who had been more responsive to psychologic stimulation than any male we have known.

4. Sources of Psychologic Stimulation. Among those females who had been aroused by psychologic stimuli, there were some who had responded to only a single sort of situation, and some who had responded to every conceivable sort of psychologic stimulation, including the observation of other persons, the observation of sexual objects or activity, fantasies of sexual objects or activity, recall of past experience, and the anticipation of new experience (Chapter 16). There was every

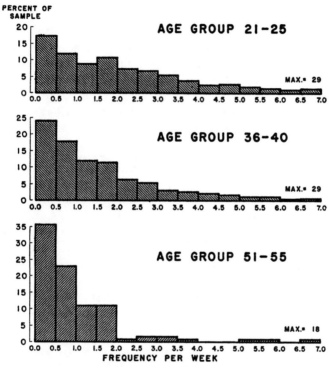

Figure 117. **Individual variation: frequency of total outlet, among married females**

For three age groups. Each class interval includes the upper but not the lower frequency. For incidences of females not reaching orgasm from any source, see Table 154.

gradation between females who had responded only occasionally to a single sort of psychologic stimulus, and females who had responded regularly to every item of psychosexual stimulation.

5. Age at First Erotic Response. We have reported female infants showing erotic responses at birth and specific responses by four months of age, we have reported the various ages at which other females first responded, and we have reported that there were some females who had not experienced their first erotic arousal until they were over thirty years of age (Table 146, Figure 98).

6. **Age at First Response to Orgasm.** We have recorded female infants reaching orgasm as early as four months of age, and we have indicated that there were some 9 per cent of the females in the sample who had lived into their late forties without reaching orgasm (Table 147, Figure 98). There were females in the sample who had reached their first orgasm at every age between these extremes, including three who had not reached their first orgasm until they were between forty-eight and fifty years of age.

7. **Frequencies of Orgasm.** Among the females who had responded to orgasm, there were some who had never responded more than once or twice in their lives. This was true even of some of the females who had been married for long periods of years. There were others who had responded in 1 or 2 per cent of their marital coitus, but there were many more who had responded much more often, including some 40 to 50 per cent who had responded to orgasm in nearly all of their coitus (Table 112).

8. **Continuity of Response.** We have shown that there were females who had been only occasionally aroused to the point of orgasm, with lapses of weeks or months and in some instances of some years between periods of arousal or orgasm. We have the record of one female who had gone for twenty-eight years between periods of coitus, with a masturbatory outlet of not more than one orgasm every two years in that long period. On the other hand, there were cases of females who had responded with high frequencies and great regularity throughout their lives, and there were cases with every conceivable pattern of discontinuity.

9. **Aging Effects.** There were females in the sample who had responded earlier in their lives but who had ceased to experience orgasm and, in some instances, to be aroused erotically after their late thirties or forties. More of the females had responded until they were in their fifties or sixties, and we have recorded the case of one ninety-year-old female who was still responding regularly.

10. **Sources of Outlet.** There were females in the sample who had derived their entire sexual outlet from a single source, which was sometimes masturbation, or petting to climax, pre-marital coitus, marital coitus, or some homosexual contact. There were a few instances of females who had never experienced orgasm except in nocturnal dreams, and instances of married females who had never experienced orgasm except in extra-marital coitus. There were females in the sample who, in the course of their lives, had utilized all six of the possible sources of outlet, and some who were utilizing all six in a single five-year period (Table 169). After the age of fifteen, something between 27 and 44

per cent of the females in the sample were depending upon a single source of outlet within each five-year period; something between 16 and 33 per cent were utilizing two sources of outlet more or less simultaneously; something between 6 and 16 per cent were utilizing three sources, and a smaller percentage was utilizing four to six sources within single age periods. There had been every conceivable combination of the possible types of sexual outlet.

11. **The Combination of Variables.** The sexual history of each individual represents a unique combination of these variables. There is little chance that such a combination has ever existed before, or ever will exist again. We have never found any individual who was a composite of all of the averages on all of the aspects of sexual response and overt activity which we have analyzed in the present volume. This is the most important fact which we can report on the sexual histories of the females who have contributed to the present study.

Table 146. Accumulative Incidence: Erotic Arousal From Any Source
By Educational Level

AGE	TOTAL SAMPLE	EDUCATIONAL LEVEL				TOTAL SAMPLE	EDUCATIONAL LEVEL			
		0–8	9–12	13–16	17+		0–8	9–12	13–16	17+
	%		Percent			Cases		Cases		
3	1	0	—	1	1	5846	176	1002	3245	1137
5	4	0	3	3	6	5799	176	1002	3245	1137
10	16	10	15	15	22	5735	176	1002	3245	1137
12	27	15	26	25	33	5711	176	1002	3245	1137
13	34	25	32	32	41	5697	176	1002	3245	1137
15	53	43	52	52	56	5648	170	1001	3245	1137
20	89	78	90	92	86	4293	124	851	2182	1136
25	95	86	96	98	94	2751	114	678	1021	938
30	97	91	98	99	97	2019	107	494	694	724
35	98	90	99	99	98	1446	89	316	483	558
40	98	92	99	100	98	933	65	188	298	382
45	98		98	100	97	557		121	161	232

Table based on total sample, including single, married, and previously married females.

Italic figures throughout the series of tables indicate that the calculations are based on less than 50 cases. No calculations are based on less than 11 cases. The dash (—) indicates a percentage or frequency smaller than any quantity which would be shown by a figure in the given number of decimal places.

Table 147. Accumulative Incidence: Orgasm From Any Source
By Educational Level

AGE	TOTAL SAMPLE	EDUCATIONAL LEVEL				TOTAL SAMPLE	EDUCATIONAL LEVEL			
		0–8	9–12	13–16	17+		0–8	9–12	13–16	17+
	%		Percent			Cases		Cases		
3	—	0	—	—	—	5873	177	1004	3269	1137
5	2	0	2	1	3	5826	177	1004	3269	1137
10	8	6	8	7	12	5762	177	1004	3269	1137
12	13	10	14	12	17	5738	177	1004	3269	1137
15	23	29	27	20	27	5675	171	1003	3269	1137
20	53	52	58	52	51	4309	125	854	2194	1136
25	77	67	80	79	72	2774	115	683	1034	942
30	86	73	87	88	85	2037	107	499	703	728
35	90	74	93	93	89	1463	89	320	491	563
40	91	79	94	95	89	945	65	192	303	385
45	91		93	95	89	568		125	165	234
48	91		91	99	88	402		98	105	164

Table based on total sample, including single, married, and previously married females.

Table 148. Sources of First Erotic Arousal and Orgasm

SOURCES	TOTAL SAM-PLE	TOTAL MARR. SAM-PLE	AGE AT MARRIAGE				DECADE OF BIRTH			
			16–20	21–25	26–30	31–35	Bf. 1900	1900–1909	1910–1919	1920–1929
FIRST EROTIC AROUSAL										
	%	%	*Percent*				*Percent*			
Masturbation	27	30	26	31	31	33	26	33	31	26
Dreams	1	1	1	1	1	3	1	1	1	1
Petting	34	30	37	30	29	28	31	31	30	33
Coitus	2	3	3	2	3	0	5	4	2	1
Homosexual	3	4	3	4	4	3	5	3	3	5
Animal contacts	1	2	2	1	2	2	2	1	2	2
Psych. stim.	32	30	28	31	30	31	30	27	31	32
Number of cases	4444	1972	515	976	351	93	284	476	720	490
FIRST ORGASM										
	%	%	*Percent*				*Percent*			
Masturbation	40	37	32	37	41	41	38	40	38	33
Dreams	5	4	5	4	4	6	7	4	4	3
Petting	24	18	18	18	17	20	10	15	18	25
Pre-mar. coitus	10	8	9	8	8	8	3	8	9	10
Mar. coitus	17	30	33	30	25	21	37	30	28	26
Homosexual	3	3	2	3	4	3	4	3	2	2
Animal contacts	—	—	1	—	0	1	1	—	—	1
Psych. stim.	1	—	—	—	1	0	0	0	1	—
Number of cases	3826	2181	568	1077	387	101	316	544	789	530

The first column is based on the total sample, single and married. All of the other calculations are based on married females, because they had had a maximum opportunity to have been aroused erotically or to have experienced orgasm.

Table 149. Accumulative Incidence: Pre-Marital Orgasm From Any Source By Educational Level

AGE	TOTAL SAMPLE	EDUCATIONAL LEVEL				TOTAL SAMPLE	EDUCATIONAL LEVEL			
		0–8	9–12	13–16	17+		0–8	9–12	13–16	17+
	%	Percent				Cases	Cases			
3	—	0	—	—	—	5864	176	1003	3264	1135
5	2	0	2	1	3	5817	176	1003	3264	1135
10	8	6	8	7	12	5753	176	1003	3264	1135
12	13	10	14	12	17	5728	176	1003	3264	1135
15	23	29	27	20	27	5662	169	1000	3264	1135
20	49	39	50	48	50	3926	95	699	2038	1094
25	62		64	62	62	1460		270	480	661
30	70		66	70	74	670		97	184	360
35	75			79	77	380			85	239
40	73				75	206				138
45	70				76	128				94

Table 150. Accumulative Incidence: Orgasm From Any Source By Age at First Marriage

AGE	NEVER MARR. BY AGE 40	TOTAL MARR. SAMPLE	AGE AT FIRST MARRIAGE			NEVER MARR. BY AGE 40	TOTAL MARR. SAMPLE	AGE AT FIRST MARRIAGE		
			16–20	21–25	26–30			16–20	21–25	26–30
	Percent		Percent			Cases		Cases		
3	1	—	—	—	0	191	2453	646	1208	432
5	4	3	2	2	3	191	2453	646	1208	432
10	13	12	10	12	12	191	2453	646	1208	432
12	16	18	18	17	18	191	2453	646	1208	432
15	21	30	31	29	29	191	2453	646	1208	432
20	37	60	82	55	46	191	2402	600	1208	432
25	51	83	89	89	66	191	2056	428	1036	432
30	64	91	93	94	91	191	1570	308	727	378
35	69	95	94	96	95	191	1138	213	514	266
40	73	96	94	96	98	191	754	144	332	180
45	70	97	95	97	98	126	442	85	193	100

In the total married sample available for the present study, 36 per cent of the females had not experienced orgasm from any source prior to marriage; 15 per cent had experienced orgasm from 1 to 25 times, and 49 per cent had experienced orgasm more than 25 times.

Table 151. Accumulative Incidence: Orgasm From Any Source
By Age at First Marriage and Educational Level

AGE	TOTAL MARRIED SAMPLE	EDUCATIONAL LEVEL 9–12	13–16	17+	TOTAL MARRIED SAMPLE	EDUCATIONAL LEVEL 9–12	13–16	17+
		AGE AT FIRST MARRIAGE 16–20						
	%	*Percent*			*Cases*		*Cases*	
5	2	1	3	3	646	225	308	71
10	10	6	15	11	646	225	308	71
15	31	26	34	34	646	225	308	71
20	82	79	84	86	600	213	279	71
25	89	87	90	97	428	165	169	60
30	93	92	96	100	308	120	106	51
35	94	95	97		213	76	79	
		AGE AT FIRST MARRIAGE 21–25						
	%	*Percent*			*Cases*		*Cases*	
5	2	3	1	3	1208	315	544	316
10	12	12	11	15	1208	315	544	316
15	29	31	27	31	1208	315	544	316
20	55	55	53	59	1208	315	544	316
25	89	90	89	90	1036	287	448	268
30	94	94	93	95	727	202	301	192
35	96	97	96	97	514	132	202	152
		AGE AT FIRST MARRIAGE 26–30						
	%	*Percent*			*Cases*		*Cases*	
5	3	4	3	3	432	104	149	163
10	12	12	11	12	432	104	149	163
15	29	29	32	28	432	104	149	163
20	46	50	46	44	432	104	149	163
25	66	66	69	64	432	104	149	163
30	91	83	92	95	378	90	130	144
35	95	94	95	97	266	51	100	103
40	98		98	99	180		66	77

Table 152. Accumulative Incidence: Orgasm From Any Source
By Parental Occupational Class

AGE	PARENTAL CLASS 2+3	4	5	6+7	PARENTAL CLASS 2+3	4	5	6+7
	Percent				*Cases*			
5	2	—	2	2	996	808	1520	2712
10	8	6	8	9	983	805	1515	2670
15	30	21	23	23	934	800	1509	2642
20	57	50	53	53	703	622	1127	1994
25	77	72	78	77	505	415	716	1208
30	85	83	88	86	375	308	501	899
35	89	88	92	91	264	194	342	685
40	89	88	90	93	170	117	205	465
45	88	89	88	94	105	70	116	280

Table based on total sample, including single, married, and previously married females.

The occupational classes are as follows: 2+3 = unskilled and semi-skilled labor. 4 = skilled labor. 5 = lower white collar class. 6+7 = upper white collar and professional classes.

Table 153. Active Incidence and Frequency: Total Outlet to Orgasm
By Age

AGE DURING ACTIVITY	ACTIVE SAMPLE Active incid. %	Median freq. per wk.	Mean frequency per wk.	TOTAL SAMPLE Median freq. per wk.	Mean frequency per wk.	CASES IN TOTAL SAMPLE
Adol.–15	22	0.5	1.4 ± 0.08	0.0	0.3 ± 0.02	5677
16–20	50	0.4	1.1 ± 0.04	0.0	0.5 ± 0.02	5649
21–25	72	0.8	1.7 ± 0.05	0.4	1.2 ± 0.04	3607
26–30	84	1.1	2.2 ± 0.07	0.8	1.8 ± 0.06	2554
31–35	88	1.1	2.1 ± 0.08	0.9	1.8 ± 0.07	1855
36–40	89	1.0	2.0 ± 0.10	0.8	1.8 ± 0.09	1301
41–45	87	0.8	1.9 ± 0.12	0.6	1.6 ± 0.11	810
46–50	84	0.6	1.4 ± 0.12	0.4	1.2 ± 0.10	469
51–55	77	0.5	1.1 ± 0.16	0.3	0.9 ± 0.13	240
56–60	66	0.4	0.7 ± 0.11	0.2	0.5 ± 0.08	121
61–65	47	*0.3*	*0.5 ± 0.11*	0.0	0.2 ± 0.06	53
66–70	*37*			0.0	*0.1 ± 0.04*	27
71–75	*30*			0.0	*0.1 ± 0.04*	10
76–80	*0*			0.0	0.0	3
81–85	*0*			0.0	0.0	2
86–90	*0*			0.0	0.0	1

While there are fewer than 11 cases in each of the oldest age groups, the data are included because of the importance of information on those groups.

Table 154. Active Incidence and Frequency: Total Outlet to Orgasm
By Age and Marital Status

AGE DURING ACTIVITY	ACTIVE SAMPLE			TOTAL SAMPLE		CASES IN TOTAL SAMPLE
	Active incid. %	Median freq. per wk.	Mean frequency per wk.	Median freq. per wk.	Mean frequency per wk.	
SINGLE FEMALES						
Adol.–15	22	0.5	1.4 ± 0.08	0.0	0.3 ± 0.02	5677
16–20	47	0.3	0.9 ± 0.04	0.0	0.4 ± 0.02	5613
21–25	60	0.4	1.1 ± 0.05	0.1	0.6 ± 0.03	2810
26–30	66	0.5	1.4 ± 0.10	0.2	0.9 ± 0.07	1064
31–35	70	0.4	1.4 ± 0.16	0.2	1.0 ± 0.12	539
36–40	71	0.5	1.7 ± 0.23	0.3	1.2 ± 0.17	315
41–45	68	0.4	1.6 ± 0.34	0.2	1.1 ± 0.24	179
46–50	61	0.5	1.4 ± 0.35	0.1	0.9 ± 0.22	109
51–55	52	0.3	1.1 ± 0.48	—	0.6 ± 0.25	58
56–60	44	0.4	0.9 ± 0.48	0.0	0.4 ± 0.23	27
MARRIED FEMALES						
16–20	78	2.2	3.6 ± 0.22	1.3	2.8 ± 0.18	578
21–25	88	1.8	2.8 ± 0.10	1.5	2.5 ± 0.09	1654
26–30	92	1.7	2.6 ± 0.09	1.5	2.4 ± 0.09	1661
31–35	94	1.5	2.4 ± 0.10	1.3	2.3 ± 0.09	1245
36–40	95	1.2	2.2 ± 0.11	1.1	2.1 ± 0.11	850
41–45	93	1.0	2.0 ± 0.15	0.9	1.9 ± 0.14	497
46–50	94	0.8	1.5 ± 0.16	0.7	1.4 ± 0.15	260
51–55	89	0.7	1.2 ± 0.19	0.6	1.1 ± 0.17	118
56–60	82	0.5	0.8 ± 0.13	0.4	0.6 ± 0.11	49
PREVIOUSLY MARRIED FEMALES						
16–20	70	0.8	2.9 ± 0.85	0.2	2.1 ± 0.62	71
21–25	76	0.9	2.3 ± 0.29	0.4	1.7 ± 0.23	238
26–30	80	0.7	2.0 ± 0.21	0.4	1.6 ± 0.18	328
31–35	85	0.6	1.9 ± 0.20	0.4	1.6 ± 0.18	303
36–40	86	0.7	1.9 ± 0.23	0.5	1.7 ± 0.20	245
41–45	84	0.5	1.8 ± 0.24	0.4	1.5 ± 0.21	195
46–50	84	0.4	1.2 ± 0.16	0.3	1.0 ± 0.14	126
51–55	73	0.4	1.1 ± 0.27	0.2	0.8 ± 0.21	82
56–60	55	0.3	0.6 ± 0.15	—	0.3 ± 0.09	53

Table 155. Active Incidence and Frequency: Total Outlet to Orgasm
By Educational Level and Marital Status

AGE	EDUC. LEVEL	SINGLE			MARRIED			PREVIOUSLY MARRIED		
		Active incid. %	Active median freq. per wk.	Cases in total sample	Active incid. %	Active median freq. per wk.	Cases in total sample	Active incid. %	Active median freq. per wk.	Cases in total sample
Adol. −15	0–8	27	0.8	162						
	9–12	27	0.4	983						
	13–16	19	0.5	3271						
	17+	27	0.5	1128						
16–20	0–8	45	0.7	143						
	9–12	51	0.4	976	73	2.1	210			
	13–16	46	0.3	3299	82	2.1	257			
	17+	48	0.4	1149	83	2.5	66			
21–25	0–8	46	0.6	67	81	1.8	72			
	9–12	60	0.4	537	89	1.7	487	74	1.1	88
	13–16	59	0.4	1204	88	1.7	727	75	0.8	89
	17+	61	0.4	1002	89	2.1	368			
26–30	0–8				84	1.5	81			
	9–12	62	0.5	181	91	1.7	489	76	1.0	100
	13–16	68	0.4	313	92	1.6	670	82	0.6	122
	17+	69	0.4	531	94	1.8	421	88	0.8	84
31–35	0–8				87	1.2	68			
	9–12	65	0.5	65	93	1.3	338	88	0.8	86
	13–16	69	0.5	139	96	1.4	479	84	0.6	106
	17+	75	0.4	309	95	1.7	360	83	0.8	88
36–40	0–8				90	1.0	50			
	9–12				92	1.4	210	87	1.0	62
	13–16	78	0.8	68	96	1.2	322	87	0.6	85
	17+	72	0.5	205	96	1.3	268	90	0.8	78
41–45	9–12				93	1.1	117			
	13–16				92	1.0	181	85	0.4	67
	17+	70	0.4	122	96	1.1	163	84	0.5	67
46–50	9–12				92	1.0	71			
	13–16				96	0.8	91			
	17+	65	0.4	80	95	0.8	76			

Table 156. Active Incidence and Frequency: Total Outlet to Orgasm
By Parental Occupational Class and Marital Status

AGE DURING ACTIVITY	PARENTAL CLASS	SINGLE			MARRIED		
		Active incid. %	Active median freq. per wk.	Cases in total sample	Active incid. %	Active median freq. per wk.	Cases in total sample
Adol.–15	2+3	29	0.6	947			
	4	21	0.4	796			
	5	22	0.4	1506			
	6+7	22	0.5	2654			
16–20	2+3	53	0.4	881	77	2.5	141
	4	45	0.3	796	67	1.6	86
	5	46	0.3	1512	78	2.1	150
	6+7	47	0.3	2648	81	2.2	225
21–25	2+3	62	0.5	460	89	1.7	309
	4	58	0.4	422	89	1.6	230
	5	57	0.4	734	90	1.8	425
	6+7	61	0.4	1283	87	1.9	732
26–30	2+3	65	0.5	195	91	1.6	307
	4	63	0.4	181	92	1.5	232
	5	67	0.4	275	92	1.6	408
	6+7	68	0.5	447	92	1.8	738
31–35	2+3	69	0.4	96	93	1.3	213
	4	62	0.4	91	94	1.2	177
	5	73	0.4	148	96	1.4	291
	6+7	73	0.5	224	95	1.6	572
36–40	2+3	61	*0.6*	62	93	1.3	134
	4				91	1.0	97
	5	72	0.5	85	96	1.2	208
	6+7	76	0.7	141	95	1.4	419
41–45	2+3				90	0.8	81
	4				90	0.7	63
	5				91	1.1	89
	6+7				95	1.1	261

The occupational classes are as follows: 2+3 = unskilled and semi-skilled labor. 4 = skilled labor. 5 = lower white collar class. 6+7 = upper white collar and professional classes.

Table 157. Accumulative Incidence: Orgasm From Any Source
By Decade of Birth

AGE		DECADE OF BIRTH					DECADE OF BIRTH		
	Bf. 1900	1900–1909	1910–1919	1920–1929	Bf. 1900	1900–1909	1910–1919	1920–1929
	Percent				*Cases*			
5	3	3	2	1	447	771	1319	3058
10	13	13	10	6	447	771	1319	3058
15	28	27	28	20	447	771	1319	3041
20	47	50	55	54	447	771	1318	1771
25	66	75	79	85	447	771	1181	373
30	78	87	89		447	771	818	
35	85	92	95		447	746	270	
40	88	94			447	497		
45	90	95			429	139		

Table based on total sample, including single, married, and previously married females.

Table 158. Accumulative Incidence: Orgasm From Any Source
By Age at Marriage and Decade of Birth

AGE	TOTAL MARR. SAMPLE	DECADE OF BIRTH				TOTAL MARR. SAMPLE	DECADE OF BIRTH		
		Bf. 1900	1900–1909	1910–1919	1920–1929		Bf. 1900	1900–1909	1910–1919	1920–1929

AGE AT FIRST MARRIAGE 16–20

AGE	%	*Percent*				*Cases*	*Cases*		
		Bf. 1900	1900–1909	1910–1919	1920–1929		Bf. 1900	1900–1909	1910–1919	1920–1929
5	2	0	4	3	1	646	64	113	185	283
10	10	13	12	10	10	646	64	113	185	283
15	31	34	34	31	29	646	64	113	185	283
20	82	75	74	82	86	600	64	113	185	239
25	89	83	85	91	95	428	64	113	177	75
30	93	88	95	93		308	64	113	131	
35	94	89	95			213	64	110		

AGE AT FIRST MARRIAGE 21–25

AGE	%	*Percent*				*Cases*	*Cases*		
		Bf. 1900	1900–1909	1910–1919	1920–1929		Bf. 1900	1900–1909	1910–1919	1920–1929
5	2	4	3	2	2	1208	145	267	475	321
10	12	12	15	12	12	1208	145	267	475	321
15	29	28	30	31	26	1208	145	267	475	321
20	55	43	52	59	57	1208	145	267	475	321
25	89	81	89	91	93	1036	145	267	457	167
30	94	88	95	96		727	145	267	315	
35	96	93	97	100		514	145	265	104	

AGE AT FIRST MARRIAGE 26–30

AGE	%	*Percent*			*Cases*	*Cases*		
		Bf. 1900	1900–1909	1910–1919		Bf. 1900	1900–1909	1910–1919
5	3	4	1	4	432	79	150	190
10	12	13	12	11	432	79	150	190
15	29	30	28	28	432	79	150	190
20	46	42	44	48	432	79	150	190
25	66	52	64	72	432	79	150	190
30	91	87	91	92	378	79	150	149
35	95	94	96		266	79	143	
40	98	97	99		180	79	101	

Table 159. Accumulative Incidence: Orgasm From Any Source
By Age at Onset of Adolescence

AGE	ADOLESCENT				At 15+	ADOLESCENT				At 15+
	By 11	At 12	At 13	At 14		By 11	At 12	At 13	At 14	
	Percent					*Cases*				
5	2	2	2	3	2	1196	1659	1735	789	346
10	10	8	8	9	7	1196	1659	1735	789	346
15	26	22	23	25	19	1176	1643	1724	784	346
20	58	52	53	51	44	868	1209	1309	640	281
25	79	77	76	75	73	508	735	840	457	232
30	85	87	87	84	84	353	483	646	361	193
35	89	94	90	89	88	248	318	467	275	155
40	90	95	91	90	87	150	196	294	188	116
45	90	92	92	92	86	93	108	172	118	76

Table based on total sample, including single, married, and previously married females.

Table 160. Active Incidence and Frequency: Total Outlet to Orgasm
By Decade of Birth and Marital Status

AGE DURING ACTIVITY	DECADE OF BIRTH	SINGLE			MARRIED		
		Active incid. %	Active median freq. per wk.	Cases in total sample	Active incid. %	Active median freq. per wk.	Cases in total sample
Adol.–15	Bf. 1900	27	0.4	436			
	1900–1909	26	0.4	760			
	1910–1919	27	0.5	1319			
	1920–1929	19	0.4	3049			
16–20	Bf. 1900	41	0.3	451	72	3.2	61
	1900–1909	44	0.4	772	72	1.5	114
	1910–1919	50	0.4	1328	76	2.2	172
	1920–1929	48	0.3	2999	84	2.3	230
21–25	Bf. 1900	45	0.3	366	80	2.2	206
	1900–1909	61	0.5	617	86	1.7	377
	1910–1919	63	0.4	985	90	1.8	624
	1920–1929	61	0.4	843	92	1.8	447
26–30	Bf. 1900	54	0.4	218	86	1.9	272
	1900–1909	69	0.5	344	91	1.6	506
	1910–1919	72	0.5	448	94	1.6	731
	1920–1929	59	0.4	54	93	1.8	153
31–35	Bf. 1900	64	0.5	151	90	1.5	290
	1900–1909	71	0.5	228	95	1.5	508
	1910–1919	74	0.4	160	96	1.4	448
36–40	Bf. 1900	66	0.5	123	93	1.4	283
	1900–1909	72	0.6	165	95	1.2	463
	1910–1919				97	1.2	105
41–45	Bf. 1900	64	0.5	116	93	1.0	273
	1900–1909	75	0.4	63	94	0.9	224

Table 161. Active Incidence and Frequency: Total Outlet to Orgasm
By Age at Onset of Adolescence and Marital Status

AGE DURING ACTIVITY	AGE AT ADOL.	SINGLE			MARRIED		
		Active incid. %	Active median freq. per wk.	Cases in total sample	Active incid. %	Active median freq. per wk.	Cases in total sample
Adol.–15	8–11	25	0.5	1203			
	12	21	0.4	1684			
	13	22	0.5	1747			
	14	24	0.5	796			
	15+	18	0.8	262			
16–20	8–11	53	0.4	1166	80	2.9	139
	12	47	0.3	1638	80	1.8	153
	13	45	0.3	1700	77	2.4	169
	14	46	0.3	777	86	2.1	76
	15+	41	0.3	345			
21–25	8–11	63	0.4	526	88	1.9	333
	12	58	0.4	769	88	1.7	458
	13	60	0.4	850	90	1.8	487
	14	58	0.5	460	87	1.8	254
	15+	57	0.4	203	83	2.2	121
26–30	8–11	69	0.5	196	92	1.7	286
	12	64	0.4	266	92	1.6	426
	13	68	0.5	323	94	1.7	526
	14	66	0.5	192	88	1.7	275
	15+	66	0.3	87	89	1.7	147
31–35	8–11	77	0.4	99	95	1.6	185
	12	76	0.4	122	96	1.3	291
	13	67	0.4	162	96	1.6	414
	14	66	0.7	109	90	1.3	224
	15+				92	1.4	131
36–40	8–11	77	0.7	65	96	1.8	133
	12	77	0.5	66	96	0.9	187
	13	66	0.5	96	97	1.3	269
	14	74	0.5	62	90	1.3	168
	15+				91	1.2	93
41–45	8–11				91	1.3	78
	12				94	0.9	102
	13				97	1.0	144
	14				93	0.8	101
	15+				87	1.2	71

Table 162. Accumulative Incidence: Orgasm From Any Source

By Rural-Urban Background

AGE	RURAL	URBAN	RURAL	URBAN
	Percent		*Cases*	
5	1	2	391	5261
10	6	9	391	5198
15	23	23	388	5122
20	45	53	325	3841
25	70	77	217	2447
30	80	86	163	1783
35	83	91	121	1274
40	84	92	82	818
45	81	92	58	484

Table based on total sample, including single, married, and previously married females.

Table 163. Active Incidence and Frequency: Total Outlet to Orgasm

By Rural-Urban Background

Age	Back-grnd.	Active incid. %	Active median freq. per wk.	Cases in total sample	Age	Back-grnd.	Active incid. %	Active median freq. per wk.	Cases in total sample
Adol. −15	Rural	23	0.5	388	26–30	Rural	57	0.4	104
	Urban	22	0.5	5132		Urban	67	0.5	915
16–20	Rural	42	0.3	386	31–35	Rural	56	0.4	64
	Urban	48	0.3	5080		Urban	72	0.5	453
21–25	Rural	54	0.3	229					
	Urban	60	0.4	2482					

Table based on single females only.

Table 164. Accumulative Incidence: Orgasm From Any Source
By Religious Background

AGE	PROTESTANT			CATHOLIC			JEWISH		
	Dev.	Moder.	Inact.	Dev.	Moder.	Inact.	Dev.	Moder.	Inact.
				Percent					
5	1	1	3	1	3	4	1	1	2
10	7	7	12	5	8	16	3	6	10
15	20	23	29	18	24	42	20	18	26
20	44	49	58	40	53	78	57	55	63
25	69	74	79	63	84	88		83	85
30	81	85	89	72	90	96		89	92
35	85	90	94	79		98		95	95
40	86	92	94						98
45	85	90	95						97
				Cases					
5	1250	1182	1083	393	156	171	112	574	993
10	1235	1168	1073	388	152	170	109	573	983
15	1199	1152	1061	382	150	170	107	571	974
20	906	920	956	282	113	145	54	331	687
25	559	587	727	176	75	106		184	408
30	418	434	544	123	51	79		134	289
35	334	314	413	77		52		75	200
40	234	211	270						125
45	149	132	150						74

Table based on total sample, including single, married, and previously married females.

Table 165. Active Incidence and Frequency: Total Outlet to Orgasm
By Religious Background and Marital Status

AGE DURING ACTIVITY	RELIGIOUS GROUP	SINGLE			MARRIED		
		Active incid. %	Active median freq. per wk.	Cases in total sample	Active incid. %	Active median freq. per wk.	Cases in total sample
Adol.–15	Protestant						
	Devout	19	0.4	1218			
	Moderate	21	0.4	1147			
	Inactive	28	0.5	1063			
	Catholic						
	Devout	17	0.4	382			
	Moderate	23	*0.4*	150			
	Inactive	41	0.8	169			
	Jewish						
	Devout	19	*0.3*	107			
	Moderate	18	0.5	571			
	Inactive	25	0.6	978			

(*Table continued on next page*)

Table 165 (continued)

AGE DURING ACTIVITY	RELIGIOUS GROUP	SINGLE			MARRIED		
		Active incid. %	Active median freq. per wk.	Cases in total sample	Active incid. %	Active median freq. per wk.	Cases in total sample
16–20	Protestant						
	Devout	41	0.3	1197	74	2.3	92
	Moderate	43	0.3	1133	77	1.5	97
	Inactive	53	0.4	1065	81	1.8	139
	Catholic						
	Devout	34	0.4	372			
	Moderate	45	0.4	139			
	Inactive	70	0.7	160			
	Jewish						
	Devout	46	0.1	107			
	Moderate	46	0.3	571			
	Inactive	56	0.4	972	79	2.4	118
21–25	Protestant						
	Devout	51	0.3	604	86	1.7	318
	Moderate	57	0.3	615	87	1.7	309
	Inactive	67	0.5	675	89	1.9	392
	Catholic						
	Devout	44	0.4	196	83	1.1	86
	Moderate	60	0.5	57	92	1.7	53
	Inactive	78	1.0	91	94	2.4	67
	Jewish						
	Moderate	62	0.4	192	88	1.8	145
	Inactive	67	0.5	396	91	1.9	319
26–30	Protestant						
	Devout	51	0.4	221	93	1.6	331
	Moderate	70	0.4	249	90	1.6	336
	Inactive	75	0.5	309	90	1.8	413
	Catholic						
	Devout	49	0.3	79	87	1.3	84
	Inactive				97	1.8	58
	Jewish						
	Moderate				91	1.9	140
	Inactive	73	0.7	110	94	1.8	274
31–35	Protestant						
	Devout	54	0.4	121	94	1.4	264
	Moderate	72	0.4	127	93	1.4	253
	Inactive	82	0.5	159	96	1.6	315
	Catholic						
	Devout				86	1.3	63
	Jewish						
	Moderate				94	1.4	102
	Inactive	78	1.1	51	94	1.6	198
36–40	Protestant						
	Devout	50	0.4	76	94	1.1	200
	Moderate	74	0.4	77	95	1.2	175
	Inactive	80	0.5	102	93	1.3	232
	Jewish						
	Moderate				95	1.5	58
	Inactive				98	1.4	131
41–45	Protestant						
	Devout				92	0.8	125
	Moderate				90	1.0	110
	Inactive				95	1.0	128
	Jewish						
	Inactive				95	1.3	79
46–50	Protestant						
	Devout				93	0.5	72
	Moderate				92	0.7	52
	Inactive				94	1.0	62

Table 166. Accumulative Incidence: Percentage of Females Ever Married By Educational Level

AGE	TOTAL SAMPLE	EDUCATIONAL LEVEL				TOTAL SAMPLE	EDUCATIONAL LEVEL			
		0–8	9–12	13–16	17+		0–8	9–12	13–16	17+
	%		Percent			Cases		Cases		
15	—	3	1	0	—	5645	175	1013	3304	1153
20	14	33	25	13	6	4364	129	860	2223	1152
25	53	63	67	60	35	2807	119	687	1048	953
30	70	75	83	76	53	2061	111	502	714	734
35	75	81	90	84	58	1483	93	323	498	568
40	78	84	92	87	64	957	68	194	305	389
45	77		91	90	59	575		127	166	236
50	76		87	88	60	311		82	77	122

Data based on age at first marriage, whether legal or common-law.

Table 167. Accumulative Incidence: Percentage of Females Ever Married By Decade of Birth

AGE	TOTAL SAMPLE	DECADE OF BIRTH				TOTAL SAMPLE	DECADE OF BIRTH			
		Bf. 1900	1900–1909	1910–1919	1920–1929		Bf. 1900	1900–1909	1910–1919	1920–1929
	%		Percent			Cases		Cases		
15	—	0	1	—	—	5645	456	784	1345	3056
20	14	14	15	14	14	4364	456	784	1344	1775
25	53	47	50	54	66	2807	456	784	1194	368
30	70	65	69	73		2061	456	784	821	
35	75	72	75	78		1483	456	754	272	
40	78	74	83			957	456	499		
45	77	75	84			575	435	139		
50	76	76				311	311			

Data based on age at first marriage, whether legal or common-law.

Table 168. Active Incidence, Frequency, and Percentage of Outlet Post-Marital Coital Experience and Orgasm

By Age

AGE DURING ACTIVITY	ACTIVE SAMPLE			TOTAL SAMPLE		CASES IN TOTAL SAMPLE
	Active incid. %	Median freq. per wk.	Mean frequency per wk.	Mean frequency per wk.	% of total outlet	
COITAL EXPERIENCE						
16–20	61	0.4	1.3 ± 0.31	0.8 ± 0.20		70
21–25	69	0.8	1.5 ± 0.14	1.0 ± 0.10		238
26–30	68	0.6	1.3 ± 0.13	0.9 ± 0.09		328
31–35	72	0.5	1.3 ± 0.12	0.9 ± 0.09		305
36–40	68	0.6	1.3 ± 0.14	0.9 ± 0.10		246
41–45	59	0.4	0.9 ± 0.12	0.5 ± 0.08		197
46–50	47	0.5	1.0 ± 0.16	0.5 ± 0.09		128
51–55	36	0.4	1.2 ± 0.41	0.4 ± 0.16		83
56–60	25	0.3	0.6 ± 0.28	0.2 ± 0.08		53
COITUS TO ORGASM						
16–20	44	0.3	1.7 ± 0.51	0.7 ± 0.24	83	71
21–25	54	0.7	1.9 ± 0.30	1.0 ± 0.17	66	237
26–30	58	0.5	1.7 ± 0.25	1.0 ± 0.15	63	328
31–35	62	0.5	1.6 ± 0.21	1.0 ± 0.14	62	303
36–40	63	0.5	1.8 ± 0.30	1.2 ± 0.20	69	245
41–45	54	0.4	1.7 ± 0.34	0.9 ± 0.19	61	195
46–50	45	0.4	1.2 ± 0.21	0.6 ± 0.11	54	126
51–55	32	0.4	1.6 ± 0.49	0.5 ± 0.17	63	82
56–60	21	0.4	0.7 ± 0.28	0.2 ± 0.07	43	53

Table 169. Number of Sources of Sexual Experience and Orgasm By Five-Year Age Groups

AGE GROUP	ANY SOURCE %	NUMBER OF SOURCES					ACTIVE SAMPLE		CASES IN TOTAL SAMPLE
		1	2	3	4	5+6	Mean number of sources	Median number of sources	

PERCENTAGE OF TOTAL SAMPLE HAVING EXPERIENCE

AGE GROUP	ANY SOURCE %	1	2	3	4	5+6	Mean number of sources	Median number of sources	CASES IN TOTAL SAMPLE
Adol.–15	57	38	15	3	1	—	1.4	1.2	5703
16–20	93	41	34	15	3	—	1.8	1.7	5672
21–25	97	32	35	23	6	1	2.0	2.0	3658
26–30	97	36	36	19	5	1	1.9	1.8	2569
31–35	96	38	35	19	4	—	1.9	1.8	1871
36–40	95	36	36	18	4	1	1.9	1.8	1313
41–45	94	40	32	20	2	—	1.9	1.8	818
46–50	90	39	31	17	3	0	1.8	1.7	474
51–55	84	43	28	11	2	0	1.7	1.5	245
56+	76	41	25	9	1	0	1.6	1.4	123

PERCENTAGE OF TOTAL SAMPLE REACHING ORGASM

AGE GROUP	ANY SOURCE %	1	2	3	4	5+6	Mean number of sources	Median number of sources	CASES IN TOTAL SAMPLE
Adol.–15	22	18	3	1	—	—	1.3	1.1	5706
16–20	50	27	16	6	1	—	1.6	1.4	5669
21–25	72	34	24	11	3	—	1.8	1.6	3654
26–30	84	40	30	12	2	—	1.7	1.6	2567
31–35	88	39	33	14	2	—	1.8	1.7	1869
36–40	89	39	33	15	2	—	1.8	1.7	1311
41–45	87	39	31	16	1	—	1.8	1.7	816
46–50	84	38	32	13	1	0	1.7	1.6	476
51–55	77	44	22	10	1	0	1.6	1.4	245
56+	66	36	21	8	1	0	1.6	1.4	123

The possible sources of sexual experience are masturbation, nocturnal sex dreams, petting, coitus, homosexual contacts, and animal contacts. For the calculation of means and medians, the data were treated as continuous data.

Table 170. Percentage of Total Outlet, by Source

In Total Active Sample

Source	Adol. −15	16–20	21–25	26–30	31–35	36–40	41–45	46–50	51–55
Masturbation	83	48	21	14	14	16	19	22	26
Nocturnal orgasm	2	3	2	2	2	2	3	6	6
Petting to orgasm	4	14	6	2	1	1	1	—	1
Coitus, any	7	31	68	78	79	77	75	71	67
Pre-marital	6	11	9	6	5	6	6	6	3
Marital	—	17	53	63	62	54	49	44	42
Extra-marital	—	1	2	4	5	7	8	9	3
Post-marital	0	2	4	5	7	10	12	12	19
Homosexual	4	4	3	4	4	4	2	1	—
Total outlet	100	100	100	100	100	100	100	100	100
Total solitary	85	51	23	16	16	18	22	28	32
Total heterosexual	11	45	74	80	80	78	76	71	68
Total homosexual	4	4	3	4	4	4	2	1	—
% with any orgasm	22	50	72	84	88	89	87	84	77
Number of cases in total sample	5677	5649	3607	2554	1855	1301	810	469	240

Table based on total sample, including single, married, and previously married females. "Active sample" refers to those experiencing orgasm from any source.

Table 171. Percentage of Total Outlet, by Source
In Active Sample, by Marital Status

Source	Adol.–15	16–20	21–25	26–30	31–35	36–40	41–45	46–50
SINGLE FEMALES								
Masturbation	84	60	46	41	42	37	45	52
Nocturnal orgasm	2	3	3	3	3	2	2	4
Petting to orgasm	4	18	18	13	8	5	4	3
Coitus	6	15	26	32	33	37	43	37
Homosexual outlet	4	4	7	11	14	19	6	4
Total outlet	100	100	100	100	100	100	100	100
Total solitary	86	63	49	44	45	39	47	56
Total heterosexual	10	33	44	45	41	42	47	40
Total homosexual	4	4	7	11	14	19	6	4
% with any orgasm	22	47	60	66	70	71	68	61
Number of cases in total sample	5677	5613	2810	1064	539	315	179	109
MARRIED FEMALES								
Masturbation		9	7	7	8	10	11	11
Nocturnal orgasm		1	1	1	2	2	2	3
Coitus, marital		85	89	85	84	78	75	73
Coitus + Petting, extra-marital		5	3	6	6	10	12	13
Homosexual outlet		—	—	1	—	—	—	—
Total outlet		100	100	100	100	100	100	100
Total solitary		10	8	8	10	12	13	14
Total heterosexual		90	92	91	90	88	87	86
Total homosexual		—	—	1	—	—	—	—
% with any orgasm		78	88	92	94	95	93	94
Number of cases in total sample		578	1654	1661	1245	850	497	260
PREVIOUSLY MARRIED FEMALES								
Masturbation		13	24	23	20	23	26	29
Nocturnal orgasm		—	4	5	5	5	6	14
Coitus + Pet. to orgasm		85	68	63	65	70	62	54
Homosexual outlet		2	4	9	10	2	6	3
Total outlet		100	100	100	100	100	100	100
Total solitary		13	28	28	25	28	32	43
Total heterosexual		85	68	63	65	70	62	54
Total homosexual		2	4	9	10	2	6	3
% with any orgasm		70	76	80	85	86	84	84
Number of cases in total sample		71	238	328	303	245	195	126

"Active sample" refers to those experiencing orgasm from any source.

Table 172. Percentage of Total Outlet, by Source
In Active Sample of Single Females, by Educational Level

Source	Adol.–15	16–20	21–25	26–30	31–35	36–40	41–45
		EDUCATIONAL LEVEL 9–12					
Masturbation	73	44	36	38	58		
Nocturnal orgasm	1	2	2	2	4		
Petting to orgasm	6	21	23	16	9		
Coitus	14	29	32	32	23		
Homosexual	6	4	7	12	6		
Total outlet	100	100	100	100	100		
Total solitary	74	46	38	40	62		
Total heterosexual	20	50	55	48	32		
Total homosexual	6	4	7	12	6		
% with any orgasm	27	51	60	62	65		
Number of cases in total sample	983	976	537	181	65		
		EDUCATIONAL LEVEL 13–16					
Masturbation	93	65	48	39	30	32	
Nocturnal orgasm	2	4	3	2	2	2	
Petting to orgasm	3	18	15	9	5	3	
Coitus	1	10	26	40	52	46	
Homosexual	1	3	8	10	11	17	
Total outlet	100	100	100	100	100	100	
Total solitary	95	69	51	41	32	34	
Total heterosexual	4	28	41	49	57	49	
Total homosexual	1	3	8	10	11	17	
% with any orgasm	19	46	59	68	69	78	
Number of cases in total sample	3271	3299	1204	313	139	68	
		EDUCATIONAL LEVEL 17+					
Masturbation	90	66	47	43	47	40	45
Nocturnal orgasm	3	4	3	3	3	2	2
Petting to orgasm	3	16	20	14	11	6	4
Coitus	2	8	23	27	21	31	42
Homosexual	2	6	7	13	18	21	7
Total outlet	100	100	100	100	100	100	100
Total solitary	93	70	50	46	50	42	47
Total heterosexual	5	24	43	41	32	37	46
Total homosexual	2	6	7	13	18	21	7
% with any orgasm	27	48	61	69	75	72	70
Number of cases in total sample	1128	1149	1002	531	309	205	122

"Active sample" refers to those experiencing orgasm from any source.

Table 173. Percentage of Total Outlet, by Source
In Active Sample of Married Females, by Educational Level

Source	16–20	21–25	26–30	31–35	36–40	41–45	46–50
EDUCATIONAL LEVEL 9–12							
Masturbation	8	5	5	7	9	9	13
Nocturnal org.	1	1	2	2	3	4	3
Coitus, marital	89	91	90	88	83	73	80
Coitus + Pet., extra-marit.	2	3	3	3	5	14	4
Homosexual	—	—	—	0	—	0	0
Total outlet	100	100	100	100	100	100	100
Total solitary	9	6	7	9	12	13	16
Total hetero.	91	94	93	91	88	87	84
Total homosex.	—	—	—	0	—	0	0
% with any orgasm	73	89	91	93	92	93	92
Number of cases in total sample	210	487	489	338	210	117	71
EDUCATIONAL LEVEL 13–16							
Masturbation	8	7	7	7	11	7	10
Nocturnal org.	—	—	1	2	1	2	4
Coitus, marital	90	89	84	82	74	82	76
Coitus + Pet., extra-marit.	1	3	7	8	13	9	10
Homosexual	1	1	1	1	1	0	0
Total outlet	100	100	100	100	100	100	100
Total solitary	8	7	8	9	12	9	14
Total hetero.	91	92	91	90	87	91	86
Total homosex.	1	1	1	1	1	0	0
% with any orgasm	82	88	92	96	96	92	96
Number of cases in total sample	257	727	670	479	322	181	91
EDUCATIONAL LEVEL 17 +							
Masturbation	10	7	9	9	10	17	13
Nocturnal org.	4	2	1	1	1	2	1
Coitus, marital	86	90	83	83	79	65	60
Coitus + Pet., extra-marit.	—	1	7	7	10	16	25
Homosexual	—	—	—	—	—	—	1
Total outlet	100	100	100	100	100	100	100
Total solitary	14	9	10	10	11	19	14
Total hetero.	86	91	90	90	89	81	85
Total homosex.	—	—	—	—	—	—	1
% with any orgasm	83	89	94	95	96	96	95
Number of cases in total sample	66	368	421	360	268	163	76

"Active sample" refers to those experiencing orgasm from any source.

Part III

COMPARISONS OF FEMALE AND MALE

Chapter 14

ANATOMY OF SEXUAL RESPONSE AND ORGASM

In our previous volume (1948) we presented data on the incidences and frequencies of the various types of sexual activity in the human male, and attempted to analyze some of the biologic and social factors which affect those activities. In the previous section of the present volume we have presented similar data for the female. Now it is possible to make comparisons of the sexual activities of the human female and male, and in such comparisons it should be possible to discover some of the basic factors which account for the similarities and the differences between the two sexes.

In view of the historical backgrounds of our Judeo-Christian culture, comparisons of females and males must be undertaken with some trepidation and a considerable sense of responsibility. It should not be forgotten that the social status of women under early Jewish and Christian rule was not much above that which women still hold in the older Asiatic cultures. Their current position in our present-day social organization has been acquired only after some centuries of conflict between the sexes. There were early bans on the female's participation in most of the activities of the social organization; in later centuries there were chivalrous and galante attempts to place her in a unique position in the cultural life of the day. There are still male antagonisms to her emergence as a co-equal in the home and in social affairs. There are romantic rationalizations which obscure the real problems that are involved and, down to the present day, there is more heat than logic in most attempts to show that women are the equal of men, or that the human female differs in some fundamental way from the human male. It would be surprising if we, the present investigators, should have wholly freed ourselves from such century-old biases and succeeded in comparing the two sexes with the complete objectivity which is possible in areas of science that are of less direct import in human affairs. We have, however, tried to accumulate the data with a minimum of pre-judgment, and attempted to make interpretations which would fit those data.

It takes two sexes to carry on the business of our human social organization; but men will never learn to get along better with women, or women with men, until each understands the other as they are and not as they hope or imagine them to be. We cannot believe that social relations between the sexes, and sexual relations in particular, can ever be improved if we continue to be deluded by the longstanding fictions about the similarities, identities, and differences which are supposed to exist between men and women.

BASIC SIGNIFICANCE OF ANATOMIC DATA

It must be emphasized again that what any animal does depends upon the nature of the stimulus which it receives, its physical structure, the capacity of that structure to respond to the given stimulus, and the nature of its previous experience. Its anatomy, its physiology, and its psychology must all be considered before one can adequately understand why the animal behaves as it does.

It is the physical structure which receives the stimulus and responds to it. Some knowledge of the anatomy of that structure and of the way in which it functions must be had before one can make adequate psychologic or social interpretations of any type of behavior. But, conversely, no knowledge of structures can completely explain the behavior unless psychologic factors are taken into account. From the lowest to the highest organism, psychologic factors—the animal's previous experience, what it has learned, and the extent to which its present behavior is conditioned by its previous activity—will determine the way in which its structures function.

Psychologic factors become most significant among those species which have the most highly developed nervous systems. Among mammals, with their highly complex brains, the psychologic aspects of behavior may become of primary importance. This is true of sexual as well as of other types of behavior. It is, of course, particularly true of man, the most complex of all the mammals; and this provides some justification for the considerable attention which has, heretofore, been given to the psychologic and social aspects of human sexual behavior. Nothing we have said or may subsequently say in this chapter or in any other part of our report should be construed to mean that we are unaware of the importance of psychologic factors in human sexual behavior.

Much of the psychologic theory about sex has, however, been developed without any adequate appreciation of the anatomy which is involved whenever there is sexual response. In the human species, there have been anatomic studies of the external genitalia, primarily of the female; there have been studies of the way in which stimulation of those genitalia may be transmitted to the lower levels of the spinal cord, and of the mechanisms by which those portions of the cord then effect genital erection, pelvic thrusts, and orgasm. There have been studies of the significance of the cerebral cortex when there is psychologic stimulation which leads to sexual response (p. 710). But this is about the limit of the sexual anatomy and physiology which has been investigated, and this represents only a small portion of the anatomy and physiology which is actually involved. There has been hardly any investigation of the physiologic changes which occur throughout the body of the animal whenever it is aroused sexually. In fact, there has been a considerable failure to comprehend that sexual responses ever involve anything more than genital responses. We no longer consider the heart the seat of love, or an inflammation of the brain the source of the drive which impels the human creature to perform carnal acts; but we fall almost as far short of the fact when we think of the genitalia as the only structures or even the primary structures which are involved in a sexual response.

Some years ago we realized that it would be impossible to make significant interpretations of our data on human sexual behavior, and especially on the sexual behavior of the human female, or to make significant comparisons of female and male sexuality, until we obtained some better understanding of the anatomy and physiology of sexual response and orgasm. We could never have understood the female's responses in masturbation, in nocturnal dreams, in petting, in coitus, and in homosexual relations, as they are presented in the previous chapters of this volume, and we could never have understood the basic similarities and differences between females and males, if we had not first become acquainted with the anatomy and physiology on which sexual functions depend.

SOURCES OF DATA

There have been six chief sources of the data which we now have on the nature of sexual response and orgasm. These have been:

1. Reports from individuals who, in the course of contributing their histories to this study, have attempted to describe and analyze their own sexual reactions.

2. An invaluable record from scientifically trained persons who have observed human sexual activities in which they themselves were not involved, and who have kept records of their observations.

3. Observations which we and other students have made on the sexual activities of some of the infra-human species of mammals, and a great library of documentary film which we have accumulated for this study of animal behavior.

4. Published clinical data and a body of unpublished gynecologic data that have been made available for the present project.

5. Published data on the gross anatomy of those parts of the body which are involved in sexual response, and some special data on the histology (the detailed anatomy) of some of those structures.

6. The published record of physiologic experiments on the sexual activities of lower mammals and, to a lesser extent, of the human animal.

It is difficult, although not impossible, to acquire any adequate understanding of the physiology of sexual response from clinical records or case history data, for they constitute secondhand reports which depend for their validity upon the capacity of the individual to observe his or her own activity, and upon his or her ability to analyze the physical and physiologic bases of those activities. In no other area have the physiologist and the student of behavior had to rely upon such secondhand sources, while having so little access to direct observation.

This difficulty is particularly acute in the study of sexual behavior because the participant in a sexual relationship becomes physiologically incapacitated as an observer. Sexual arousal reduces one's capacities to see, to hear, to smell, to taste, or to feel with anything like normal acuity, and at the moment of orgasm one's sensory capacities may completely fail (Chapter 15). It is for this reason that most persons are unaware that orgasm is anything more than a genital response and that all parts of their bodies as well as their genitalia are involved when they respond sexually. Persons who have tried to describe their experiences in orgasm may produce literary or artistic descriptions, but they rarely contribute to any understanding of the physiology which is involved.

The usefulness of the observed data to which we have had access depends in no small degree upon the fact that the observations were made in every instance by scientifically trained observers. Moreover, in the interpretation of these data we have had the cooperation of a considerable group of anatomists, physiologists, neurologists, endocrinologists, gynecologists, psychiatrists, and other specialists. The materials are still scant and additional physiologic studies will need to be made.

STIMULATION THROUGH END ORGANS OF TOUCH

Among the mammals, tactile stimulation from touch, pressure, or more general contact is the sort of physical stimulation which most often brings sexual response. In some other groups of animals, sexual responses are more often evoked by other sorts of sensory stimuli.

Among the insects, for instance, the organs of smell and taste are most often involved. In such vertebrates as the fish it becomes difficult to distinguish between responses to sound and responses to pressure. It is true that among mammals, sexual responses may also be initiated through the organs of sight, hearing, smell, and taste; but tactile stimuli account for most mammalian sexual responses.

It has long been recognized that tactile responses are akin to sexual responses,[1] but we now understand that sexual responses amount to something more than simple tactile responses. A sexual response is one which leads the animal to engage in mating behavior, or to manifest some portion of the reactions which are shown in mating behavior. Mating behavior always involves a whole series of physiologic changes, only a small portion of which ordinarily develop when an individual is simply touched. The matter will become apparent as we analyze the data in this and the next chapter.

The organs which make the animal aware that it has been touched, and which at times may lead it to make more specifically sexual responses, are the end organs of touch (nerve endings) that are located in the skin, and some of the deeper nerves of the body. Certain areas of the body which are richly supplied with end organs have long been recognized as "erogenous zones."[2] In the petting techniques which many females and males regularly utilize (Chapter 7), the sexual significance of these areas is commonly recognized. It is, in consequence, surprising to find how many persons still think of the external reproductive organs, the genitalia, as the only true "sex organs," and believe that arousal sufficient to effect orgasm can be achieved only when those structures are directly stimulated.

The data that are given here on the sensitivity of certain structures must, therefore, be considered in relation to the fact that the tactile stimulation of all other surfaces which contain end organs of touch may also produce some degree of erotic arousal.

Penis. In the female and male mammal the external reproductive organs, the genitalia, develop embryologically from a common pattern.[3] They are, therefore, homologous structures in the technical meaning of the term. In spite of considerable dissimilarities in the gross anatomy of the adult female and male genitalia, each structure in the one sex

---

[1] As examples of this association of tactile and sexual responses, see: Bloch 1908:30. Havelock Ellis 1936(I,3):3–8. Freud 1938:599–600.

[2] As examples of such lists of erogenous zones, see: The Kama Sutra of Vatsyayana [between 1st and 6th cent. A.D., Sanskrit]. The Ananga Ranga [12th cent. A.D.?, Sanskrit]. Van de Velde 1930:45–46. Havelock Ellis 1936(II,1):143. Haire 1951:304–306.

[3] For the embryology of the genitalia, see, for instance: Arey 1946:283–308. Patten 1946:575–607. Hamilton, Boyd, and Mossman 1947:193–223.

is homologous to some structure in the other sex. During the first two months of human embryonic development, the differences between the male and the female structures are so slight that it is very difficult to identify the sex of the embryo. In any consideration of the functions of the adult genitalia, and especially of their liability to sensory stimulation, it is important and imperative that one take into account the homologous origins of the structures in the two sexes.

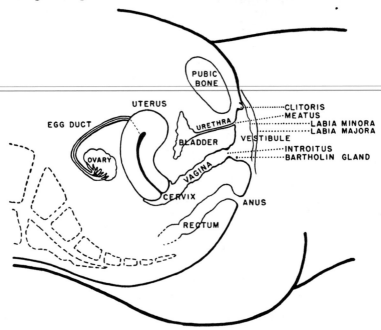

Figure 118. Female reproductive and genital anatomy

The embryonic phallus becomes the penis of the male or the clitoris of the female (Figures 118, 119). The adult structures in both cases are richly supplied with nerves which terminate in what seem to be specialized sorts of end organs of touch—some of which are called Meissner's corpuscles, and some Krause's genital corpuscles.[4]

It is commonly understood that the lower edge of the head (the glans) of the penis, or what is technically known as the corona of the glans, is the area that is most sensitive to tactile stimulation. The area on the under surface of the penis directly below the cleft of the glans

---

[4] Recent authors are inclined to discard the concept of Krause's corpuscles as end organs of touch and emphasize other end organs, especially Meissner's. See: Fulton 1949:3–6, 17. Ruch in Howell (Fulton edit.) 1949:304–306. Houssay et al. 1951:836. Blake and Ramsey 1951:29. For data on Krause's corpuscles, see: Eberth 1904:249. Jordan 1934:147, 149, 438, 467. Wharton 1947:10. Bailey 1948:585, 630. Dickinson 1949:58. Fulton 1949:6. Maximow and Bloom 1952:181, 524.

is similarly sensitive to stimulation. This latter area lies directly beneath the longitudinal fold (the frenum) by which the loose foreskin is attached if it has not been removed by circumcision.

It appears, however, that there is a minimum of sensation in the main shaft of the penis or in the skin covering it. When the frenum is moved to one side or cut away, pressure on the original point shows that it is as sensitive as it was before the removal of the skin. Stimulation applied by inserting a probe into the urethra similarly shows that the sensory nerves are not located between the urethra and the under surface of the organ, but between the urethra and the spongy mass which forms the shaft of the penis. It remains for the neurologist and the student of male anatomy to identify the exact nerves which are involved, but the present evidence seems to show that they end deep in the shaft of the penis and not in its epidermal covering.

In addition to reactions to sensory stimulation, mechanical reactions also appear to be involved in the erection of the penis. It has been generally assumed that the increased flow of blood into the organ during sexual arousal depends entirely upon circulatory changes which are effected by tactile or other sorts of erotic stimulation; but the possibility that mechanical effects may have something to do with the erection needs consideration. Forward pressures exerted on the corona of the glans not only effect sensory stimulation but, just as in stripping a wet piece of sponge rubber, also help crowd blood into the glans. Similarly, downward pressures on the upper (the distal) ends of the two spongy bodies (the corpora cavernosa) which constitute the shaft of the penis, and especially pressures on the upper edges of the spongy bodies at the point where they meet directly under the frenum, may stimulate the deep-seated nerves; but they may also have some mechanical effect in crowding blood into the corpora.

The effects of any direct stimulation of the penis are so obvious that the organ has assumed a significance which probably exceeds its real importance. The male is likely to localize most of his sexual reactions in his genitalia, and his sexual partner is also likely to consider that this is the part of the body which must be stimulated if the male is to be aroused. This overemphasis on genital action has served, more than anything else, to divert attention from the activities which go on in other parts of the body during sexual response. It has even been suggested that the larger size of the male phallus accounts for most of the differences between female and male sexual responses, and that a female who had a phallus as large as the average penis might respond as quickly, as frequently, and as intensely as the average male. But this is not in accord with our understanding of the basic factors in sex-

ual response. There are fundamental psychologic differences between
the two sexes (Chapter 16) which could not be affected by any genital
transformation. This opinion is further confirmed by the fact that
among several of the other primates, including the gibbon and some
of the monkeys, the clitoris of the female is about as large as the penis
of the male, but the basic psychosexual differences between the female
and male are still present.[5]

Clitoris. The clitoris, which is the phallus of the female, is the
homologue of the penis of the male (Figures 118, 119). The shaft of
the clitoris may average something over an inch in length. It has a
diameter which is less than that of a pencil. Most of the clitoris is em-
bedded in the soft tissue which constitutes the upper (*i.e.*, the anterior) wall of the vestibule to the vagina. The head (glans) of the
clitoris is ordinarily the only portion which protrudes beyond the body.
In many females the foreskin (the hood) of the clitoris completely
covers the head and adheres to it, and then no portion of the clitoris is
readily apparent. Because of the small size of any protrudent portions,
no localizations of sensitive areas on the corona or on other parts of
the clitoris have been recorded.

Also because of its small size and the limited protrusion of the
clitoris, many males do not understand that it may be as important a
center of stimulation for females as the penis is for males. However,
most females consciously or subconsciously recognize the importance
of this structure in sexual response. There are many females who are
incapable of maximum arousal unless the clitoris is sufficiently stimulated.

In connection with the present study, five gynecologists have co-
operated by testing the sensitivity of the clitoris and other parts of the
genitalia of nearly nine hundred females. The results, shown in Table
174, constitute a precise and important body of data on a matter which
has heretofore been poorly understood and vigorously debated. The
record shows that there is some individual variation in the sensitivity
of the clitoris: 2 per cent of the tested women seemed to be unaware
of tactile stimulation, but 98 per cent were aware of such tactile stimulation of the organ. Similarly, there is considerable evidence that most
females respond erotically, often with considerable intensity and im-
mediacy, whenever the clitoris is tactilely stimulated.

We have already noted (p. 158, Table 37) that a high percentage of
all the females who masturbate use techniques which involve some

---

[5] For data on the large clitoris of some primates, see: Hooton 1942:175, 231, 252,
   273. Ford and Beach 1951:21. In examining spider monkeys we find the
   clitoris may be as long as or longer than the penis, although not as large in
   diameter.

sort of rhythmic stimulation of the clitoris, usually with a finger or several fingers or the whole hand. Such techniques often involve the stimulation of the inner surfaces of the labia minora as well, but then each digital stroke usually ends against the clitoris. When the technique includes rhythmic pressure on those structures, the effectiveness of the action may still depend upon the sensitivity of the clitoris and of the labia minora. Even direct penetrations of the vagina during mas-

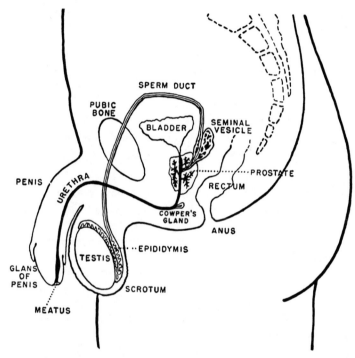

SPERM DUCT

PUBIC BONE

BLADDER

SEMINAL VESICLE

PENIS

URETHRA

PROSTATE

RECTUM

COWPER'S GLAND

ANUS

EPIDIDYMIS

GLANS OF PENIS

TESTIS

SCROTUM

MEATUS

Figure 119. Male reproductive and genital anatomy

turbation may depend for their effectiveness on the fact that the base of the clitoris, which is located in the anterior wall of the vagina, may be stimulated by the penetrating object.

Whenever female homosexual relations include genital techniques, the clitoris is usually involved (Table 130). This is particularly significant because the partners in such contacts often know more about female genital function than either of the partners in a heterosexual relation. While there are more females than males who achieve orgasm through the stimulation of some area other than their genitalia, certainly there are no structures in the female which are more sensitive than the clitoris, the labia minora, and the extension of the labia into the vestibule of the vagina.

The male who comprehends the importance of the clitoris regularly provides manual or other mechanical stimulation of that structure during pre-coital petting. In coitus, he sees to it that the clitoris makes contact with his pubic area, the base of his penis, or some other part of his body. Oral stimulation of the female genitalia is most often directed toward the labia minora or the clitoris.

Some of the psychoanalysts, ignoring the anatomic data, minimize the importance of the clitoris while insisting on the importance of the vagina in female sexual response (p. 582).

Urethra and Meatus. The penis of the male is normally penetrated for its full length by the urethra (Figure 119). The urethra of the female does not penetrate the clitoris; instead it lies in the soft tissues which constitute the upper (the anterior) wall of the vagina (Figure 118). The opening of the urethra of the male (the meatus) is normally on the tip of the head of the penis (Figure 119); in the female it is located between the clitoris and the entrance to the vagina (Figure 118).

A few persons, females and males, employ masturbatory techniques which include insertions of objects into the urethra.[6] The urethral lining which has not become accustomed to such penetration is so sensitive that most individuals, upon initial experimentation, desist from further activity. More experienced persons claim that they receive some erotic stimulation from the penetrations. The recorded satisfactions may include reactions to pain, and sometimes they may be wholly psychologic in origin. Urethral insertions may also stimulate the nerves which lie in the tissue about the urethra of the female, or deep in the shaft of the penis of the male.

There is evidence that the area surrounding the meatus is supplied in some individuals with an accumulation of nerves.[7] Consequently, direct stimulation of the meatus is sometimes included in the masturbatory procedures. This may happen more often in the female than in the male.

Labia Minora. The inner lips of the female genitalia, the labia minora (Figure 118), are homologous with a portion of the skin covering the shaft of the penis of the male. Both the outer and the inner surfaces of the labia minora appear to be supplied with more nerves than most other skin-covered parts of the body, and are highly sensi-

---

[6] Urethral insertions are also noted, for instance, in: Bloch 1908:411. Havelock Ellis 1936(I,1):171–173. Dickinson 1949:63, 69. Grafenberg 1950:146. Haire 1951:147.

[7] For nerve endings located about the urethral meatus of the female, see, for instance: Lewis 1942:8. Dickinson 1949:62.

tive to tactile stimulation. The gynecologic examinations made for this study showed that some 98 per cent of the tested women were conscious of tactile stimulation when it was applied to *either* the out-

Table 174. Responses to Tactile Stimulation and to Pressure, in Female Genital Structures

Structures	% Responding	Total no. of cases	Typical variation in response in 15 cases														
			1	2	3	4	5	6	7	8	9	10	11	12	13	14	15
AREA OF TACTILE STIMULATION																	
Labia majora																	
Right	92	854	x	x	√	√	x	x	√		√	√	√		√	√	x
Left	87	854	x	x	√	√	√	x	√		x	√	√		√	√	x
Clitoris	98	879	√	x	√	√	√	√	√	√	x	√	√	x	√	√	√
Labia minora																	
Right, outer surf.	97	879	x	x	√	√	x	√	√	√	√	√	√	x	√	√	√
Right, inner surf.	98	879	x	x	√	√	x	√	√	√	√	√	√	x	√	√	√
Left, outer surf.	95	879	√	x	√	√	x	√	√	√	x	√	√	x	√	√	√
Left, inner surf.	96	879	√	x	√	√	x	√	√	√	√	√	√	x	√	√	√
Vestibule																	
Anterior surf.	92	650	x	x	x	√	√			√	√	√	√	√	√	√	√
Posterior surf.	96	879	√	√	√	√	√	√	√	√	√	√	x	x	√	√	√
Right surf.	98	879	√	√	√	√	√	√	√	√	√	√	x	√	√	√	√
Left surf.	98	879	√	√	√	√	√	√	√	√	√	√	x	√	√	x	√
Vagina																	
Anterior wall	11	578	x	x	x	√	x	x	x	√	x	√	x	x	√	x	
Posterior wall	13	578	x	x	x	√	x	x	x	√	x	x	x	x	√	x	
Right wall	14	578	x	x	x	√	√	x	x	√	x	√	x	x	√	x	
Left wall	14	578	x	x	x	√	√	x	x	√	x	√	x	x	√	x	
Cervix	5	878	x	x	x	√	√	√	x	x	x	x	x	x	x	x	x
AREA OF PRESSURE																	
Vagina																	
Anterior wall	89	878	x	√	√	√	√	√	√	√	√	√	x	x	x	x	√
Posterior wall	93	878	√	√	√	√	√	√	√	√	√	√	√	x	x	√	√
Cervix	84	878	x	√	√	√	√	√	√	√	√	√	x	√	x	x	√

The check (√) shows response, x shows lack of response to stimulation in the designated area. The tests were made by five experienced gynecologists, two of them female, on a total of 879 females. In all of the tests, the vagina was spread with a speculum, and all testing of internal structures was done with especial care not to provide any simultaneous stimulation of the external areas. The tests of tactile responsiveness were made with a glass, metal, or cotton-tipped probe with which the indicated areas were gently stroked. Awareness of pressure was tested by exerting distinct pressure at the indicated points with an object larger than a probe. Finer standardizations of the tests proved impractical under the office conditions. It should be noted that awareness of tactile stimulation or of pressure does not demonstrate the capacity to be aroused erotically by similar stimuli; but it seems probable that any area which is not responsive to tactile stimulation or pressure cannot be involved in erotic response.

side or inside surfaces of the labia, and about equally responsive to stimulation of either the left or right labium (Table 174).

As sources of erotic arousal, the labia minora seem to be fully as important as the clitoris. Consequently, masturbation in the female usually involves some sort of stimulation of the inner surfaces of these labia. Sometimes this is accomplished through digital strokes which may be confined to the labial surfaces; usually the strokes extend to the clitoris which is located at the upper (the anterior) end of the genital area where the two labia minora unite to form a clitoral hood. However, manipulation of the labia may also involve nerves that lie deep within their tissues. This is suggested by the fact that digital stimulation of the labia is often effected while the thighs are tightly pressed together so pressure may be exerted on the labia. Sometimes the technique may involve nothing more than a tightening of the thighs or a crossing of the legs, without digital stimulation. Sometimes the labia are rhythmically pulled in masturbation. During coitus, the entrance of the male organ into the vagina may provide considerable stimulation for the labia minora. All of these techniques are also significant in producing the muscular tensions which are of prime importance in the development of sexual responses (p. 618).

Labia Majora and Scrotum. The solid, outer lips (the labia majora) of the female genitalia are (at least in part) homologues of the skin which forms the scrotum of the male (Figures 118, 119). Both structures develop from the two swollen ridges which lie on the sides of the genital area in the developing embryo. They form the lateral limits of the adult female genitalia. The homologous ridges in the male embryo become hollow during their development. Finally their two cavities unite to form a single cavity in a single sac, the scrotum of the male. In the human male, shortly before the completion of embryonic development, the testes, which have previously been located in the body cavity, descend through the inguinal canals (which lie in the groins) and take up their permanent locations in the scrotum.

The labia majora of the female are sensitive to tactile stimulation. This was so in some 92 per cent of the women who were tested by the gynecologists (Table 174). But there is a difference between the capacity to respond to tactile stimulation and the capacity to respond erotically. Although it is improbable that any area which is insensitive to tactile manipulation could be stimulated erotically, some areas which are tactilely sensitive (e.g., the backs of the hands, the shoulders) are of no especial importance as sources of erotic response. We do not yet have evidence that the labia majora contribute in any important way to the erotic responses of the female.

Neither do we have any evidence that the skin of the scrotum is more sensitive than any other skin-covered surface of the body, and the scrotum does not seem an important source of erotic arousal.[8] There are quite a few males who react erotically, and a small number who

---

[8] Havelock Ellis 1936(II,1):123 also states that the scrotum "is not the seat of any voluptuous sensation."

may respond to the point of orgasm when there is some stimulation or more active manipulation of the testes; but this response is not due to the stimulation of the skin of the scrotum.

Vestibule of Vagina. The labia minora continue inward to form a broad, funnel-shaped vestibule which leads to the actual entrance (the orifice or introitus) of the vagina (Figure 118). The structure represents an in-pocketing of external epidermis which is well supplied with end organs of touch. Nearly all females—about 97 per cent according to the gynecologic tests (Table 174)—are distinctly conscious of tactile stimulation applied *anywhere* in this vestibule, and only a very occasional female out of the 879 who were tested proved to be entirely insensitive in the area. For nearly all women the vestibule is as important a source of erotic stimulation as the labia minora or the clitoris. Since the vestibule must be penetrated by the penis of the male in coitus, it is of considerable importance as a source of erotic stimulation for the female.

The hymen of the virgin female is a more or less thin membrane which lies at the inner limits of the vestibule. It is attached by its outer rim, partially blocking the entrance to the vagina. It usually has a natural opening of some diameter located in the center of the membrane, but this may, of course, be enlarged either by coitus or by the insertion of fingers, tampons, or other objects. An unusually thick or tough hymen which might cause considerable pain if it were first stretched or torn in coitus, may be more easily stretched or cut by the physician who makes a pre-marital examination. Remnants of the hymen almost always persist even after long years of coital experience, and the remnant of the tissue is sometimes sensitive. It is not clear whether the sensitivity depends on nerves in the remnants of the tissue or on the fact that movements of the tissue may stimulate the underlying nerves.

A ring of powerful muscles (the levator muscles) lies just beyond the vaginal entrance. The cavity of the vagina extends beyond this point. The female may be very conscious of pressure on the levators. The muscles may respond reflexly when they are stimulated by pressure, and most females are erotically aroused when they are so stimulated.

Interior of Vagina. The vagina is the internal cavity which lies beyond the external genitalia of the female (Figure 118). Unlike its vestibule, the vagina is derived embryologically from the primitive egg ducts which, like nearly all other internal body structures, are poorly supplied with end organs of touch. The internal (entodermal) origin of the lining of the vagina makes it similar in this respect to the rectum and other parts of the digestive tract. There is no functional homologue of the vagina in the male.

In most females the walls of the vagina are devoid of end organs of touch and are quite insensitive when they are gently stroked or lightly pressed. For most individuals the insensitivity extends to every part of the vagina. Among the women who were tested in our gynecologic sample, less than 14 per cent were at all conscious that they had been touched (Table 174). Most of those who did make some response had the sensitivity confined to certain points, in most cases on the upper (anterior) walls of the vagina just inside the vaginal entrance. The limited histologic studies of vaginal tissues confirm this experimental evidence that end organs of touch are in most cases lacking in the walls of the vagina, although some nerves have been found at spots in the vaginal walls of some individuals.[9]

This insensitivity of the vagina has been recognized by gynecologists who regularly probe and do surface operations in this area without using anesthesia.[10] Under such conditions most patients show little if any awareness of pain. There is some individual variation in this regard, and clinicians are aware of this, for they ordinarily stand prepared to administer a local anesthetic if the patient does register pain.

The relative unimportance of the vagina as a center of erotic stimulation is further attested by the fact that relatively few females masturbate by making deep vaginal insertions (p. 161, Table 37). Fully 84 per cent of the females in the sample who had masturbated had depended chiefly on labial and clitorial stimulation. Although some 20 per cent had masturbated on occasion by inserting their fingers or other objects into the vagina, only a small portion had regularly used that technique. Moreover, the majority of those who had made insertions did so primarily for the sake of providing additional pressure on the introital ring of muscles, or to stimulate the anterior wall of the vagina at the base of the clitoris, and they had not made deeper insertions. As we shall note below, there is satisfaction to be obtained from deeper penetration of the vagina by way of nerve masses that lie outside of the vaginal wall itself, but all the evidence indicates that the vaginal walls are quite insensitive in the great majority of females.[11]

---

[9] Relative lack of nerves and end organs in the vaginal surface is noted by: Dahl acc. Kuntz 1945:319. Undeutsch 1950:447. Dr. F. J. Hector (Bristol, England) and Dr. K. E. Kranz (University of Vermont) have furnished us with histologic data on this point. Vaginal sensitivity applying primarily to the area on the anterior wall at the base of the clitoris is also noted in: Lewis 1942:8. Grafenberg 1950:146, 148.

[10] From our gynecologic consultants, we have abundant data on the limited necessity of using anesthesia in vaginal operations. See also Döderlein and Krönig 1907:88.

[11] Simone de Beauvoir 1952:373 says vaginal pleasure certainly exists, and pro-

In most of the homosexual relations had by females, there is no attempt at deep vaginal insertions (Table 130). Once again, the insertions that are made are usually confined to the introitus or intended to stimulate the anterior wall of the vagina at the base of the clitoris. Occasionally there are deeper penetrations in order to reach the perineal nerves (p. 584). This restriction of so much of the homosexual technique is especially significant because, as we have already noted, homosexual females have a better than average understanding of female genital anatomy.

On the other hand, many females, and perhaps a majority of them, find that when coitus involves deep vaginal penetrations, they secure a type of satisfaction which differs from that provided by the stimulation of the labia or clitoris alone. In view of the evidence that the walls of the vagina are ordinarily insensitive, it is obvious that the satisfactions obtained from vaginal penetration must depend on some mechanism that lies outside of the vaginal walls themselves.

There is a parallel situation in anal coitus. The anus, like the entrance to the vagina, is richly supplied with nerves, but the rectum, like the depths of the vagina, is a tube which is poorly supplied with sensory nerves. However, the receiving partner, female or male, often reports that the deep penetration of the rectum may bring satisfaction which is, in many respects, comparable to that which may be obtained from a deep vaginal insertion.

There may be six or more sources of the satisfactions obtainable from deep vaginal penetrations, and several or all of these may be involved in any particular case. The six sources are:

1. Psychologic satisfaction in knowing that a sexual union and deep penetration have been effected. The realization that the partner is being satisfied may be a factor of considerable importance here.

2. Tactile stimulation coming from the full body contact with the partner, and from his weight. This may result in pressures on various internal organs which can produce "referred sensations." These may be incorrectly interpreted as coming from surface stimulation.

3. Tactile stimulation by the male genitalia or body pressing against the labia minora, the clitoris, or the vestibule of the vagina. This alone would provide sufficient stimulation to bring most females to orgasm. The location of this stimulation may be correctly recognized, or it may be incorrectly attributed to the interior of the vagina (see p. 584).

poses (without specific data) that vaginal masturbation seems more common than we have indicated.

4. Stimulation of the levator ring of muscles in coitus. Such stimulation may bring reflex spasms which may have distinctly erotic significance.

5. Stimulation of the nerves that lie on the perineal muscle mass (the so-called pelvic sling), which is located between the rectum and the vagina (see the discussion of the perineum, below).

6. The direct stimulation, in some females, of end organs in the walls of the vagina itself. But this can be true only of the 14 per cent who are conscious of tactile stimulation of the area. There is, however, no evidence that the vagina is ever the sole source of arousal, or even the primary source of erotic arousal in any female.

Some of the psychoanalysts and some other clinicians insist that only vaginal stimulation and a "vaginal orgasm" can provide a psychologically satisfactory culmination to the activity of a "sexually mature" female. It is difficult, however, in the light of our present understanding of the anatomy and physiology of sexual response, to understand what can be meant by a "vaginal orgasm." The literature usually implies that the vagina itself should be the center of sensory stimulation, and this as we have seen is a physical and physiologic impossibility for nearly all females. Freud recognized that the clitoris is highly sensitive and the vagina insensitive in the younger female, but he contended that psychosexual maturation involved a subordination of clitoral reactions and a development of sensitivity within the vagina itself; but there are no anatomic data to indicate that such a physical transformation has ever been observed or is possible.[12]

---

[12] For Freud's interpretation of the relative importance of the clitoris and vagina, and the adoption of this interpretation by many of the psychoanalysts, see, for instance: Freud 1933:161 (". . . in the phallic phase of the girl, the clitoris is the dominant erotogenic zone. But it is not destined to remain so; with the change to femininity, the clitoris must give up to the vagina its sensitivity, and, with it, its importance, either wholly or in part"). Freud 1935:278 ("The clitoris in the girl, moreover, is in every way equivalent during childhood to the penis. . . . In the transition to womanhood very much depends upon the early and complete relegation of this sensitivity from the clitoris over to the vaginal orifice"). Hitschmann and Bergler 1936:15 (". . . the girl . . . must undertake a *removal of the leading sexual zone* from the clitoris to the vagina. . . . If this transition is not successful, then the woman cannot experience satisfaction in the sexual act. . . . *The first and decisive requisite of a normal orgasm is vaginal sensitivity*"). Deutsch 1945(2):80 (". . . the clitoris preserves its excitability during the latency period and is unwilling to cede its function smoothly, while the vagina for its part does not prove completely willing to take over both functions, reproduction and sexual pleasure"). Fenichel 1945:82 ("The significance of the phallic period for the female sex is associated with the fact that the feminine genitals have two leading erogenous zones: the clitoris and the vagina. In the infantile genital period the former and in the adult period the latter is in the foreground. The change from the clitoris as the leading zone to the vagina is a step that definitely occurs in or after puberty only"). Kroger and Freed 1950:528 ("Hence, in the child the clitoris gives sexual satisfaction, while in

The concept of a vaginal orgasm may mean, on the other hand, that the spasms that accompany or follow orgasm involve the vagina; and in much of the psychoanalytic literature there is an implication that the vagina must be chiefly involved before one may expect any maximum and "mature" psychosexual satisfaction.[13] This is an equally untenable interpretation for, as we shall see in Chapter 15, most parts of the nervous system, and all parts of the body which are controlled by those parts of the nervous system, are involved whenever there is sexual response and orgasm. In some individuals the spasms or convulsions that follow orgasm are intense and prolonged, and in others they are mild and of short duration. The individual differences in patterns of response are quite persistent throughout an individual's lifetime, and probably depend upon inherent capacities more than upon learned acquirements. Those females who have extensive spasms throughout their bodies when they reach orgasm are the ones who are likely to have vaginal convulsions of some magnitude at the same time. Those who make few gross body responses in orgasm are not likely to show intense vaginal contractions. No question of "maturity" seems to be involved, and there is no evidence that the vagina responds in orgasm as a separate organ and apart from the total body. Whether a female or male derives more or less intense sensory or psychologic satisfaction when the vaginal spasms are more or less extreme is a matter which it would be very difficult to analyze.

---

the normal adult woman the vagina is supposed to be the principal sexual organ . . . . in frigid women the transference of sexual satisfaction and excitement from the clitoris to the vagina, which usually occurs with emotional maturation, does not take place). See also: Chideckel 1935:39. Ferenczi 1936: 255–256. Knight 1943:28. Abraham (1927) 1948:284.

[13] The shift from clitoris to vagina is sometimes stated to be psychologic rather than physiologic, and in psychoanalytic theory the failure to effect this change is frequently considered the chief cause of frigidity. For example, see: Freud 1935:278 ("In those women who are sexually anaesthetic, as it is called, the clitoris has stubbornly retained this sensitivity"). Hitschmann and Bergler 1936:20 ("Under frigidity we understand the incapacity of woman to have a *vaginal orgasm*. . . . The sole criterion of frigidity is the absence of the vaginal orgasm"). Deutsch 1944(1):233 ("The competition of the clitoris, which intercepts the excitations unable to reach the vagina, and the genital trauma then create the dispositional basis of a permanent sexual inhibition, *i.e.*, frigidity"). Abraham (1927) 1948:359 ("In . . . frigidity the pleasurable sensation is as a rule situated in the clitoris and the vaginal zone has none"). Kroger and Freed 1950:526 ("However, as a general rule, the question of what constitutes true frigidity depends on whether clitoric or vaginal response is achieved. It is believed that the clitoris does not often come into contact with the male organ during intercourse, and, if a transfer in sensation occurs from the clitoris to the vagina, it is purely psychologic and unconscious. In completely frigid women this psychologic transmission is always disturbed. Therefore, the problem of frigidity is reduced to a psychologic basis"). See also: Lundberg and Farnham 1947:266. Stokes 1948:39. Bergler 1951:216 (transfer is purely psychologic). Beauvoir 1952:372 ("the clitorid orgasm . . . is a kind of detumescence . . . only indirectly connected with normal coition, and it plays no part in procreation").

This question is one of considerable importance because much of the literature and many of the clinicians, including psychoanalysts and some of the clinical psychologists and marriage counselors, have expended considerable effort trying to teach their patients to transfer "clitoral responses" into "vaginal responses." Some hundreds of the women in our own study and many thousands of the patients of certain clinicians have consequently been much disturbed by their failure to accomplish this biologic impossibility.

Cervix. The cervix is the lower portion of the uterus (Figure 118). It protrudes into the deeper recesses of the vaginal cavity, and stands out from the vaginal wall as a rounded and blunt tip about as large or larger than the tip of a thumb. It has been identified by some of our subjects, as well as by many of the patients who go to gynecologists, as an area which must be stimulated by the penetrating male organ before they can achieve full and complete satisfaction in orgasm; but most females are incapable of localizing the sources of their sexual arousal, and gynecologic patients may insist that they feel the clinician touching the cervix when, in reality, the stimulation had been applied to the upper (anterior) wall of the vestibule to the vagina near the clitoris. All of the clinical and experimental data show that the surface of the cervix is the most completely insensitive part of the female genital anatomy. Some 95 per cent of the 879 women tested by the gynecologists for the present study were totally unaware that they had been touched when the cervix was stroked or even lightly pressed (Table 174). Less than 5 per cent were more or less conscious of such stimulation, and only 2 per cent of the group showed anything more than localized and vague responses.

Histologic studies show that there are essentially no tactile nerve ends in the surfaces of the cervix. This is further confirmed by gynecologic experience, for the surfaces are regularly cauterized and operated upon in other ways without the use of any anesthesia—and nearly all such patients are unaware that they have been touched. Cutting deeper into the tissue of the cervix may lead an occasional patient to register pain, and the dilation of the cervical canal causes most patients to feel intense pain. In none of these instances, however, is there any evidence of erotic response.[14]

Perineum. The perineal area includes and lies between the lower portions of the digestive and the reproductive tracts. The area is essentially the same in the female and the male. The surface of the perineum includes the anal and genital areas and all of the space between.

---

[14] Our data on the insensitivity of the cervix come from the abundant experience of our gynecologic consultants. See our Table 174. See also: Döderlein and Krönig 1907:88. Malchow 1923:183. Lewis 1942:8. Dickinson ms.

This surface is highly sensitive to touch, and tactile stimulation of the area may provide considerable erotic arousal.

The perineal area is occupied for the most part by layers of muscles, the so-called pelvic sling. Within and on this muscular mass there are nerves, and the area is, in consequence, definitely sensitive when it is stimulated with any sufficiently strong pressure. Many males are quickly brought to erection when pressure is applied on the perineal surface at a point which is about midway between the anus and the scrotum. In the case of the female, strong pressure applied from inside on the posterior (lower) wall at the back of the vagina may stimulate these same nerves, and this is one of the sources of the satisfaction which many females experience when the vagina is penetrated in coitus. Deep penetrations of the rectum may stimulate the same perineal nerves, and prove to be similarly erotic.

Anus. The anal area is erotically responsive in some individuals. In others it appears to have no particular erotic significance even though it may be highly sensitive to tactile stimulation. As many as half or more of the population may find some degree of erotic satisfaction in anal stimulation, but good incidence data are not available. There are some females and males who may be as aroused erotically by anal stimulation as they are by stimulation of the genitalia, or who may be more intensely aroused.

The erotic sensitivity of the anal area depends in part upon the fact that there are abundant end organs of touch throughout the anal surfaces, and in part upon the fact that reactions of the muscles (the anal sphincters) which normally keep the anus closed may be erotically stimulating. Some persons are psychologically aroused while others respond negatively to the idea of anal intercourse, and psychologic factors may have a good deal to do with the erotic or non-erotic significance of such contacts. These generalizations apply to both females and males.

Penetration of the anus may cause pain, and this may intensify the sexual responses for some persons.

The anus is particularly significant in sexual responses because the anal and genital areas share some muscles in common, and the activity of either area may bring the other area into action. Stimulation of the genitalia, both in the female and the male, may cause anal constrictions. Gynecologists frequently observe that stimulation of the clitoris, or of the areas about the clitoris and the urethral meatus, may cause contractions of the anus,[15] the hymenal ring, and the vaginal and peri-

---

[15] An early reference to anal contractions during coitus is found in Aristotle [4th cent. B.C.]: Problems, Bk.IV:879b.

neal muscles. As rhythmic muscular movements develop during sexual responses, and particularly in the spasms that follow orgasm, the anal sphincter may rhythmically open and close. The incentives for anal insertions of various sorts and for anal intercourse lie partly in the significance of these rhythmic responses of the anal sphincters.

Conversely, contractions of the anal sphincter, whether they be voluntary or initiated by erotic stimulation, may bring contractions of the muscles that extend into the genitalia and produce erection in the male or movement of the genital parts in the female. As a matter of fact, contractions of the anal sphincter appear to produce contractions of muscles in various remote parts of the body, including areas as far away as the throat and the nose. Anal contractions may cause the sides of the nose to flare, and the subject is inclined to inhale deeply—both of which are characteristic aspects of an individual who is responding sexually. When the clinician has difficulty in bringing a patient out of anesthesia, he may start deep breathing by inserting a gloved finger into the anus of the patient. Anal contractions may also be associated with contractions of still other muscles elsewhere in the body; for the contraction of any muscle, anywhere in the body, may develop tensions in every other muscle in the body (p. 618).

Apparently the abdominal diaphragm is involved when there is anal contraction, for it is very difficult for an individual to exhale as long as the anal sphincters are under any considerable tension.

In brief, anal contractions, perineal responses, genital responses, and nasal and oral responses are so closely associated that one may believe that some sort of simple and direct reflex arc is involved. We do not yet understand the neural bases of such a connection.

Breasts. The breasts of both the male and the female may be more sensitive than some other parts of the body. Because of the greater size of the female breast its reactions to tactile stimulation are quite generally known. Among the infra-human mammals, the breast rarely plays a part in sexual activity,[16] but among human animals there may be a considerable amount of manual or oral stimulation of the female breast. In American patterns of sexual behavior, breast stimulation most regularly occurs among the better educated groups where nearly all of the males (99%) manually manipulate the female breast, and about 93 per cent orally stimulate the female breast during heterosexual petting or pre-coital play (Tables 73, 100). All of this usually stimulates the male erotically, but the significance for the female has probably been

---

[16] The relative infrequency of breast stimulation in the sexual activity of the infra-human mammal is also noted by Ford and Beach 1951:48. See also p. 255 of the present volume.

overestimated. There are some females who appear to find no erotic satisfaction in having their breasts manipulated; perhaps half of them do derive some distinct satisfaction, but not more than a very small percentage ever respond intensely enough to reach orgasm as a result of such stimulation (Chapter 5).[17] Some females hold their own breasts during masturbation, coitus, or homosexual activities, evidently deriving some satisfaction from the pressure so applied; but there were only 11 per cent of the females in our sample who recorded any frequent use of breast stimulation as an aid to masturbation (Table 37).

Because of their smaller size the sensitivity of the male breasts has not so often been recognized except among some of the males who have had homosexual experience. This is undoubtedly not due to differences between heterosexual and homosexual males, but to the fact that relatively few females ever try to stimulate the breasts of their male partners, whereas such behavior is rather frequent in male homosexual relations. Our homosexual histories suggest that there may be as many males as there are females whose breasts are distinctly sensitive. A few males may even reach orgasm as a result of breast stimulation.[18]

Mouth. The lips, the tongue, and the whole interior of the mouth constitute or could constitute for most individuals an erogenous area of nearly as great significance as the genitalia. Erotic arousal always involves the entire nervous system, and there appears to be some oral response whenever there is sexual arousal. We have already noted (p. 586) the apparently reflex connections between contractions of the genital, anal, nasal, and oral musculatures. Most uninhibited individuals become quite conscious of their oral responses, particularly if they accept deep kissing, mouth-breast contacts, or mouth-genital contacts as part of their sexual play.

The significance of the mouth depends, of course, upon the fact that all of its parts are richly supplied with nerves. This is true throughout the class Mammalia, and also true of some of the other vertebrates. Fish, lizards, many birds, and practically all of the mammalian species which have been studied are likely to place their mouths on some part of the partner's body during pre-coital play or actual coitus (see p. 229). Mouth-to-mouth contacts between some of the birds and mammals may occasionally be continued for hours. The sexual significance

---

[17] Records of females reaching orgasm from breast stimulation alone are rare. In addition to our own data (p. 161), we note: Moraglia 1897:7–8. Rohleder 1921:44. Eberhard 1924:246. Kind and Moreck 1930:156. Dickinson 1949: 66–67. Grafenberg 1950:146.

[18] The sensitivity of male breasts is recorded in our 1948:575. It is also recorded in: Van de Velde 1930:164. Féré 1932:105. Pillay 1950:81.

of the mouth is obviously of long phylogenetic standing.[19] The human animal testifies to its origin when it engages in mouth activity during sexual relationships. The human is exceptional among the mammals when it abstains from oral activities because of learned social proprieties, moral restraints, or exaggerated ideas of sanitation.

In the course of mammalian oral activity, the lips may be pressed against the mouth or against some other portion of the partner's body; tongues are brought into contact or used to lick the body; the lips and the teeth may nibble, or the teeth may bite more severely somewhere on the partner's body. In some of the lower mammals (e.g., the mink) such biting may penetrate the skin of the partner,[20] and while this is the means by which the males of these species hold the females in position for coitus, the immediate source of such behavior probably lies in the sensory satisfaction which the male secures by making these contacts. Even such large animals as stallions, which are ill-equipped to make contacts with their mouths, keep their lips in constant motion during sexual activity and constantly attempt to nibble or bite over the surface of the body of the mare. In many mammalian species, where the tip of the nose may be as sensitive as the lips of the mouth, the lips and the nose are used indiscriminately to make contacts with all parts of the body.[21] All of these things also happen during sexual activity in an uninhibited human animal.

It is not surprising that the two areas of the body which are most sensitive erotically, namely the mouth and the genitalia, should frequently be brought into direct contact. The high incidence of such mouth-genital contacts in all species of mammals, from one to the other end of the class Mammalia, and the high frequency of mouth-genital contacts in the human animal have been detailed in our volume on the male (1948: especially pages 368–373, 573–578), and in the present volume (pp. 229, 257).

Ears. The lobe of the ear and the inner cavity leading from it are points of especial sensitivity for at least some persons.[22] The ear lobes

---

[19] For our own data on oral activity among the lower mammals, see p. 229 including footnotes 3 and 4. Also see: Yerkes 1929:296. Zuckerman 1932:123, 147, 227, 230. Yerkes and Elder 1936a:10–11. Havelock Ellis 1936(I,2):85. Reed 1946:passim. Beach in Hoch and Zubin 1949:55–62. Ford and Beach 1951:46–55.

[20] Biting through the skin of the sexual partner's neck is recorded for the mink, ferret, marten, and sable. See also: Ford and Beach 1951:58–59.

[21] Manipulation of the sexual partner's body with the nose and lips has occurred among nearly all of the species of mammals which we have observed (see p. 229). See also: Stone 1922 (rat). Louttit 1927 (guinea pig). Shadle 1946 (porcupine). Reed 1946 (various mammals).

[22] The erotic sensitivity of the ear is quite common knowledge. In the literature see: Van de Velde 1930:45. Havelock Ellis 1936(II,1):143. Stone and Stone 1937:221; 1952:182. Beach 1947b:246.

become engorged with blood during sexual arousal, and become increasingly sensitive at that time. An occasional female or male may reach orgasm as a result of the stimulation of the ears.

Buttocks. Tactile stimulation and heavier pressure on the buttocks may elicit unusually strong responses from the gluteal muscles.[23] These are the largest and the chief muscles on the back surface of the upper part of the leg. Contractions of these muscles reflect, more than any other one factor, the development of the nervous and muscular tensions which are involved in erotic arousal. Some persons, both female and male, deliberately contract these gluteal muscles to build up their erotic responses. Movements of vertebral muscles in conjunction with contractions of the gluteal muscles are the chief means by which the pelvis is thrown forward in rhythmic thrusts during copulation.

Thighs. The thin-skinned, inner surfaces of the thighs, particularly on their midlines, are richly supplied with nerves. Any tactile stimulation of these areas may contribute to erotic arousal.[24] Such stimulation may bring responses of the adductor muscles on the inner faces of the thighs, and of the abductor muscles which are located on the outer surfaces of the legs, and as a result the thighs may be rolled together or thrown apart with distinctive movements that are characteristic of much coital, masturbatory, and other sexual activity. These movements of the adductors and abductors are also involved in the build-up of nervous tensions throughout the whole body (see p. 618).

Other Body Surfaces. The areas just listed are the ones which are most often concerned in erotic stimulation and response. Other parts of the body may, however, be involved, and for some individuals these other parts may be as significant as any of the particular areas listed.

The nape of the neck, the throat, the soles of the feet, the palms of the hand, the armpits, the tips of the fingers, the toes, the navel area, the midline at the lower end of the back, the whole abdominal area, the whole pubic area, the groin, and still other parts have been recognized as areas which may be erotically sensitive under tactile stimulation. Even non-sensitive, non-living structures like teeth and hair may sometimes become sources of erotic stimulation, because movements of those structures stimulate the sensitive nerves that lie at their bases. There are females in our histories who have been brought to orgasm

---

[23] Contraction of the buttocks was once considered the mechanism by which semen was expelled during coitus. See Aristotle [4th cent. b.c.]: Problems, Bk.IV: 876b.

[24] There are abundant data in our records on the erotic sensitivity of the inner surfaces of the thigh. There are few references in the published studies, but see: Van de Velde 1930:45. Havelock Ellis 1936(II,2):113. Kahn 1939:70. Fulton 1949:140 records that manipulation of this area can cause penile erection in dogs, cats, monkeys, and humans whose spinal cords are severed.

by having their eyebrows stroked, or by having the hairs on some other part of their bodies gently blown, or by having pressure applied on the teeth alone. This may be a factor in the biting which often accompanies sexual activities. Such stimuli are most effective when there is some accompanying and significant psychologic stimulation. Acting alone, these unusual sorts of stimuli would rarely be sufficient to effect any considerable arousal. Acting in conjunction with other physical and psychologic stimuli, they may, on occasion, provide the additional impetus which is necessary to carry the individual on to orgasm.

STIMULATION THROUGH OTHER SENSE ORGANS

Many persons are conscious of the fact that they are sexually stimulated by things that they see, smell, taste, or hear. Oriental, Islamic, Classical Greek and Roman, Medieval, Galante, and more modern European sex literatures regularly refer to odors and to other chemical stimuli which are "aphrodisiacs" capable of exciting sexual responses.

But it is not certain that stimuli received through these other sense organs effect erotic arousal in the same direct fashion that tactile stimuli do. It is conceivable that great intensities of light, lights of particular colors, or movements of lights of different intensities or of different colors, do have some direct effect upon an animal's sexual responses; but specific investigations have not yet been made.

It is also not impossible and even probable that strong odors, spices, and loud noises have direct effects upon the nervous system and thus start the physiologic changes which constitute sexual response. It is rather clear that particular sorts of rhythms (e.g., march and waltz time), variations of tempo, particular sequences of pitch (e.g., continued repetitions of one note or chord, alternations of tones which lie an exact octave apart), and variations in volume (e.g., crescendos and diminuendos, sudden and sharp notes, heavy chords) produce physiologic effects on the human and on some other animals which have some affinity to sexual responses. But on none of these points have there been sufficient scientific studies to warrant further discussion here.

It is more certain that stimuli received through these other sense organs operate primarily through the psychic associations which they evoke (p. 647). The arousal which the average male and some females experience in seeing persons with whom they have had previous sexual contacts, in seeing potential new partners, or in seeing articles of clothing or other objects which have been connected with some previous experience, must depend largely upon the fact that they are reacting, consciously or subconsciously, to memories of the past events. In the same fashion one may be aroused by entering a room, seeing a beach, a mountain top, or a sunset which, by its similarity to some situation associated with the previous sexual contact, brings arousal through recall of that experience. Sexual arousal from the taste of some particular food, through particular odors, or from hearing particular birds sing, particular tones of ringing bells, certain words spoken, the tones of particular voices, particular musical themes sung or played, or any other particular sounds, appears to depend largely upon associations of those things with previous sexual experience.

All of this is a picture of psychologic learning and conditioning, and not one of direct mechanical stimulation of the sort which tactile stimuli provide. In considering the erotic significances of things that are seen, smelled, tasted, or heard, the possibility that both psychologic and sensory factors are involved should, therefore, always be taken into account.

SUMMARY AND COMPARISONS OF FEMALE AND MALE

The physical differences between the genitalia and reproductive functions of the mammalian female and male have been the chief basis for the longstanding opinion that there must be similarly great differences in the sexual physiology and psychology of the two sexes. In view of the considerable significance of such concepts in human social relations, it has, therefore, been important to reexamine and reevaluate the similarities and differences in the anatomic structures which are involved in sexual responses and orgasm in the two sexes. Specifically we have found that:

1. In both sexes, end organs of touch are the chief physical bases of sexual response. There seems to be no reason for believing that these organs are located differently in the two sexes, that they are on the whole more or less abundant in either sex, or that there are basic differences in the capacities of the two sexes to respond to the stimulation of these end organs.

2. The genitalia of the female and the male originate embryologically from essentially identical structures and, as adult structures, their homologous parts serve very similar functions. The penis, in spite of its greater size, is not known to be better equipped with sensory nerves than the much smaller clitoris. Both structures are of considerable significance in sexual arousal. The chief consequence of the larger size of the penis is the fact that sexual stimulation is more often directed specifically toward it. Its larger size accounts in part for the greater psychologic significance of the penis to the male.

3. The labia minora and the vestibule of the vagina provide more extensive sensitive areas in the female than are to be found in any

homologous structure of the male. Any advantage which the larger size of the male phallus may provide is equalled or surpassed by the greater extension of the tactilely sensitive areas in the female genitalia.

4. The larger and more protrudent and more extensible phallus of the male, and the more internal anatomy of the female, are factors which may determine the roles which the two sexes assume in coitus. The female may find psychologic satisfaction in her function in receiving, while the male may find satisfaction in his capacity to penetrate during coitus, but it is not clear that this could account for the more aggressive part which the average male plays, and the less aggressive part which the average female plays in sexual activity. The differences in aggressiveness of the average female and male appear to depend upon something more than the differences in their genital anatomy (see Chapters 16 and 18).

5. The vagina of the female is not matched by any functioning structure in the male, but it is of minimum importance in contributing to the erotic responses of the female. It may even contribute more to the sexual arousal of the male than it does to the arousal of the female.

6. The perineal area provides a considerable source of stimulation for both the female and the male. In the male, nerves located among the perineal muscles may be stimulated through direct pressure on the external surface of the perineum or through the rectum; in the female, the area may be stimulated in exactly the same fashion or through vaginal penetrations.

7. Because female breasts are larger, they may appear to be more significant to the female than male breasts are to the male. But since most males are aroused by seeing female breasts, and because most females are, in actuality, only moderately aroused by having their breasts tactilely stimulated, female breasts may be more important sources of erotic stimulation to males than they are to females.

8. The mouth, which is one of the most important erotic areas of the mammalian body, appears to be equally sensitive in the female and the male.

9. Tactile stimulation of the buttocks and of the inner surfaces of the thighs may play a significant part in the picture of sexual response. Responses to such stimulation seem to be essentially the same in the female and the male.

10. All of the other body surfaces which respond to tactile stimuli seem to function, as far as the specific evidence goes, identically in the two sexes.

11. There are no data to indicate that there are any differences in female and male responses which depend directly on the senses of sight, smell, taste, or hearing.

12. In brief, we conclude that the anatomic structures which are most essential to sexual response and orgasm are nearly identical in the human female and male. The differences are relatively few. They are associated with the different functions of the sexes in reproductive processes, but they are of no great significance in the origins and development of sexual response and orgasm. If females and males differ sexually in any basic way, those differences must originate in some other aspect of the biology or psychology of the two sexes (see Chapters 16 and 18). They do not originate in any of the anatomic structures which have been considered here.

Chapter 15

PHYSIOLOGY OF SEXUAL RESPONSE AND ORGASM

The responses which an animal makes when it is stimulated sexually constitute one of the most elaborate and in many respects one of the most remarkable complexes (syndromes) of physiologic phenomena in the whole gamut of mammalian behavior. The reactions may involve changes in pulse rates, blood pressure, breathing rates, peripheral circulation of blood, glandular secretions, changes in sensory capacities, muscular activity, and still other physiologic events which are described in the present chapter. As a climax to all these responses, the reacting individual may experience what we identify as sexual orgasm. There is every reason for believing that most of the physiologic changes which are described in the present chapter take place even in the mildest sexual response, even though the gross movements of the body may be limited and the individual fails to reach orgasm.

The gross aspects of sexual response and orgasm may differ considerably in different individuals. The stimuli which initiate the response may vary in intensity, continuity, and duration, and the animal's responses may depend not only on such variation in the nature of the stimuli, but upon its physiologic state and psychologic background. There is nothing more characteristic of sexual response than the fact that it is not the same in any two individuals. On the other hand, the most obvious variations lie in the gross body movements which are part of the response, and particularly in the spasms or convulsions which follow orgasm; and while these variations are striking and sometimes very prominent, the basic physiologic patterns of response are remarkably uniform among all the mammals, including both man and the infra-human species. Even more significant is the fact that the basic physiology of sexual response is essentially the same among females and males, at least in the human species.

The record in the present chapter emphasizes the physical reality of any sexual response. Whatever the poetry and romance of sex, and whatever the moral and social significance of human sexual behavior, sexual responses involve real and material changes in the physiologic functioning of an animal. The present chapter describes the gross physiologic changes which occur whenever there is sexual response;

594

the neural and hormonal mechanisms which may be responsible for these changes are discussed in Chapters 17 and 18.

PHYSIOLOGIC CHANGES DURING SEXUAL RESPONSE

1. **Tactile and Pressure Responses.** One of the most characteristic qualities of living matter, plant or animal, is its capacity to respond to touch. The normal, first reaction of an organism is to press against any object with which it may come into contact. One-celled animals mass against objects. Multicellular bodies like cockroaches crowd into corners. Infants and small children spontaneously snuggle against other human bodies. Uninhibited human adults do the same thing whenever the opportunity affords.

When the contact causes pain, or subjects the animal to extreme temperatures, the organism may respond negatively and pull away from the stimulus. Higher animals become conditioned by experience and may learn to react negatively to the mere possibility of repeating such a contact. Moral codes and social custom aid in this conditioning, and the human adult, in consequence, often reacts adversely to contacts with other living bodies. It is probable, however, that most negative responses are learned responses, and they do not represent the innate qualities of uneducated protoplasm. Sexual difficulties in marriages, and personal maladjustments, are not infrequently the product of this sort of perversion of the biologically normal reactions to tactile stimuli.

If an animal pulls away from the stimulating object, little else may happen to it physiologically. If it responds by pressing against the object, a considerable series of physiologic events may follow. If the tactile stimulation becomes rhythmic, or the pressure is long-continued, the level of response may increase and build up neuromuscular tensions which become recognizable as sexual responses.

2. **Pulse Rate.** A rise in pulse rate is one of the most obvious and widely recognized results of a mammal's reaction to tactile stimulation or to any sort of erotic stimulation.[1] Such changes in pulse rate are probably an invariable outcome of sexual arousal, although we cannot be certain of this until more extensive studies are made. Unfortunately, exact measurements of these changes are rare in the literature, although they could easily be obtained with automatic recording devices.

---

[1] Such an increase in pulse rate is widely recognized in the literature, as in: Roubaud 1876:16. Caufeynon 1903:57. Bloch 1908:49. Talmey 1912:61; 1915:92. Krafft-Ebing 1922:40. Kisch 1926:288, 344. Bauer 1927(1):154. Van de Velde 1930:245. Dickinson 1933, 1949:fig. 126. Havelock Ellis 1936 (II,1):149, 151. Haire 1937:200. Reich 1942:82. Podolsky 1942:49. Sadler 1944:41. Negri 1949:97. Faller in Hornstein and Faller 1950:234, 237. Ford and Beach 1951:244–246. Stone and Stone 1952:173.

The few records which are available on the human animal indicate that a pulse which normally runs at something between 70 and 80 per minute may be raised to as much as 150 or more when there is erotic arousal, and particularly if the reaction proceeds to the point of orgasm (Figures 120–125).[2] This may approach the pulse rate of an athlete

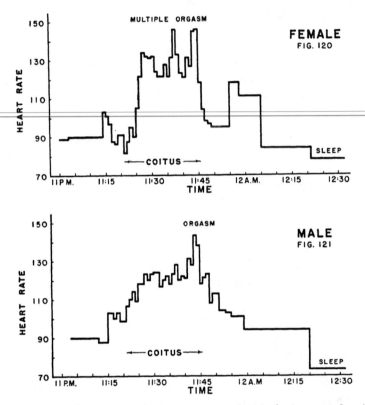

Figures 120–121. Heart rate in human female and male during sexual activity

Showing erotic responses before, during, and after coitus between husband and wife. The female had four orgasms. Data from Boas and Goldschmidt 1932.

[2] Precise measurements on the increase in pulse rate during arousal are few. Mendelsohn 1896:381–384 reproduces graphs of the pulse of human males and females before, during, and after coitus, showing a maximum pulse of 150. Boas and Goldschmidt 1932:99–100 record pulses in human coitus reaching points between 128 and 146 in four consecutive orgasms in a human female, and a pulse of 143 at orgasm in a male. Gantt 1944:128–129, and Gantt in Hoch and Zubin 1949:44, graph the pulses of three male dogs, showing increases at orgasm more marked than those found under other emotional situations. Klumbies and Kleinsorge 1950a:953–956; 1950b:61–66, report on one human male and one female who masturbated to orgasm, finding a maximum pulse of 142 at orgasm in the male, while the female's normal pulse of 63 rose to something between 85 and 97 in five successive orgasms. Polatin and Douglas (ms. 1951) record spontaneous orgasm in a female with a pulse rising from 60 to 140 at orgasm.

during his maximum effort, or that of a man involved in heavy labor.[3] But sometimes the rise is less than this. There is considerable individual variation in this regard, and there may be variation in the same individual on different occasions. We have records which show that an individual whose pulse ordinarily rises to 150 or more during maximum arousal, and who normally does not experience orgasm unless his pulse reaches that height, may sometimes reach orgasm after a less intense

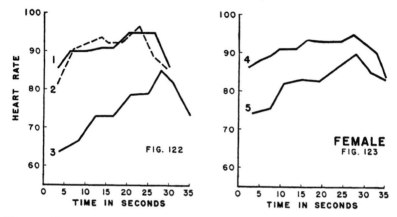

Figures 122–123. Heart rate in human female in five consecutive orgasms

The female masturbated to orgasm by fantasy alone, without genital manipulation. Data from Klumbies and Kleinsorge 1950.

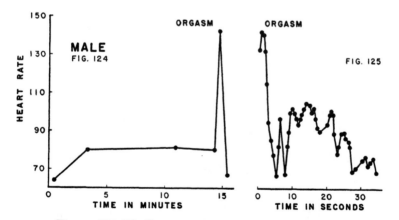

Figures 124–125. Heart rate in human male in orgasm

Orgasm reached in manual masturbation; Figure 125 shows the detail of the after-effects following orgasm. Data from Klumbies and Kleinsorge 1950.

---

[3] Pulse rates during violent exertion are given in: Boas and Goldschmidt 1932:83, 91. Robinson 1938:253–266. Hoff in Howell (Fulton edit.) 1949:661–663. Houssay et al. 1951:475, 481. Pulses in violent exercise varied between 160 and 195 among boys, and 160 to 170 in adults. The maximum pulses of 10 young men ranged from 182 to 208 while running on a treadmill.

experience during which his pulse does not rise above 100. Fatigue, starvation, ill health, psychologic blockages and distractions, and other factors may account for the occasionally lower rates at the time of orgasm. There are similar data on the rise in pulse rate among some other animals, including dogs, during sexual activity (Figures 126–128).

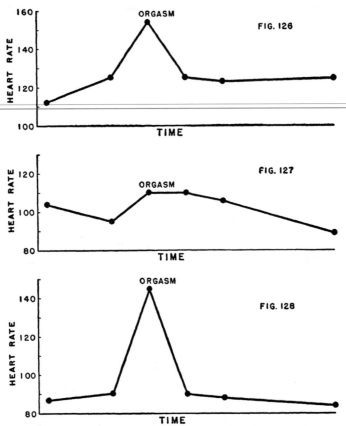

Figures 126–128. Heart rate in three male dogs during sexual activity
Data from Gantt 1944, and Gantt in Hoch and Zubin 1949.

This rise in pulse rate is one of the products of sexual response which many persons may observe in themselves and in their sexual partners. Consequently this has provided one means for determining the nature of the experience of the subjects of the present study. There has been doubt on the part of some persons whether all females are able to recognize erotic arousal, and whether all of them could correctly report their experience for the present study. To those of us who have done the interviewing, this apprehension has seemed for the most part unnecessary. It is impossible to believe that any male would ever be unconscious of the fact that he was aroused sexually, even if he did not have a penis to bear testimony to that effect; and neither does there appear to be any uncertainty in the minds of most women as to whether or when or where they have responded to sexual stimulation. It is true that some younger girls and some of the less experienced older women do, on occasion, hesitate to record their experience until the nature of sexual arousal has been defined for them. It has, therefore, been our standard procedure to explain that arousal is ordinarily accompanied by the realization that "one's pulse is beating faster, one's heart is thumping, and one's breathing has become deeper." This description has almost invariably brought an instantaneous reply from the subject to the effect that she had or had not had such experience during a sexual contact. We venture the opinion that normally intelligent persons who have never become conscious of at least the circulatory disturbances which occur during sexual contacts have never, in actuality, been very much aroused.

3. **Blood Pressure.** The blood pressure of an animal may rise materially when it is sexually aroused. There are few measurements on the human female or male, but these indicate that diastolic blood pressures which have normally been as low as 65 may be raised to 160, and systolic pressures may be raised from 120 to 250 or more at the time of orgasm (Figures 137, 138). There are similar findings in an excellent study on dogs (Figures 129–136).[4] However, even these few records indicate that there is considerable variation among different individuals, and it is probable that there is variation within the history of any single individual. This is another matter on which it would be highly desirable and quite possible to secure further data.

Hemiplegia, stroke, or death may occasionally occur, although not at all frequently, during coital and other sexual activities. The average physician may see a few such cases during a lifetime of practice. In such instances the fatality may be the consequence of this rise in blood pressure which accompanies sexual activity. The increased pressure may lead to a rupture of blood vessels, especially in older individuals

---

[4] A rise in blood pressure during sexual arousal is also mentioned in: Mendelsohn 1896:383–384 (in human). Urbach 1921:126. Bauer 1927(1):154. Hirschfeld 1928(2):242. Van de Velde 1930:245. Sadler 1944:41. Faller in Hornstein and Faller 1950:237. Ford and Beach 1951:244–245. Exact measurements are in: Pussep 1922:61 ff. (in male dogs, 215 mm. at orgasm, in female dogs, 195 mm. at moment of intromission). Scott 1930:97 ff. (nearly all of 100 medical students responded to an erotic moving picture with increases in blood pressure of 5 to 15 mm., and in some cases of 45 mm.). Havelock Ellis 1936(II,1):151–152 (a rise from 65 mm. to 160 mm. in a human female during erotic arousal). Klumbies and Kleinsorge 1950a:953, 955; 1950b:61–62, 64 (in masturbation, an increase in systolic pressure from 120 to 250 mm. in the human male, and from 110 to 160 mm. (and once to 200 mm.) in the female; in diastolic pressure an increase from 80 to 120 in the male, and 80 to 105 in the female).

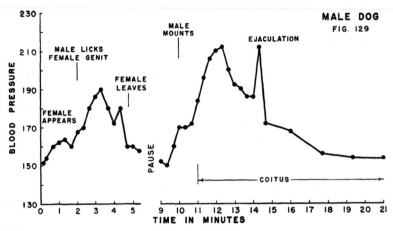

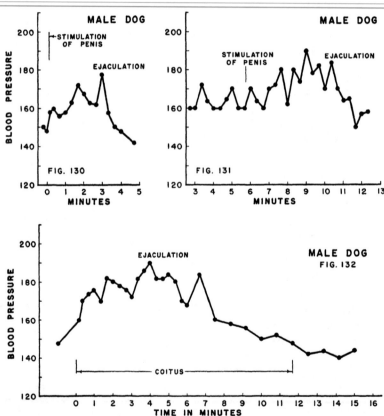

Figures 129–132. Blood pressure in three male dogs during sexual activity

Figures 129, 130 represent two different experiences for the same dog. Figures 129, 132 represent responses in coitus; Figures 130, 131 represent responses in masturbation. Data from Pussep in Weil 1922.

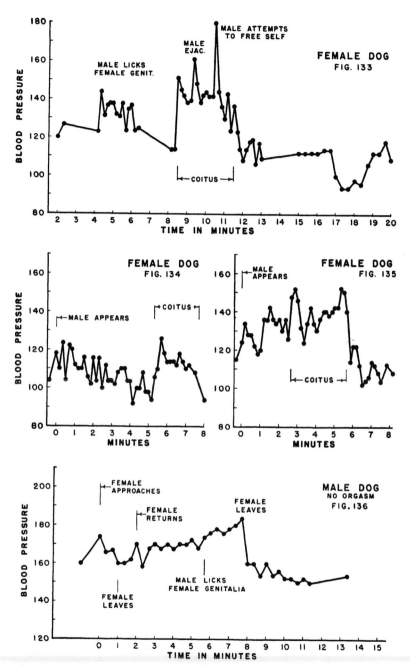

Figures 133–136. Blood pressure in three female dogs and one male dog during sexual activity

The recession from the peak of response in the female bears a striking resemblance to orgasm in the male. Data from Pussep in Weil 1922.

who are handicapped by high pressures to begin with.[5] There would be considerable value in having additional records on this point.

4. **Increased Peripheral Flow of Blood.** Probably as a result of the direct neural stimulation of arterioles, blood is forced into the peripheral areas of the whole body during sexual response. For instance, the

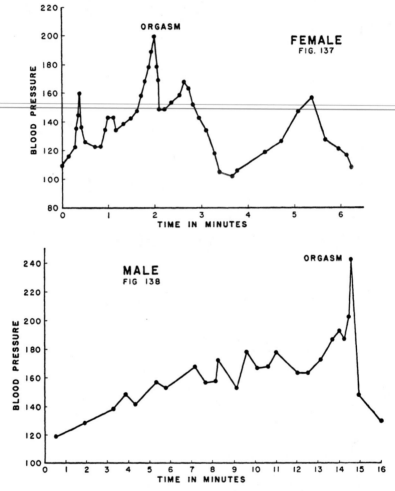

Figures 137–138. Blood pressure in human female and male during sexual activity

The female fantasied, the male masturbated to orgasm. Data from Klumbies and Kleinsorge 1950.

---

[5] Death during sexual activity has been reported from Pliny [23–79 A.D.]: Natural History, Bk. 7, LIII, 184, to the following, for instance: Mendelsohn 1896:384. Rohleder 1907(1):372. Hirschfeld 1928(2):242. Van de Velde 1930:245. Havelock Ellis 1936(II,1):168–169. Klumbies and Kleinsorge 1950a:957–958; 1950b:63, 66.

face of the sexually aroused individual usually becomes flushed. Such changes in coloration may be marked enough to be noticeable to other persons, even when the responses represent nothing more than the mildly erotic reactions which one may experience in meeting a friend. At maximum arousal, the faces of some individuals radically change color, the whole chest and throat may become brilliant or dark red or deep reddish purple, and the genital area may become deeply colored.

One may become conscious of an increase in temperature in his own or the sexual partner's body surfaces, partly due to this peripheral circulation of blood, and perhaps in part due to the neuromuscular tensions which develop when there is any sexual response. Even very cold feet may become warm during sexual activity. The identification of sexual arousal as a fever, a glow, a fire, heat, or warmth, testifies to the widespread understanding that there is this rise in surface temperatures [6]; but there seem to be no precise measurements of the extent of these temperature changes. There seem to be no data which indicate whether there is any rise in the temperature of the body as a whole, as a result of sexual arousal, even though there are these changes in surface temperatures.

The peripheral circulation of blood which occurs during sexual response is paralleled in other emotional situations. The face of the person who is embarrassed, angry, or excited in some other way (but not frightened) may become flushed and red. Since the parasympathetic nervous system is known to be the source of this vasodilation in the other emotional situations, it is probable that the same system is involved in sexual responses. Sexual flushing cannot be the outcome of adrenal or sympathetic nervous stimulation, for both of these agencies effect the sort of vasoconstriction which is responsible for the white and cold body surfaces of most frightened persons.

5. **Tumescence.** In consequence of this increase in the peripheral circulation of blood, all distensible parts of the body become swollen (tumescent) in an individual who is sexually stimulated. This tumescence provides some of the most obvious evidences of sexual arousal.

Almost instantly, or within a matter of seconds or a minute or so after the initiation of a sexual contact, certain areas of the body may become swollen, enlarged, and stiff with an excess of blood. This is equally true of the human and lower mammalian species, both female and male. In some of the most distensible areas, such as the penis, blood is forced in by the arteries faster than the capillaries and the

---

[6] Records of an increase in skin temperature during sexual arousal are also given, for instance, in: Roubaud 1876:16. Bloch 1908:49. Urbach 1921:133. Bauer 1927(1):157. Hirschfeld 1935:44. Havelock Ellis 1936(II,1):149, 167. Haire 1937:200.

veins can carry it away. This engorgement may in itself aid in the process by closing the veins so they cannot carry blood away as rapidly as they normally do.[7] The penis, clitoris, some of the tissues near the entrance to the vagina, the nipples of the breast, and the side walls of the nose contain a spongy *erectile* tissue which makes those structures especially liable to enlargement during sexual arousal. Blood is carried by small arterioles into the spaces of the spongy tissue, or actually forced through the walls of the capillaries into the cavities, and the whole structure consequently enlarges. The penis, for instance, may become from half again to double its usual length, and become turgid and stand forward or up in erection when it is tumescent. The clitoris may become swollen and erect. The labia minora, which are usually limp and folded, may become swollen and prominently protrudent. In both the female and the male, the nipples of the breast may become enlarged, hard, and erect. The soft parts of the nose, the alae, may become swollen and the nostrils in consequence become expanded.

While the phenomenon of tumescence is commonly recognized in connection with the penis, and sometimes with these other erectile organs, it is not so often realized that a considerable tumescence may also occur during sexual arousal in all other parts of the body, even though these other parts do not contain erectile tissues. The surface outlines of the whole body of one who is sexually aroused become quite different from the outlines of one who is not so aroused. The lobes of the ears may become thickened and swollen. The lips of the mouth may become filled with blood and, in most individuals, more protrudent than under ordinary circumstances. The whole breast, particularly of the female, may become swollen, enlarged, and more protrudent, and the general outline of the breast may become more rounded.[8] The anal area may become turgid. The arms and the legs may have their outlines altered. The tumescence is so apparent everywhere over the body that it alone is sufficient evidence of the presence of erotic arousal. Women who pretend arousal when there is none may, to some degree, simulate the motions of coitus; but they cannot voluntarily produce the peripheral circulation of blood and the consequent tumescence of the lips, the breasts, the nipples, the labia minora, and

---

[7] Entrapment of blood in the penis as a result of venous compression in erection is also noted in: Eberth 1904:253. Dickinson 1933, 1949:78–80. Hirsch 1949: 66 (locked). Hooker in Howell (Fulton edit.) 1949:1203 (veins possess funnel-like valves). Maximow and Bloom 1952:487.

[8] Tumescence of various parts of the body during arousal has also been noted by: Rohleder 1907(1):372. Bauer 1927(1):157. Van de Velde 1930:245. Dickinson 1933, 1949:65–68, figs. 103–104 (recognized the tumescence of the whole breast). Havelock Ellis 1936(II,1):144 ff. Sadler 1944:37. Faller in Hornstein and Faller 1950:237. Stone and Stone 1952:173.

the whole body contour which are the unmistakable and almost invariable evidences of erotic arousal.

In some males the erection of the penis may occur in a matter of seconds—even three or four seconds—and in some females the clitoris and the labia minora may respond as quickly. In many species of lower mammals the reactions are even more nearly instantaneous. Stallions, bulls, rams, rats, guinea pigs, porcupines, cats, dogs, apes, and males of other species may come to full erection almost instantaneously upon contact with a sexual object.[9] Rapid tumescence in the human animal occurs most frequently among vigorous, younger persons in whom erectile capacities far exceed those of most older persons, although some older females and males may still retain their capacities for rapid response.

6. **Respiration.** As a correlate of the increase in pulse rate and blood pressure which occurs in the sexually responding individual, there is an increase in breathing rate. In the earlier stages of arousal the breathing becomes deeper and faster, but with the approach of orgasm the respiration becomes interrupted. Inspiration is then effected with prolonged gasps, and expiration follows with a forceful collapse of the lungs (Figure 139). There is some popular understanding of this, as is evidenced by the fact that the panting of the actor in the old-time melodrama became a stylized representation of sexual passion. Some of the gasping and sucking sounds and some of the more specific vocalizations which may occur during the climax of sexual activity result from this forced type of breathing.[10] The tortured facial expressions of persons who are sexually aroused, and particularly of those who are near the point of orgasm, usually include expanded nostrils, an open mouth, and pursed lips which suggest that the individuals are struggling to secure air to satisfy the demands of the increased pulse rates and high blood pressures.

7. **Anoxia.** The facial conspectus of a sexually responding individual, especially at the time of orgasm, suggests that he or she may be suffering from a shortage of oxygen—an anoxia. This sort of shortage is

---

[9] We have observed almost instantaneous erection in all of the animals listed above. Root and Bard 1947:82 mention this in the cat, and other authors imply rapid erection in their descriptions of the brevity of coitus—for example, in: Ford and Beach 1951:35–37.

[10] Modifications in respiratory rate as a result of sexual arousal are matters of common knowledge, and are also recorded by various authors: Roubaud 1876:17. Rohleder 1907(1):372. Talmey 1912:61. Kisch 1926:288, 345. Bauer 1927 (1):154, 157. Van de Velde 1930:245. Havelock Ellis 1936(II,1):150–151. Reich 1942:82. Podolsky 1942:49. Negri 1949:97. Brown and Kempton 1950: 207. Faller in Hornstein and Faller 1950:237. Stone and Stone 1952:173. Actual measurements have been made less often, but see: Gantt in Hoch and Zubin 1949:37, 42–43. Klumbies and Kleinsorge 1950a:954; 1950b:61 ff.

characteristic of the athlete at the peak of his performance, or of the person involved in heavy labor; and the face of the human female or male who is approaching sexual climax often bears a striking resemblance to that of the runner who is making a supreme effort to finish his race.[11]

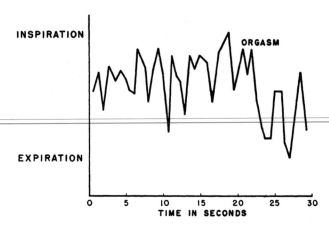

Figure 139. Respiration in human female during sexual activity

Record on the same female shown in Figures 122, 123, 137. Data from Klumbies and Kleinsorge 1950.

The face of the individual who is approaching orgasm similarly and for the same reason presents the traditional aspect of a person who is being tortured. Prostitutes who attempt to deceive (jive) their patrons, or unresponsive wives who similarly attempt to make their husbands believe that they are enjoying their coitus, fall into an error because they assume that an erotically aroused person should look happy and pleased and should smile and become increasingly alert as he or she approaches the culmination of the act. On the contrary, an individual who is really responding is as incapable of looking happy as the individual who is being tortured.

It would not be difficult to investigate the possibility that an anoxia is actually involved during sexual response. Such an anoxia could be detected with simple mechanical devices, and this is one of the first

---

[11] The possibility of an anoxia during sexual response is further suggested by the following: Cannon 1920:205, 209 states that great exertion and strong emotions may cause "asphyxia." Van de Velde 1930:245 mentions an excess of CO_2 in the blood at orgasm. Rossen, Kabat, and Anderson 1943:513–515 list the following symptoms of anoxia: fixation of the eyes, narrowed and blurring of vision, loss of consciousness, mild convulsion—all of which we have found in the sexual syndrome. Shock in Reymert 1950:279 speaks of hyperventilation leading to "local tissue anoxia" and cerebral vasoconstriction. Klumbies and Kleinsorge 1950a:956; 1950b:63, show that sexual activity increases oxygen consumption.

aspects of the physiology of sexual response which might well be studied.

8. **Bleeding.** There are limited data which indicate that bleeding from cut blood vessels is much reduced during sexual arousal. Skin abrasions, accidental cuts and even cuts on the genitalia and on other tumescent parts of the body, and the injuries that may be incurred during sado-masochistic activity, seem to be remarkably free from extensive bleeding. We have records of menstrual flow being slowed up when there is erotic stimulation, although the physical activity in coitus might be expected to induce an increased flow.[12] The records include data on an increase in the flow of blood from wounds after an individual experiences orgasm and returns to a normal physiologic state.

9. **Female Genital Secretions.** The glands connected with the female reproductive tract increase their activity during sexual arousal. The Bartholin glands, which open in the vestibule just outside the entrance to the vagina (Figure 118), are the source of a clear, quite liquid, and somewhat slippery secretion. This secretion should not be confused with the usually thicker and often more colored secretions which frequently come from vaginal or cervical infections, or which constitute the so-called uterine discharges. During sexual activity, an increase in Bartholin secretions provides one of the best indicators of erotic response. Of this fact many observant participants in sexual activities are well aware.[13] The absence of such a secretion is ordinarily evidence that there is no arousal, except among some older women in whom all secretions may be limited, and in occasional instances of anatomic abnormalities. In addition to providing lubrication, the alkaline Bartholin secretions may neutralize the normal acidity of the vagina and prevent that acidity from killing the sperm which are ejaculated in coitus.[14]

The cervix, which is the tip of the uterus which projects into the vagina (Figure 118), also secretes a mucus, and when there is arousal this secretion may become even more copious than the Bartholin secretions. In clinical practice, when it has been necessary to remove the

---

[12] Assertions that the menstrual flow is increased during coitus, are also in: Malchow 1923:136. Van de Velde 1930:292. Stone and Stone 1952:241–242.

[13] Secretions of the Bartholin glands during sexual arousal are commonly recognized. In the literature, they are mentioned, for instance, by: Rohleder 1907 (1):310. Bloch 1908:50. Talmey 1912:60. Moll 1912:25. Malchow 1923: 133. Kisch 1926:290–291. Hirschfeld 1928(2):225, 231. Van de Velde 1930: 195. Dickinson 1933, 1949:48. Havelock Ellis 1936(II,1):145. Kahn 1939:74, 85. Sadler 1944:14, 39. Negri 1949:82. Brown and Kempton 1950:19. Faller in Hornstein and Faller 1950:236. Stone and Stone 1952:61, 173.

[14] The function of the Bartholin secretions in reducing the acidity of the vagina, thus giving the sperm greater longevity, was suggested some years ago (e.g., Talmey 1912:58), and is treated more fully by Siegler 1944:223.

Bartholin glands, it is found that the cervical glands still supply enough mucus for vaginal lubrication. On the other hand, extirpation of the cervical glands, even while the Bartholin glands are still functioning, may so reduce the vaginal secretions as to interfere with coitus.[15] This indicates that the cervical secretions are more important than they have sometimes been considered.

The cervical secretions are also important because they may loosen the mucous plug which ordinarily lies in the opening (the os) of the cervix.[16] Unless the cervical canal is opened, sperm which have been deposited in the vagina cannot move into the uterus and egg ducts (Fallopian tubes), and fertilization may therefore be prevented.[17]

There is considerable variation in the quantity of the vaginal secretions among different females. There may also be variation in the quantity of the secretion at different times in the same individual. This depends upon the intensity of her sexual response, upon her physiologic state, and, interestingly enough, upon the timing of the activity within the menstrual month. About 59 per cent of our sample of women with coital experience recognized such a monthly fluctuation in their vaginal secretions during erotic arousal. About 69 per cent of those who recognized such a fluctuation reported that the mucus was most abundant when sexual activity occurred one to four (or more) days before the onset of menstruation. Some 39 per cent reported that the maximum secretion during arousal occurred soon after the menstrual flow had ceased. About 10 per cent reported that it occurred in the course of the menstrual flow itself, and 11 per cent reported that it occurred in the middle of the month near the time of ovulation. These percentages total more than one hundred because some of the women reported that the increased secretion had occurred both before and after menstruation.

In interviewing for the present study, we have inquired about fluctuations in vaginal mucous secretions before we have discussed fluctuations in erotic responsiveness. The data so obtained indicate that the time of maximum mucous secretion and the time of maximum erotic responsiveness are almost always the same. The record of fluctuations in erotic response is therefore of especial significance, since it

---

[15] Data from Dr. Sophia Kleegman *in litt.*

[16] Cervical secretions are discussed in: Bloch 1908:50. Talmey 1912:60; 1915:91. Urbach 1921:133. Kisch 1926:287, 296. Bauer 1927(2):159. Hirschfeld 1928 (2):225. Van de Velde 1930:194. Dickinson 1933, 1949:fig. 102. Havelock Ellis 1936(II,1):162. Brown and Kempton 1950:18. In an unpublished ms., Dickinson estimates a cervical secretion of 1 to 4 cc. during arousal.

[17] That the cervical os may be opened to the passage of sperm, through cervical secretory activity, is discussed by: Dickinson 1933, 1949:92–94. Weisman 1941:111–113, 130–131. Siegler 1944:229–231. Gardner in Howell (Fulton edit.) 1949:1179.

originated in questions concerning a physical reality, the vaginal secretions, which could be precisely identified. If it had been based on questions concerning erotic arousal, the answers might have represented more subjective judgments.

However, these records on the human female do not fit the laboratory data on the periods of sexual responsiveness in some other mammals. These periods of response, the so-called periods of heat or estrus, occur periodically, and in many of the mammals (but not in all of them) the females rarely accept the males for coitus except during these periods of heat.[18] Ovulation, the release of an egg from the ovary preparatory to its passage down the egg ducts where fertilization may occur (Figure 118), takes place during this period of heat, and at no other time in the female's cycle. This means that the infra-human female has coitus near the time of ovulation, and therefore at the period in which coitus is most likely to lead to fertilization and reproduction. Among the monkeys and apes, which are practically the only animals besides the human which menstruate, the period of maximum sexual arousal may also come just before or concurrently with the time of ovulation, which is about midway between the periods of menstrual flow.[19]

The location of the period of heat near the time of ovulation is, obviously, advantageous to the propagation of any species. The occurrence of the period of maximum sexual responsiveness in the human female just before the onset of menstruation and therefore at her most sterile period, does not appear so advantageous for the accomplishment of fertilization. To those who believe that the evolutionary origins of new structures and physiologic characters are determined by their advantage or disadvantage (i.e., by Darwinian adaptation and selection), it seems proper and correct that a period of arousal should occur close to the time of ovulation. Therefore some laboratory students,

---

[18] Since heat or estrus is often defined as a period in which the female accepts the male, the acceptance of copulation outside of estrus is not often mentioned. Williams 1943:125 says "Copulation in female domestic animals is physiologically limited to the period of estrum . . . outside these periods the female absolutely refuses the sexual advances of the male." But also see: Miller 1931: 384, 387. McKenzie and Terrill 1937:10. Andrews and McKenzie 1941:7, 11, 20–22. Hartman 1945:23 ff. Asdell 1946:*passim*. Roark and Herman 1950: 7–11. Whitney; Farris; and Bissonette in Farris 1950b:199, 246–247, 264. A general summary of the data on periodicity is to be found in Marshall 1936. See also Chapter 18, footnote 39.

[19] The close relationship between ovulation and estrus is discussed by: Ball and Hartman 1935. Marshall 1936:447–448. Corner 1942:70. Yerkes 1943:62–66. Hartman 1945:23 ff. Asdell 1946:13–25. Beach 1947b:272–274, 292, 293. Gardner in Howell (Fulton edit.) 1949:1170. Ford and Beach 1951:201–204, 273. Houssay 1951:639. There may be some slight discharge of blood at the time of ovulation in some human females and in the females of some other species. This *Mittelschmerz* is more pronounced in the dog, but it is an ovulatory discharge and not menstruation.

working with lower mammals, have been loath to accept the human data.[20] On the other hand, most of the research on human subjects has produced data which accord with our own.[21] They are confirmed by observations which have been made by a number of the husbands who have contributed histories to the present study. Some of the women who masturbate only once in a month do so in the period just before or immediately after menstruation. Evidently the human female, in the course of evolution, has departed from her mammalian ancestors and developed new characteristics which have relocated the period of maximum sexual arousal near the time of menstruation.

10. **Male Genital Secretions.** In the male the so-called Cowper's glands connect with the urethra near the base of the penis (Figure 119). These glands in the male and the Bartholin glands in the female originate from the same embryonic tissues. They are, therefore, actual homologues. The clear and slippery Cowper's secretions supply a pre-coital or pre-ejaculatory mucus (the glad-come of the Negro vernacular), which may exude from the urethral opening at the tip of the penis of some males when they are sexually aroused.[22] In many species of mammals the Cowper's glands are well developed and in some species (e.g., the boar) they are enormous.[23] Stallions, rams, boars, male goats, and some others may dribble or run continuous streams of such secretions as soon as they approach a female in which they are sexually interested. In the human species, the Cowper's glands are much more poorly developed. Most human males do not secrete more than a drop of mucus during sexual activity, and perhaps a third, especially at older ages, do not secrete enough mucus to have it ever exude from the urethra. Another third, however, may develop enough of the Cow-

---

[20] Reluctance to believe that the human female's maximal arousal does not coincide with ovulation is noted or discussed in: Tinklepaugh 1933:335. Hartman 1936:77–86. Yerkes and Elder 1936a:38. Stone in Allen et al. 1939:1228–1230, 1258. Benedek and Rubenstein 1942:4. Ford and Beach 1951:210–213.

[21] Others who find that the period of maximum responsiveness in the human female is close to the time of menstruation, are: Sturgis ca.1908:23. Schbankov acc. Weissenberg 1924a:10. Robie 1925:124. Hamilton 1929:197–198. Hamilton and Macgowan 1929:91. Hoyer 1929:20–21. Davis 1929:220–229. Kelly 1930:222. Van de Velde 1930:285–286. Kopp 1933:98–100. McCance et al. 1937:597–599, 609. Terman 1938:351. Popenoe 1938:17. Dickinson ms. Stone and Stone 1952:239. The outstanding exception is Benedek and Rubenstein 1942 and Benedek 1952:6–12, 144–159. This latter work, which combines psychoanalytic and endocrine data, concludes that sexual desire in the human female is highest at mid-month, immediately preceding ovulation. No understandable interpretation is given as to why other studies fail to concur.

[22] That there is a secretion from Cowper's glands as a result of sexual arousal is, of course, commonly known, and is noted in: Roubaud 1876:26. Rohleder 1907(1):310. Moll 1912:22. Hirschfeld 1928(2):231. Dickinson 1933, 1949:81. Havelock Ellis 1936(II,1):153. Sadler 1944:4, 14. Faller in Hornstein and Faller 1950:92, 236.

[23] The boar possesses enormous Cowper's glands measuring in some cases 4 to 5 inches in length and 1 to 2 inches in diameter, according to McKenzie, Miller, and Bauguess 1938:6–7, 37–41.

per's secretion to wet the head of the penis during sexual arousal, and there are a few human males who secrete Cowper's mucus in such quantity that it may pour copiously from the urethra whenever they are erotically stimulated. When there are pre-coital mucous secretions they do provide evidence of erotic arousal; but the absence of such secretions in a human male cannot be taken as evidence that he is not aroused.

Cowper's secretion usually appears before orgasm. It is, therefore, a pre-coital mucus, although there is an occasional male in whom secretion does not develop until after orgasm. Since these secretions are alkaline, it is commonly pointed out that they may neutralize the acidity of urine in the urethra. Sperm should in consequence have a better chance to survive their passage through the urethra. The secretions may also provide lubrication for genitalia which are in copulo.[24] But the fact remains that few human males secrete enough of this mucus to accomplish either of these ends, and most males do manage to copulate and to effect fertilization without especial difficulty.

In addition to the Cowper's secretions, the liquids which exude from the male's penis during sexual arousal may contain secretions from the lining of the urethra.

The largest gland contributing to the male genital secretions is the prostate. It is located at and around the base of the penis (Figure 119). It provides the major portion of the liquid semen which is ejaculated by the male at the time of orgasm.

Most of the remainder of the semen comes from two seminal vesicles, which are expanded portions of the two ducts (vas deferens) which carry sperm from the testes to the seminal vesicles (Figure 119). Sperm are continually moving up from the testes and into the seminal vesicles. There they are stored until there is sexual arousal and orgasm and then, along with other sperm which have been accumulating in the upper ends of the sperm ducts, they are thrown out as a microscopic part of the ejaculate. While the cavities of the seminal vesicles serve as storehouses for the sperm, their spongy walls are glandular, and these are the structures which contribute their secretions to the semen. Both the prostate and seminal vesicles are probably stimulated into secretion as soon as erotic arousal starts, and by the time orgasm occurs a considerable quantity of liquid, averaging about 3 cc. or the equivalent of a teaspoonful in volume, is ready for ejaculation.

---

[24] That Cowper's secretion may neutralize the urethral acidity is suggested in: Dickinson 1933, 1949:72. Hotchkiss 1944:55. Farris 1950b:14. On the other hand, Hooker in Howell (Fulton edit.) 1949:1202 feels that the prostatic secretion is the neutralizing agent and that Cowper's secretion is merely a lubricant.

There is a popular opinion that the testes are the sources of the semen which the male ejaculates. The testes are supposed to become swollen with accumulated secretions between the times of sexual activity, and periodic ejaculation is supposed to be necessary in order to relieve these pressures. Many males claim that their testes ache if they do not find regular sources of outlet, and throughout the history of erotic literature and in some psychoanalytic literature the satisfactions of orgasm are considered to depend upon the release of pressures in the "glands"—meaning the testes.[25] Most of these opinions are, however, quite unfounded. The prostate, seminal vesicles, and Cowper's are the only glands which contribute any quantity of material to the semen, and they are the only structures which accumulate secretions which could create pressures that would need to be relieved. Although there is some evidence that the testes may secrete a bit of liquid when the male is erotically aroused, the amount of their secretion is too small to create any pressure. The testes may seem to hurt when there is unrelieved erotic arousal (the so-called stone-ache of the vernacular), but the pain probably comes from muscular tensions in the perineal area, and possibly from tensions in the sperm ducts, especially at the lower ends (the epididymis) where they are wrapped about the testes (Figure 119).[26] Such aches are usually relieved in orgasm because the muscular tensions are relieved—but not because of the release of any pressures which have accumulated in the testes. Exactly similar pains may develop in the groins of the female when sexual arousal is prolonged for some time before there is any release in orgasm.

In both the female and the male, a considerable congestion of the whole pelvic area may be the consequence of sexual arousal. This depends both upon the tumescence and upon the muscular tensions which have developed. Whenever there is prolonged or repeated arousal without orgasm, this pelvic congestion may become chronic (as in the "engagement pelvis") and lead to the continual discomfort or more acute pain which the clinician sometimes meets in his practice.

The prostate gland and seminal vesicles of the male do have embryonic equivalents in the female embryo, but they never develop in the adult female and do not produce any secretions equivalent to those of the male.

---

[25] For examples of the carry-over of the popular concept of full male "glands" accounting for sexual interest, see such material as: Bauer 1929:234. Haire 1937:150. Freud 1938:608–609. Deutsch 1945(2):85.

[26] Pain in the genital region as a result of unrelieved sexual arousal is, of course, common knowledge. Curiously enough, it is mentioned infrequently in the literature, but see: Malchow 1923:252. Haire 1948:109. Weisman 1948:137–138. Negri 1949:114–115.

11. Nasal and Salivary Secretions. When there is sexual arousal, the membranes which line the nostrils may secrete more than their usual amounts of mucus. This mucus may contribute to the generally swollen state of the nose during sexual activity.

In the mouth, the salivary glands may also increase their secretions during sexual arousal. This is particularly true at the approach of orgasm. Then the glands may spurt quantities of saliva into the mouth. This is often sufficient to provide an abundant lubrication when there is uninhibited kissing or mouth-genital contact. The secretions are often so copious that one's mouth may in actuality "water" in anticipation of a sexual relationship, and one who is erotically aroused may have to swallow repeatedly to clear his mouth of the overabundant supply of saliva. If one's mouth is open when there is a sudden upsurge of erotic stimulation and response, saliva may be spurted some distance out of the mouth. Such behavior becomes especially characteristic at the approach of orgasm. Some of this may be due to difficulty in swallowing because of the muscular tensions which develop in the throat during erotic arousal.[27]

12. Reduction in Sensory Perception. There is a general impression that one who is aroused erotically becomes more sensitive to tactile and other types of sensory stimuli.[28] Quite on the contrary, all of our evidence indicates that there is a considerable and developing loss of sensory capacity which begins immediately upon the onset of sexual stimulation, and which becomes more or less complete, sometimes with complete unconsciousness, during the maximum of sexual arousal and orgasm. At orgasm some individuals may remain unconscious for a matter of seconds or even for some minutes.[29] There are French terms,

---

[27] We have abundant data from recorded observations on the increase in salivation with sexual arousal, but the phenomenon has not often been reported in the literature. See, however, Van de Velde 1930:244. Havelock Ellis 1936(II,1): 153, 166. Lashley 1916:487–488, on the contrary, found that in three human subjects sexual arousal inhibited parotid salivation; but swallowing was also inhibited so that drooling resulted despite the decreased salivation. Our data, however, apply to the increased secretions from the sublingual glands.

[28] The idea that sexually aroused individuals are more sensitive to stimuli is a common one, appearing in such references as: Roubaud 1876:12. Caufeynon 1903:56. Rohleder 1907(1):310. Urbach 1921:127. Malchow 1923:181. Bauer 1927(1):155. Havelock Ellis 1936(I,2):236. Van de Velde 1930:178, 246–247. But the latter author adds: "There is here a profound contradiction. In spite of this magnified receptiveness to sensory impressions during sexual excitement, an individual under its immediate impact, will pay no attention to extraneous things which would otherwise rouse most violent reactions. He is deaf and blind to the world."

[29] Loss of sensory capacity or even of consciousness during extreme emotion or sexual arousal is also noted by: Roubaud 1876:16–17. Caufeynon 1903:57. Talmey 1912:61–62; 1915:92–93. Prince 1914:491. Kantor 1924(1):103. Bauer 1927(2):159–160. Hirschfeld 1928(2):241. Van de Velde 1930:247. Hirschfeld 1935:44. Havelock Ellis 1936(II,1):149–150. Kahn 1939:86. W. Reich 1942:83. Negri 1949:97. Brown and Kempton 1950:207.

"La petite mort" (the little death), and "La mort douce" (the sweet death), which indicate that some persons do understand that unconsciousness may enter at this point. Most persons, however, including technically and professionally trained persons, have failed to comprehend the considerable loss of sensitivity which actually occurs, and the matter therefore needs discussion.

This loss of sensory capacity was first brought to our attention by prostitutes who were contributing histories to this study. Many of the prostitutes rob (roll) their patrons during their sexual contacts. They well understand that their confederates (the creepers) can move about a room without the victim hearing or seeing them if they do not pass directly in front of him; and they can touch him without his being conscious of their presence—providing they confine their activities to the period when he is erotically aroused.

The situation may involve some psychologic distraction. The attention of the individual may be so centered on the sexual activities that he is not consciously aware of other sensory stimulation; but there is some evidence that an actual anesthesia of the sensory structures may occur. It is possible and not improbable that both distraction and anesthesia may be involved. Similar situations are recognized in anger, in fear, and in epilepsy—all of which are phenomena that are physiologically related to sexual response. It is popularly understood that a person may become "too mad to see straight"; "so excited he did not hear the train coming"; or "too angry to know what was happening to him." Love too appears to be blind, and probably more blind than the poets realized when they wrote about it.[30]

Specific observations and experimental data indicate that the whole body of the individual who is sexually aroused becomes increasingly insensitive to tactile stimulation and even to sharp blows and severe injury. Any stimulation which is maintained at a constant level becomes ineffective; and it is quite usual for the participants in a sexual relationship, even though they are unaware of the physiology which is involved, to progressively increase the speed and the force of their techniques. Toward the peak of sexual arousal there may be considerable slapping and heavier blows, biting and scratching, and other activities which the recipient never remembers and which appear to have a minimum if any effect upon him at the time they occur. Not only does

---

[30] The physiology of such a sensory loss, and of the loss of consciousness, is ill understood and probably very complex. One factor may be the reduction of the blood supply to the brain which may result from excessive vasodilation (Engel 1950:19) or hyperventilation (Engel 1950:76; Shock in Reymert 1950:279). Hyperventilation, according to Houssay et al. 1951:287, also may cause dizziness with muscular hyperexcitability, muscular contractions, and tetany. These symptoms resemble some of those in the sexual syndrome.

the sense of touch diminish, but the sense of pain is largely lost. If the blows begin mildly and do not become severe until there is definite erotic response, the recipient in flagellation or other types of sado-masochistic behavior may receive extreme punishment without being aware that he is being subjected to more than mild tactile stimulation.

There is also evidence that even the genitalia, contrary to the general opinion, become anesthetic as the sexual relationships progress. It is not impossible that precise measurements might show that the genital structures preserve their sensitivity longer than some other parts of the body, but there are no data to establish that point. Since so much of the stimulation during a sexual contact is directed toward the genitalia, and since their pronounced turgidity may increase one's awareness of those structures, it is understandable that there should be some concentration of attention on those organs during sexual arousal.

It may take only mild stimulation and sometimes only the slightest touch of the genitalia at the peak of sexual arousal to precipitate orgasm. This, however, does not seem to be evidence of increasing sensitivity, but evidence of the high level to which the physiologic changes may have proceeded before the final touch brings the individual to the point of climax. A drop of water cannot fill an empty cup, but a single drop can make an already filled cup overflow. The effectiveness of the minor tactile stimulation which may precipitate orgasm may emphasize nothing more than the fact that there has been a previous build-up of physiologic changes which are now ready to culminate in orgasm.

Not only the sense of touch but all of the other senses become increasingly ineffective during erotic arousal. The sense of sight is considerably contracted during sexual activity.[31] The pupils dilate,[32] and the range of vision is so narrowed that the individual loses his capacity to observe things that lie to the side and can see only those objects which are directly ahead (p. 614). Some persons become so blind at the peak of sexual arousal that they do not see lights which are moved directly in front of them or recognize other sorts of visual stimuli. Such a loss of visual capacity is one of the known effects of anoxia; and since an anoxia may be involved in sexual response, this is one possible explanation of the sensory loss here.

---

[31] The reduction of visual capacity during arousal is also noted by several authors, including: Roubaud 1876:17. Talmey 1912:61. Bauer 1927(2):159. Goldstein and Steinfeld 1942:44.

[32] Dilation of the pupil of the eye during arousal is also described by: Van de Velde 1930:247. Havelock Ellis 1936(II,1):166–167. Haire 1937:200. Sadler 1944:42. Faller in Hornstein and Faller 1950:237. Stone and Stone 1952:173. Note that the injection of adrenaline may also produce such a dilation (Cannon 1920:37).

The sense of hearing is similarly impaired during erotic arousal. Minor sounds are completely overlooked by the sexually responding individual. The wife who hears the shade flapping or the baby crying after coitus has begun, simply registers the fact that she is not responding erotically. The occasional records of persons being apprehended by police or other intruders when they are engaged in sexual activity may depend on the fact that sexually occupied persons do not hear with their normal acuity. The sense of hearing may be so completely lost during maximum sexual response that very loud noises close at hand and voices at their maximum are not heard.[33]

The senses of smell and taste may similarly be reduced and ultimately more or less lost whenever there is real sexual arousal. Contacts which would be offensive to an individual who was not sexually stimulated, no longer offend. Data on the reception of semen in the performance of fellatio indicate that there may be considerable consciousness and some offense when the sexual relation is not sufficiently stimulating, but that there is a reduction of consciousness of both the contact and the taste when the sexual relation is had between persons who are erotically aroused.

The temperature sense is similarly diminished and may become quite lost during sexual activity. In the earlier stages of arousal there is, as we have already noted, a considerable recognition of the surface warmth of the body which develops as a result of the peripheral circulation of blood. But the sexual arousal may progress to a point at which most persons become unconscious of the extreme temperatures of summer or of winter, of an overheated or a very cold room, or even of objects like cigarettes which may actually burn them.

Since few persons are aware of these losses of sensory capacities, they do not comprehend that they are incapable of giving a coherent account of the physiologic events which transpire during their sexual activities. There is usually some recognition of the fact that one becomes "hot" when there is sexual arousal (because of the peripheral flow of blood), "bothered" (because of the neuromuscular tensions), and "aroused" (by the change in physiologic state). But that is usually the limit of awareness of one's own reactions. Persons whose sexual reactions regularly involve the most vigorous sorts of body movements, such as rhythmic or propulsive extensions of the arms and legs and movements and torsions of the back and neck, are amazed to learn from other persons who have observed them that they have ever behaved so.

---

[33] The impairment of hearing during arousal is also noted in: Bauer 1927(2):159. Van de Velde 1930:247 manufactures a paradox by deciding that while hearing becomes more acute, the aroused person is nevertheless "deaf and blind to the world."

Some insight into the effects of sexual arousal comes from our histories of amputees who have lost arms or lower limbs. It is well known that such individuals frequently experience what have always been considered to be phantom pains in their non-existent limbs. It is, however, sometimes possible to stop such phantom pains by blocking the spinal centers which would control those areas in a normal limb. The spinal centers are still connected with the remnants of the nerves that belong to those areas, and the localization of the pains in the non-existent portion of the limb is due to the fact that no one, amputee or non-amputee, is able to localize the point of origin of any stimulation when the stimulus is applied to the basal portion of a nerve.

From the standpoint of the sexual physiology which is involved, it is notable that our data indicate that amputees rarely if ever experience phantom pains when they are aroused sexually. On the other hand, such pains may suddenly return after orgasm, thereby emphasizing the fact that they were absent during the period of arousal. This appears to be another instance of a loss of sensory perception, or of the distraction of perception during sexual arousal.

13. **Central Nervous System.** There is evidence that the whole central nervous system is affected whenever there is sexual arousal. The data are fragmentary, but sufficient to indicate some of the points which the neurophysiologist should investigate.

During sexual arousal, inhibitions and psychologic blockages are relieved or completely eliminated. A considerable series of histories indicate that stutterers are not likely to stutter when they are with a companion to whom they are sexually responsive. Similarly, gagging may be eliminated, even among individuals who are quite prone to gag when objects are placed deep in their mouths. If sufficiently aroused, these individuals do not gag even when they perform fellatio, although that may involve deep penetration of the mouth. Some persons have their hay fever and sinus congestions relieved when they are sexually aroused, and this may depend in part on the relief of some psychologic state by the arousal. Some persons break out into nervous sweating when they are aroused sexually, although it is more usual for the skin to stay dry or normally moist during sexual activity.[34]

Interferences with the free action of the musculatory system may be relieved during sexual arousal. We have histories which indicate that spastics are able to move more freely when they are sexually aroused, and they may be surprisingly capable in coitus. The improved condition does not persist, however, after the cessation of the sexual arousal.

---

[34] Increased perspiration as a result of sexual arousal is also noted by: Van de Velde 1930:244. Havelock Ellis 1936(II,1):153. Sadler 1944:42. Negri 1949:97.

There is also some information to the effect that some partial paraplegics may be improved when they are aroused sexually (p. 700).

Most persons display unusual muscular strength during sexual arousal, and may become capable of performing feats that require abilities which they do not ordinarily exhibit. This is not because they actually acquire strength, but because they are released from the inhibitions which normally prevent them from utilizing their full capacities. The possibility of a rapist doing unusual damage may depend on this fact; and many another male becomes unexpectedly strong and handles his partner roughly during sexual activity. When there is arousal, many persons become capable of bending and distorting the body to an extent which would be impossible if there were no arousal. The doubling of the body which is necessary in self-fellation, for instance, may be impossible before arousal, but may become possible for some males as they approach orgasm.

14. **Movements in Buttocks and Pelvis.** One of the most striking aspects of a sexual performance is the development of neuromuscular tensions throughout the body of the responding individual, female or male. From head to toe, the muscles contract and relax, involuntarily, in steady or more convulsive rhythms. The movements may vary at various times in the experience of each individual, but they may vary even more between different individuals. Sometimes the muscular action is sufficient to effect major movements of the limbs and of still other parts of the body. Sometimes the movements are violent. Sometimes they are so limited that they are hardly noticeable; but in even the most quiescent individuals, whenever there is sexual response there is likely to be some evidence that muscles are rhythmically tensing and relaxing, everywhere in the body. There may be occasional moments when the movements cease and the muscles are held in continuous tension; but in an uninhibited and responding individual there is usually a flow of continuous muscular movement, from the first moment of arousal to the moment of orgasm.

The most prominent of these muscular activities effect rhythmic movements of the buttocks and of the whole pelvis, and the consequent rhythmic pelvic thrusts during sexual activity are among the distinctive characteristics of the class Mammalia. Without the capacity to make these rhythmic movements, mammalian coitus could not occur. Simple intromission of a male organ into a vagina occurs among a few of the birds and reptiles and among some of the insects, and this is copulation, but it is not coitus in the mammalian sense. The mammals are nearly the only animals in which there are rhythmic alternations of in-and-out movements of the male copulatory organ and, in at least some cases, correspondingly rhythmic movements of the pelvis of the female partner.

Coital movements depend in part upon the gluteal muscles, which are the large muscles in the buttocks, working in conjunction with certain lower vertebral muscles. Even when there is no gross movement of the pelvis, the gluteal muscles may rhythmically contract during erotic arousal. More extended movements of these muscles may crowd or crush the two halves of the buttocks together. Since tensions in these gluteal muscles are correlated with or contribute to a rise in neuromuscular tensions throughout the body, some persons begin their sexual activities by voluntarily moving their buttocks, and thus they may build up their erotic responses.

The erotic significance of gluteal contractions may depend in part upon the fact that movements of the buttocks can stimulate the perineal area, with its abundant supply of nerves. Movements of the buttocks may also stimulate the anal area, which is erotically sensitive in many individuals. In addition, gluteal contractions may stimulate a flow of blood into the genitalia, thereby contributing to the erection of the genital structures. Some males are able to effect full erection by voluntarily tensing and moving their buttocks, and may occasionally reach orgasm without the genitalia being touched. Not a few females have also learned that voluntary contractions of their buttocks and movements of the pelvis may develop their erotic reactions and even effect orgasm in masturbation, petting, coitus, and homosexual activities.

The muscular movements which occur during sexual response are usually involuntary. While most of the muscles which are involved are ordinarily under voluntary control, no such control could effect the coordinated flow of movement which occurs throughout the body of one who is sexually aroused, or the extraordinary rapidity of some of the rhythmic movements which may occur during sexual activity.

15. **Movements of Thighs.** Coordinated with the gluteal contractions, there may be movements of the adductor and/or abductor muscles in the upper portions of the legs. The adductor muscles lie toward the front, on the inner surfaces of the thighs, and the abductors lie toward the sides. Rhythmic contractions of the adductors during sexual activity may pull the upper halves of the legs together rhythmically or, sometimes, quite convulsively. Contractions of the abductors may, on the contrary, roll the legs out. The coordination of the gluteal and adductor contractions may bring the buttocks and thighs together simultaneously in rhythmic sweeps of movement. Many persons intensify the force with which the upper halves of the legs come together by crossing their feet, or by placing objects between their legs to in-

crease the muscular tensions. The sexual partner, placed between the legs of an intensely reacting individual, may be caught in a vise of considerable force.

16. **Movements of Feet and Toes.** Outside of the movements of gluteal, adductor, and abductor muscles, the muscular reactions which are next most noticeable during sexual activity may involve the feet and the toes. The whole foot may be extended until it falls in line with the rest of the lower leg, thereby assuming a position which is impossible in non-erotic situations for most persons who are not trained as ballet dancers. The toes of most individuals become curled or, contrariwise, spread when there is erotic arousal. Many persons divide their toes, turning their large toes up or down while the remaining toes curl in the opposite direction. Such activity is rarely recognized by the individual who is sexually aroused and actually doing these things, but the near universality of such action is attested by the graphic record of coitus in the erotic art of the world. For instance, in Japanese erotic art curled toes have, for at least eight centuries, been one of the stylized symbols of erotic response.

Some persons tense so severely during sexual activity that their feet and toes develop cramps as soon as they have experienced orgasm. They rarely recognize such cramps during the sexual activity itself, but upon the sudden release of the tensions at orgasm, they may have to rise and shake their legs to rid themselves of the cramps.[35] Dramatic instances of the development of such tensions are found in the histories of some of the amputees who have contributed to the present study. Although these persons may have nothing but remnants of the nerves that would have served the muscles of their lower limbs, they may build up neuromuscular tensions in the non-existent portions of those limbs which are quite like those of persons who have complete limbs. Consequently the amputee may also have to rise after orgasm and shake the cramps out of his non-existent toes.

17. **Movements of Arms and Legs.** The legs of a person who is involved in sexual activity may be thrust out in a straight line and held there with considerable tension.[36] The arms may be thrust into similar positions. Or the arms and legs may be bent at the elbows and knees in angular positions which may be rigidly maintained throughout the period of maximum arousal. Contractions of the individual muscles

---

[35] Cramps in the feet or limbs are noted in: Roubaud 1876:17. Van de Velde 1930:246.

[36] Roubaud 1876:17 (who is also quoted by Kisch 1926:288 and Havelock Ellis 1936(II,1):149–150) describes the limbs at the approach of orgasm as becoming "stiff like iron bars." See also Faller in Hornstein and Faller 1950: 238.

usually become distinctly visible, standing out from the body surfaces as they do in an athlete who is demonstrating his muscular capacities.

The upper halves of the arms and of the legs may move rhythmically, sometimes in a slow rhythm, but often with an increase in speed as orgasm is approached. In amputees, the remaining portions of the arms and legs, being unimpeded by the weight of the limbs, may move in an even more distinctive fashion.

18. **Movements of Hands and Fingers.** In the midst of intense sexual activity the hands and fingers may move and curl in a manner which is comparable to the movements of the feet and toes. In some cases the fingers curl under and clench into a tight fist. In other cases the fingers spread in somewhat the same fashion that the toes may spread. Many persons clasp some object when they approach the point of maximum arousal. This may be the side of a chair, some part of the bed, the bed covers, or some other solid object which is available. It may be the body of the sexual companion. Usually the hands and the fingers move rhythmically. Often the movements become spasmodic as the tensions increase, and then the hand may grab objects with considerable force. The tensions are often so great that the sexual partner who is caught in such a situation may be bruised or cut by the fingernails of the re-acting individual.

19. **Abdominal Muscles.** During sexual arousal the abdominal muscles may contract with considerable force. The "stomach" is pulled in, sometimes being held in continuous tension, but more often contract-ing with spasmodic jerks which may rock the whole body. In persons with well trained muscles, such as weight-lifters and dancers, the con-tractions may be phenomenal in their magnitude and intensity. Some-times the contractions build up to the very moment of orgasm with increasingly prolonged periods of continuous tension. In other cases, the contractions cease to be spasmodic and become more evenly rhyth-mic at the approach of orgasm. Sometimes the final contractions are slow, but often they become faster. In a few persons, both female and male, the movements immediately before, during, and after orgasm become so fast that the individual elements are no longer discernible to the human eye. The speed of some of these abdominal movements at the approach of orgasm may approach the most rapid rhythmic movements of which the human body is capable. Such rapid abdominal and pelvic movements regularly occur among many human females and males, and in a number of infra-human mammalian species during coitus.

20. **Thoracic Muscles.** Because of the fixed skeleton which lies un-der the thoracic muscles, their movements during sexual arousal are

not as prominent as the muscular movements in some other parts of the body. Nevertheless, the tensions of the muscles on the chest and on the sides often become apparent enough to outline the ribs with a typically washboard effect. That these muscles are tensed during sexual response becomes most apparent when they are relaxed after orgasm. Then the thorax assumes a smoothed-out appearance which it did not have during sexual arousal.

The pectoral muscles, on the chest, may protrude prominently under erotic tension. They become so prominent on some males that they may assume something of the appearance of female breasts. The increased prominence of the female breast during sexual arousal may also depend on some protrusion of the underlying pectorals, as well as upon the increased tumescence of the tissue in the breast itself.

21. **Neck Muscles.** In most persons the neck becomes rigid during sexual arousal. The tensions may cause the muscles and the tendons to stand out prominently.[37] Under maximum tensions, at the approach of orgasm, the neck may shift in position, moving the head either forward or backward, but often poising it in a fixed position which is maintained until the tensions are released in orgasm. The prolonged maintenance of such a rigid position may account for some of the "rheumatic neck pains" which an occasional patient takes to the physician for treatment.

22. **Facial Muscles.** Some degree of muscular tension is usually apparent in the face of the person who is reacting sexually, and in some individuals such tensions become extreme at the moment of orgasm. This, combined with the fact that the mouth is open to secure air, may cause the face to take on a drawn, tense, and tortured expression which is paralleled only in the facial expressions of persons who are suffering intense pain and agony (see p. 606).

23. **Eye Muscles.** The eyes of sexually aroused persons acquire a distinctive glare, particularly at the moment of orgasm. This glare cannot be mistaken, and is one of the things that the sexual partner may sometimes report; but it is difficult to analyze exactly what is involved. The pupil of the eye becomes dilated. The lids are held in a fixed position, and the eyes stare without being focused. This gives them something of the blankness which is evident in the eye of a blind person. The eyeball appears to protrude and the eye glistens to a greater degree than usual, in part because there is an increased lachry-

---

[37] During extreme arousal the muscular tensions in the neck become quite prominent, especially in the sternocleidomastoid muscle. This phenomenon was also described by Roubaud 1876:17.

mal secretion.[38] There may be some tumescence of the tissues about the eye. Often, however, the eyelids are kept closed.

24. Scrotum and Testes. During sexual arousal, and especially at the moment of maximum tension, the testes are usually pulled up by their supporting cremaster muscles. The walls of the scrotum also contract and in many males the testes are pulled tight against the shaft of the penis, against the perineal surfaces, or into the groins. In a few human males in whom the inguinal canals are pathologically open, and among some other species of mammals in which the inguinal canals are normally open, the testes may be pulled high enough to enter the canals or even to enter the abdominal cavity. This accounts for the near (or more rarely complete) disappearance of the testes of some human males when they are engaged in sexual activity.[39]

25. Other Structures. Sexual responses obviously involve a great deal more than genital structures. In actuality, every part of the mammalian body may be involved whenever there is sexual response, and many parts of the body may respond as notably as the genitalia during sexual contact. The activities of any of these other parts of the body may be as useful as the genital activities for the identification of the onset of the response, the continuity or discontinuity of the response, the gradual rise in the level of response, the sudden approach of the ultimate peak in orgasm, and the moment of release in orgasm. The progress of the response may be as obvious in the neck tensions, the aspect of the eye, the behavior of the fingers or toes, or the movements of the buttocks or abdominal muscles, as it is in the genitalia themselves.

THE APPROACH TO ORGASM

The Build-up. With continued and uninterrupted stimulation, the physiologic changes which characterize sexual response may progress and build up in intensity until they approach some maximum point of departure from their normal physiologic states. While exact measurements of these progressive developments are available on only a few of the phenomena which are involved—most specifically on pulse rates and blood pressures, as we have already shown (Figures 120–138)—the more general observations on tumescence, the development of muscular and nervous tensions, the loss in sensory perception, the

---

[38] Changes in the appearance of the eye during sexual arousal have also been noted by: Anon. 1772: The Virgin's Dream, 143–144 ("My Body was all Pulse, my Breath near gone:/ My Cheeks inflam'd, distorted were my eyes . . ."). Roubaud 1876:17. Talmey 1912:61. Moll 1912:164. Kisch 1926:288, 297. Bauer 1927(1):157. Havelock Ellis 1936(II,1):167. Sadler 1944:43.

[39] The drawing up of the testes in coitus was noted as long ago as Aristotle [4th cent. B.C.]: Problems, Bk. 4, 879a and Bk. 27, 949a.

rates of respiration, glandular secretions, and still other phenomena make it apparent that in all of its aspects sexual response may be an accumulative phenomenon which is most effectively concluded by the occurrence of orgasm.

In most human females and males, sexual responses usually develop irregularly, with sudden upsurges of arousal and the development of preliminary peaks and periods of regression in the course of an over-all rise to continually higher levels of response (again see Figures 120–138). The maximum peak may be reached only after a sudden rise which is more abrupt than any of the previous rises, and the level of the final peak may be much higher than the level of any of the preceding peaks of response.

Some persons, particularly younger males, may respond instantaneously to sexual stimulation and proceed quite immediately, with sharp peaks and regressions which follow rapidly upon each other, until they suddenly surge to the maximum level of response at the moment of orgasm. Some persons, particularly older persons, experience more even and more steady rises in their responses, and reach their maxima without the abrupt developments which characterize the approach to orgasm in many other individuals. There are, however, some older persons, including both females and males, who continue to react in an abrupt fashion throughout their lives.[40]

The speed of reaction to sexual stimuli, the speed with which physiologic changes progress toward the peak of arousal, the presence or absence of preliminary peaks and regressions, the abruptness of the approach toward orgasm, the level at which orgasm occurs, and the pattern of the muscular reactions which may follow orgasm are likely to remain more or less uniform throughout most of an individual's life. This is established by specific records which we have on the nature of the responses of a limited number of persons who had been observed over long periods of years—over as many as sixteen years in one case, and in some cases from pre-adolescence into the late teens or twenties. These individual patterns of response may depend at least to some extent on the physiologic equipment with which an individual is born, for clinicians report striking individual variation in the responses of even very young infants to general contact, to pressure, to specific tactile stimuli, and to genital stimulation. There is, however, considerable reason for believing that some aspects of the behavioral pattern

---

[40] For examples of variation in the build-up to orgasm, see the diagrams and charts in: Boas and Goldschmidt 1932:99. Dickinson 1933, 1949:figs. 126–127. Klumbies and Kleinsorge 1950a:952–958; 1950b:64–66. Ford and Beach 1951:245–248 (Fig. 16 is theoretic and inadequate).

represent learned behavior which has become habitual after early experience (pp. 643 ff.).

Many persons may exert some deliberate control over the normal course of their sexual responses in order to modify or prolong the activity or particularly pleasurable aspects of it. By regulating the frequency of sexual contacts, by controlling the breathing rate, by holding muscles in continuous tension, by avoiding continuous stimulation, by avoiding fantasies or other controllable sources of psychosexual stimulation, and by still other means, it is possible to extend the period of preliminary activity and delay the upsurge of response which carries the physiologic developments to climax. The sexual literature for at least four thousand years, from early Sanskrit to current marriage manuals, has recommended that the male in particular delay his responses in order that the female may reach orgasm simultaneously with him. There are, however, many persons who, at least on occasion, find greater satisfaction in sexual relations which proceed directly to the point of orgasm.

Coitus reservatus, the *Karezza* of the Sanskrit and Hindu literature, represents a maximum sophistication of such deliberate control. It has been practiced by whole communities, like the Oneida Colony in the nineteenth century in New York State. In this technique it is common for the individual to experience as many as a dozen or twenty peaks of response which, while closely approaching the sexual climax, deliberately avoid what we should interpret as actual orgasm. Persons who practice such techniques commonly insist that they experience orgasm at each and every peak even though each is held to something below full response and, in the case of the male, ejaculation is avoided. We now interpret the supposed orgasms as preliminary peaks of arousal. The possibility of prolonging this sort of experience, especially at any high level of response, apparently depends on the very fact that there is no orgasm.[41]

Speed of Response. There is a longstanding and widespread opinion that the female is slower than the male in her sexual responses and needs more extended stimulation in order to reach orgasm. This opinion is accepted and is the basis of much clinical practice today.[42]

---

[41] Coitus reservatus, or *Karezza*, is the subject of some literature, including two volumes: Stockham 1901(?). Lloyd 1931. For the history of the Oneida colony, see Parker 1935.

[42] The idea that the female is inherently slower than the male in sexual response occurs repeatedly in the literature, as in: Rohleder 1907(1):312. Moll 1912: 26. Talmey 1912:63; 1915:94. Urbach 1921:124 ff. Malchow 1923:164, 231. Hirschfeld 1928(2):230. Stopes 1931:74. Rutgers 1934:135–136. Havelock Ellis 1936(I,2):236. Wright 1937:85. Clark 1937:46–49. Kahn 1939:fig.31. Butterfield 1940:94. Podolsky 1942:52. Magoun 1948:216–220. Hirsch 1949: 134. Faller in Hornstein and Faller 1950:238–239. Stone and Stone 1952:180.

Certain it is that many males reach orgasm before their wives do in their marital coitus, and many females experience orgasm in only a portion of their coitus (Chapter 9). On the other hand, a high proportion of the males could ejaculate soon after coitus begins. These facts seem to substantiate the general opinion that the female is slower than the male, but our analyses now make it appear that this opinion is based on a misinterpretation of the facts.

This becomes apparent when we examine the time which the average female needs to reach orgasm in masturbation. Apparently many females, even though they may be slow to respond in coitus, may masturbate to orgasm in a matter of a minute or two. Masturbation thus appears to be a better test than coitus of the female's actual capacities; and there seems to be something in the coital technique which is responsible for her slower responses there.

The crux of the matter seems to lie in the fact that the female in masturbation usually proceeds directly to orgasm and is not interrupted or distracted as she often is in coitus. The record indicates that the average (median) female ordinarily takes a bit less than four minutes to reach orgasm in masturbation, although she may need ten or twenty minutes or more to reach that point in coitus. Similar records show that the average male needs something between two and four minutes to reach orgasm in masturbation although, in order to increase his pleasure, he often delays his performance. The record indicates, therefore, that the female is not appreciably slower than the male in her capacity to reach orgasm.

Actually there are some females who regularly reach orgasm within a matter of fifteen to thirty seconds in their petting or coital activities. Some regularly have multiple orgasms (p. 375) which may come in rapid succession, with lapses of only a minute or two, or in some instances of only a few seconds between orgasms. Such speed is found in only a small percentage of the females, but it is found, similarly, in only a small percentage of the males. Of the 2114 females in our sample who supplied data on the time usually taken to reach orgasm in masturbation, some 45 per cent had regularly done so in something between one and three minutes, and another 24 per cent had averaged four to five minutes. About 19 per cent had averaged something between six and ten minutes, and only 12 per cent regularly took longer than that to reach orgasm. In all of these groups there were, of course, females who had deliberately taken longer than necessary to reach orgasm, in order to prolong the pleasure of the experience.

The slower responses of the female in coitus appear to depend in part upon the fact that she frequently does not begin to respond as promptly as the male, because psychologic stimuli usually play a more important role in the arousal of the average male, and a less important role in the sexual arousal of the average female (Chapter 16). The average male is aroused in anticipation of a sexual relationship, and he usually comes to erection and is ready to proceed directly to orgasm as soon as or even before he makes any actual contact. The average female, on the contrary, is less often aroused by such anticipation, and sometimes she does not begin to respond until there has been a considerable amount of physical stimulation.

Moreover, because she is less aroused by psychologic stimuli, the female is more easily distracted than the male in the course of her coital relationships (Chapter 9). The male may be continuously stimulated by seeing the female, by engaging in erotic conversation with her, by thinking of the sexual techniques he may use, by remembering some previous sexual experience, by planning later contacts with the same female or some other sexual partner, and by any number of other psychologic stimuli which keep him aroused even though he may interrupt his coital contacts. Perhaps two-thirds of the females find little if any arousal in such psychologic stimuli. Consequently, when the steady build-up of the female's response is interrupted by the male's cessation of movement, changes of position, conversation, or temporary withdrawal from the genital union, she drops back to or toward a normal physiologic state from which she has to start again when the physical contacts are renewed. It is this, rather than any innate incapacity, which may account for the female's slower responses in coitus.[43]

ORGASM

As the responding individual approaches the peak of sexual activity, he or she may suddenly become tense—momentarily maintain a high level of tension—rise to a new peak of maximum tension—and then abruptly and instantaneously release all tensions and plunge into a series of muscular spasms or convulsions through which, in a matter of seconds or a minute or two, he or she returns to a normal or even subnormal physiologic state (Figures 120–138).

This explosive discharge of neuromuscular tensions at the peak of sexual response is what we identify as orgasm. The spasms into which the individual is thrown as a result of that release, we consider the after-effects of that orgasm. Many psychologists and psychiatrists, emphasizing the satisfactions that may result from sexual experience,

---

[43] We have had access to a considerable body of data on the continuous nature of male response and the discontinuous nature of female response. Although this situation has rarely been recognized in the literature, it is noted in Urbach 1921:132.

suggest that the after-effects of this release from sexual tensions may be a chief source of those satisfactions. They are, therefore, inclined to extend the term orgasm to cover both the release from tensions and the after-effects of that release. There are, however, several advantages in restricting the concept of orgasm to the sudden and abrupt release itself, and it is in that sense that we have used the term throughout the present volume.[44]

Some, and perhaps most persons may become momentarily unconscious at the moment of orgasm, and some may remain unconscious or only vaguely aware of reality throughout the spasms or convulsions which follow orgasm. Consequently few persons realize how they behave at and immediately after orgasm, and they are quite incapable of describing their experiences in any informative way.

Sometimes the recession from the high peak of orgasm is accomplished in a single great sweep. Sometimes there are momentary pauses in the recession, or some brief resurgence to a subsidiary peak or peaks which are, however, always lower than the maximum at which orgasm occurred (Figures 120–138). The return to the physiologically normal or subnormal state is, however, usually accomplished in a short period of time.

The abrupt cessation of the ofttimes strenuous movements and extreme tensions of the previous sexual activity, and the peace of the resulting state, provide in their contrast the most obvious evidence that orgasm has occurred. It is the best means of identifying orgasm in the human female, and it is the best means of identifying orgasm in the human male in instances in which ejaculation cannot be observed. Because this sudden release is not often seen among females of infra-human species of mammals, we conclude that they are not reaching orgasm. In coital relations, the neuromuscular tensions of the responding infra-human female are usually maintained after the cessation of coitus, and they disappear only gradually, without the explosive discharge which characterizes orgasm. Moreover, after coitus the females of the infra-human species are usually as responsive as they were before coitus, and this is not generally true of animals that have experienced orgasm.

On the other hand, the statement that orgasm never occurs among the females of any of the infra-human species does not seem to be

---

[44] The concept of orgasm as the period of time during which the spasms (and, in the male, ejaculation) occur is implicit in various works, as in: Adler 1911: 30 ff. Van de Velde 1930:173, 184–188. Kuntz 1945:312–313, 323. Hardenbergh 1949:226. Stone and Stone 1952:185–186. On the other hand, the concept of orgasm as a sudden release of neuromuscular tensions is supported by such measurements as those in: Pussep 1922:61 ff. Boas and Goldschmidt 1932:99. Klumbies and Kleinsorge 1950a:955, 957; 1950b:61–66.

entirely correct, although it apparently applies to most individuals of most of the species. Reference to Figures 133 to 135 will show that blood pressures among female dogs build up in essentially the same way as they do among male dogs in coitus (Figures 129–132), and the females similarly show a sudden return from the high peak of tension to normal levels of blood pressure. If measurements of other aspects of the physiology of responding female dogs parallel the data shown for blood pressure, it is difficult to understand why this should not be identified as a record of orgasm.

The difficulty has originated in the fact that there have not been well understood criteria for interpreting orgasm when it appears in these infra-human females. However, looking at the specific record which some of the students of mammalian behavior have been able to provide, we find descriptions of what appear to be good cases of orgasm among rabbits and among chimpanzees. There is a record of masturbation in a female chimpanzee which has apparently been reaching orgasm regularly over a considerable period of years. We ourselves have had the opportunity to observe another female chimpanzee in a physiologic reaction which may well have been orgasm.

While most cows who mount other cows give evidence of erotic arousal, they do not ordinarily show any build-up to peaks of tension, from which there is the sort of sudden release which is characteristic of the bull at the time of orgasm. On the other hand, we have reports of mounting cows giving a sudden lunge at the peak of response, and then dropping back into inactivity as though they had experienced orgasm.[45]

The very strenuous activities which the females of all species of cats, including the house cat, the tiger, and the lion, regularly display after coitus, may involve a most strenuous series of convulsive movements which turn the body in a wild succession of circles for some moments or some minutes after the cessation of coitus. While we have

---

[45] Dr. Marc Klein of the Université de Strasbourg informs us that he believes orgasm occurs in the female rabbit. In a letter of February, 1952 he states: "I have very often observed a quite definite peak of response with climax, from which the female falls back abruptly into a quiet state . . . . the two partners fall together on one side; . . . the female is shrieking and sometimes even the male is shrieking himself. The female thrusts the male aside by a sudden movement of the pelvis and the two animals at once come back to the normal stature on the four limbs . . . . it is an individual response of the female which is far from appearing in all females even of a definite strain . . . . there are enormous individual differences in this behavior." Our chimpanzee data come from the observations of Dr. Henry Nissen of the Yerkes Primate Laboratory at Orange Park, Florida, and concern primarily the adult female Alpha. We have observed an additional female in what seemed to be orgasm. The data on completed thrusts in cows mounting other cows come from Dr. Albert Shadle of the University of Buffalo.

PHASE 1

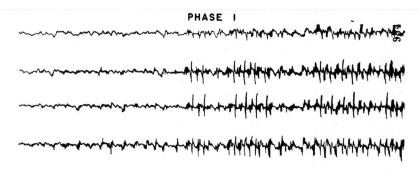

PHASE 2

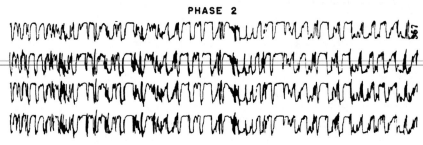

PHASE 3

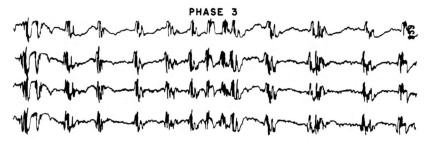

Figure 140. Electroencephalogram of human subject in sexual activity

From unpublished data provided by courtesy of Dr. Abraham Mosovich. Four electrodes, attached to the head at different points, recorded four lines simultaneously. The three sets of lines are not continuous, but represent samples from three phases of the record. For example, the sample of phase 2 begins approximately half a minute after the end of the sample of phase 1. Read from left to right.

PHASE 1: With the beginning of sexual stimulation, rapid, low voltage (small) waves develop in the brain. Such waves are known to occur during excitement. These are subsequently obscured by spikes resulting from muscle contraction and tremor.

PHASE 2: With orgasm, very high voltage, slow (large) waves develop in the brain. These appear to be flat-topped only because the machine was not set to record waves of such height. Interspersed with and partially obscuring the large slow waves are spikes resulting from muscle spasm. This phase resembles petit mal epilepsy or the later stages of grand mal when the subject is passing from tonic rigidity to clonic spasms.

PHASE 3: Later, the large slow waves vanish and the low voltage, rapid (small) waves reappear. These are periodically interrupted by spikes from rhythmic clonic muscular spasm.

previously been uncertain of the interpretation of this phenomenon, it may represent the after-effects of a true orgasm.

It is true, however, that there are no records of orgasm among the females of most of the infra-human species of mammals, and apparently most individuals among most of the species just cited do not appear to reach orgasm in any of their sexual activities, even though some individual females may do so. The matter will not be finally settled until extensive physiologic measurements are made on these infra-human species; but as far as our knowledge yet goes, the human female is unique among the mammals in her capacity to reach orgasm with some frequency and regularity when she is aroused sexually.

Sexual orgasm constitutes one of the most amazing aspects of human behavior. There is only one other phenomenon, namely sneezing, which is physiologically close in its summation and explosive discharge of tension. Sneezing is, however, a localized event, while sexual orgasm involves the whole of the reacting body. While the summation of neuromuscular tensions is a phenomenon with which neurophysiologists are well acquainted, the mechanism of the explosive discharge of such tension is not understood.

AFTER-EFFECTS OF ORGASM

Muscular Spasms and Convulsions. Muscular spasms or more intense convulsions are the usual product of the sudden release of tension in orgasm. Sometimes these spasms are localized and mild, amounting to nothing more than slight movements of some particular part of the body. Usually they are more pronounced and extend to all parts of the body. Sometimes they may involve the entire body in extreme convulsions.[46]

In some persons the spasms which follow orgasm may subside within a moment or two. They may be so mild and pass so quickly that it would be difficult to observe that there had been any spasms at all. The average individual, however, may be in spasm for a half minute or a minute or more before subsiding into a normal physiologic state. Some persons may experience spasms or more extreme convulsions for two or three minutes, or in rare cases for as long as five minutes after the moment of orgasm.

The more extreme convulsive movements into which some individuals are thrown after orgasm bear a striking resemblance to those which may be observed in epilepsy (Figure 140). There may be some

---

[46] We have had access to a considerable body of observed data on the involvement of the entire body in the spasms following orgasm. The situation was graphically described by Roubaud 1876:17, whose description has not been surpassed by later authors.

common physiologic mechanism which accounts for the resemblances between the spasms or convulsions in epilepsy and those following orgasm, but the mechanisms are not yet understood. The convulsions following orgasm also resemble those which follow an electric shock. This makes it all the more amazing that most persons consider that sexual orgasm with its after-effects may provide one of the most supreme of physical satisfactions.

In the most extreme types of sexual reaction, an individual who has experienced orgasm may double and throw his whole body into continuous and violent motion, arch his back, throw his hips, twist his head, thrust out his arms and legs, verbalize, moan, groan, or scream [47] in much the same way as a person who is suffering the extremes of torture. In all of these respects, human females and males may react in essentially the same way. In some individuals the whole body may be thrown, or tossed, or rolled over a distance of several feet or yards. On occasion the sexual partner may be crushed, pounded, violently punched, or kicked during the uncontrolled responses of an intensely reactive individual. The movements are obviously involuntary, and they are for the most part beyond voluntary control. Some persons whose responses are mild can control their movements if there is some social advantage in reacting without attracting attention; but for those whose responses are more extreme, any deliberate control is nearly impossible.

Genital Spasms and Convulsions. The genitalia of both the female and the male are usually included in the spasms or convulsions which follow orgasm. In the female the perineal muscles may go into convulsion, and the levator muscles of the vagina may also move convulsively and sometimes may grab the male's penis or any other object which has been inserted into the entrance to the vagina.[48] The fact that some women experience vaginal spasms or convulsions may provide some basis for the references in the psychiatric literature to a "vaginal orgasm." These vaginal spasms are, however, simply an extension of the spasms which may involve the whole body after orgasm.

While the vaginal contractions may prove a source of considerable pleasure both for the female and for her male partner, it is a more

---

[47] Involuntary vocalization at orgasm is, of course, a matter of common knowledge. It is also reported in: Roubaud 1876:17. Bloch 1908:50. Talmey 1912:62; 1915:93. Bauer 1927(1):157. Van de Velde 1930:246. Havelock Ellis 1936(II,1):150, 166. Haire 1937:200. Brown and Kempton 1950:207.

[48] Genital spasms at or immediately following orgasm are also noted by: Roubaud 1876:13. Rohleder 1907(1):310–311. Bloch 1908:50. Moll 1912:25. Urbach 1921:125–126. Malchow 1923:135. Kisch 1926:295–296. Bauer 1927(1):156. Dickinson 1933, 1949:98. Van de Velde 1930:199. Havelock Ellis 1936(II,1): 159. W. Reich 1942:83. Hardenbergh 1949:228. Faller in Hornstein and Faller 1950:238.

difficult matter to determine whether the lack of vaginal spasms represents any loss of pleasure for a female. The absence of vaginal contractions in a woman who customarily has them may, however, provide some evidence that she is not responding in that particular relationship.

Strong rhythmic contractions of the abdominal muscles sometimes push the uterus and the attached walls of the vagina closer to the vaginal entrance (Figure 118) if it is not blocked by the inserted penis. Recently the action of the uterus has been studied with electrical instruments which can follow the movements of fine particles that are inserted in the cavity of the organ. These studies show that the upper end of the uterus goes into rhythmic contractions of considerable frequency whenever there is sexual arousal. The rate of movement, however, decreases in the body of the uterus, and the cervix at the lower end of the uterus shows a minimum of movement. There is, in consequence, an actual sucking effect which may pull semen through the cervix into the uterus.[49]

In the male, spasms or convulsions following orgasm may involve both the penis and the scrotum. The scrotal walls may suddenly relax, or expand and contract in rhythmic movements; and the testes, which have been drawn up into the groins during sexual arousal, may undergo considerable movement within the scrotum as a specific aftereffect of orgasm.

In most males there are only limited gross movements of the penis following orgasm, but in some individuals the movements may become spasmodic jerks of some magnitude. The most pronounced movements appear to depend upon contractions of the muscular attachments of the two spongy bodies (the corpora cavernosa) which constitute the main shaft of the penis. For most of their length the two spongy bodies are joined to form a single body, but at the base of the penis the two corpora separate into divergent roots which are fastened at their tips (the crura of the penis) to the pelvic bones. Movements of either the flaccid or erect penis may be effected by voluntary contractions of these crura, and at orgasm there may be strong and violently convul-

---

[49] The descent and/or sucking of the uterus is mentioned in: Rohleder 1907(1): 531. Talmey 1912:56–57; 1915:87–89. Urbach 1921:133. Malchow 1923: 135–136. Kisch 1926:295–300. Hirschfeld 1928(2):226. Dickinson 1933, 1949:90–93. Havelock Ellis 1936(II,1):160–162. Haire 1937:199, 202, 283. Kahn 1939:85, fig. 17. Weisman 1941:109–110 (a good summary of opinions). Hutton 1942:71. Huhner 1945:454–455. Gardner in Howell (Fulton edit.) 1949:1179. Faller in Hornstein and Faller 1950:238. Such uterine action has been described in animals (e.g., Marshall 1922:173–174, and Baker 1926:32), and data on its occurrence in human females are brought together in Colmeiro-Laforet 1952:125–126. Extensive work done under the direction of Carl Hartman is not yet published.

sive contractions of the crura. In consequence, the whole penis may sometimes move in strong jerks; but sometimes there is hardly any gross movement of the penis even though the crural contractions may be strong. Ejaculation may depend in part upon these contractions of the corpora cavernosa, as well as upon contractions of the urethra.

Any contraction of the muscles in the crura simultaneously contracts the anal sphincters, and the anus may open and close in violent convulsion as an after-effect of orgasm. Most persons are unconscious of this anal action unless they have had anal intercourse or utilized anal insertions as a source of erotic stimulation.

Ejaculation. Ejaculation is one of the most characteristic products of the spasms or convulsions which follow orgasm in the male. The prostate and seminal vesicles are thrown into spasms which press their liquid secretions into the urethra, and urethral contractions and contractions of muscle tissues in the crura of the penis may propel the semen with a force which is sufficient to carry it at least to the meatus (the opening of the urethra).

It is generally believed that semen is usually ejaculated with a force which is sufficient to propel it for some distance beyond the tip of the penis. Some clinicians have considered that the force with which the semen is thrown against the cervix in vaginal coitus may be a factor in determining whether fertilization occurs.[50] However, a considerable body of data to which we have had access, based on observations of some hundreds of males, indicates that there is considerable individual variation in this regard and some lesser variation in the experience of any single individual. In perhaps three-quarters of the males the semen merely exudes from the meatus or is propelled with so little force that the liquid is not carried more than a minute distance beyond the tip of the glans of the penis. In other males the semen may be propelled for a matter of some inches, or a foot or two, or even as far as five or six (or rarely eight) feet. This variation in function may depend upon anatomic or physiologic variations, for the pattern of ejaculation is largely fixed for each individual except as fatigue and, ultimately, age may reduce the intensity of all the physiologic responses.

Since the prostate gland and seminal vesicles are only vestigial structures in the female, she does not actually ejaculate. Muscular contractions of the vagina following orgasm may squeeze out some of the genital secretions, and in a few cases eject them with some force.[51] This

---

[50] Ejaculation against or into the cervix is mentioned by: Roubaud 1876:45. Rohleder 1907(1):311. Friedlaender 1921:27–28. Forel 1922:57. Hirschfeld 1928(2):226. This idea is, however, denied by Dickinson 1933, 1949:93–94.
[51] The expulsion of genital secretions by the female at orgasm, which is the so-called "female ejaculation," is popularly known and talked about. In the

is frequently referred to, particularly in the deliberately erotic litera-ture, as an ejaculation in the female; but the term cannot be strictly used in that connection. Ejaculation is, in fact, the only phenomenon in the physiology of sexual response which is not identically matched in the male and the female, or represented by closely homologous functions.

Because ejaculation is almost invariably and immediately associated with orgasm, it is often considered as the orgasm of the male. This interpretation is not acceptable, for the following reasons:

1. The data already presented show that sexual arousal and orgasm involve the whole nervous system and, therefore, all parts of the body. Ejaculation is only one of the events that may follow the release of nervous tensions at orgasm.

2. Orgasm in the female matches the orgasm of the male in every physiologic detail except for the fact that it occurs without ejaculation.

3. Pre-adolescent boys may experience orgasm, duplicating the experience of the adult male in every respect except for the fact that they do not ejaculate. This simply depends on the fact that the prostate gland and seminal vesicles of the younger boy are not sufficiently developed to secrete seminal fluids.

4. Adult males who are capable of multiple orgasm may have several experiences with ejaculation and then, when the secretions of the pros-tate and seminal vesicles are exhausted, they may have further or-gasms without semen. The later orgasms may be duplicates of the earlier ones, except that they do not lead to ejaculation. Physiologically and psychologically they may be as satisfactory as those in which ejaculation occurred.

5. There are some males in whom ejaculation does not occur until some seconds after orgasm, and in whom, therefore, it is possible to distinguish orgasm and ejaculation as two separate events.

6. There are a few adult males (perhaps one in four thousand) who are anatomically incapable of ejaculation, although they may experi-ence orgasms which are in all other respects similar to those which are accompanied by ejaculation.

7. Males who have had their prostate glands or a portion of the sympathetic system removed by surgical operation (a prostatectomy or a sympathectomy) are no longer capable of ejaculation, although they may still be capable of orgasm if no other complications have

literature it is described, for instance, in: Van de Velde 1930:195–196. Have-lock Ellis 1936(II,1):145. Grafenberg 1950:147.

been involved in the operation. Although the males of some of the lower mammalian species (*e.g.*, rats) may have their capacity to reach orgasm stopped by such a prostatic operation, they may again become capable of reaching orgasm if they are given male hormones (androgens); and such orgasms may be typical in every respect of orgasm in the normal rat except that they occur without ejaculation.

Because of this mistaken identification of ejaculation as orgasm, many persons have concluded that orgasm in the female is something different from orgasm in the male. On the contrary, we find that orgasm in the female is, physiologically, quite the same as orgasm in the male. Ejaculation may constitute a spectacular and biologically significant event which is unique to the male, but it is an event which depends on relatively simple anatomic differences, rather than upon differences in the basic physiology of sexual response in the female and male.

Other Readjustments. In addition to the physiologic readjustments noted above, all other functions which have been distorted during sexual response are quickly restored by orgasm to their normal or even subnormal states.

Pulse rates which had reached 150 or more at the moment of orgasm are, within a matter of seconds or a few minutes at the most, returned to their normal 75 or 80 (Figures 120–128).[52] Blood pressures similarly drop (Figures 129–138). The increase in the peripheral circulation of the blood quickly subsides, and tumescence may be abruptly reduced, although the speed of detumescence may vary considerably in different individuals. Younger males may maintain erection for several minutes after orgasm; most older males begin to lose erection immediately. If there is any continuing excitation, some younger individuals and even some older males may maintain full erection for five or ten minutes or even for a half hour or more after orgasm. It is sometimes possible for such a male to renew sexual activity and attain another orgasm without having experienced flaccidity between performances.

After the cessation of the respiratory convulsions which are the immediate after-effects of orgasm, the respiratory rate in most individuals quickly drops to normal or, in case the individual is already fatigued, to something lower than normal frequency (Figure 139).

Wounds which were acquired during the sexual activity, and which exhibited a surprisingly scant flow of blood at that time, may begin to

---

[52] The rapidity of the decrease in pulse following orgasm is shown in Boas and Goldschmidt 1932:99, where the female's pulse fell from 146 to 117 in a matter of seconds, and the male's from 143 to 117 within a minute. In Klumbies and Kleinsorge 1950a:953, 955; 1950b:62, the male's pulse fell from 142 to 67 in four seconds.

flow more freely after orgasm. The menstrual flow, which may have been slowed up during erotic response, is resumed at its normal rate.

Even more remarkable is the sudden return of sensory acuity (or the reorientation of sensory perception) which may follow orgasm. While the increased capacity may amount to nothing more than a return to normal sensitivity, the contrast with the previous insensitive state may be enough to cause the individual considerable discomfort. Many males—perhaps a majority of them—become so sensitive after orgasm that they may experience considerable pain if there is any additional stimulation of the penis at that time. Marriage manuals frequently recommend that the male who reaches orgasm before his female partner does, should continue coital movements until she has been satisfied; but it should be recognized that such continued activity would be excruciatingly painful to many males, and for many of them a physical impossibility. There is some record of a similar hypersensitivity among females, although the specific data are more limited.

Various parts of the body may itch after orgasm. Many individuals become especially conscious of a full bladder after orgasm.[53] Muscles, such as those of the groin, the legs, the calf, the fingers, and the neck, may go into cramps during sexual activity, but one may not be conscious of that fact until after he or she has experienced orgasm (p. 620).

Some persons become conscious of the fact that they are hungry and thirsty after sexual activity. Sometimes such hunger or thirst simply reflects a return to normal perceptive capacities; but hunger, thirst, a general restlessness, a desire to get away from other people, a desire to smoke, and still other nervous types of behavior may, on occasion, be the product of the psychologic disturbance which is engendered by doubts over the moral propriety or the social acceptability of the act in which one has just engaged. Sometimes there are increased movements of the digestive tract following sexual activity, and an occasional individual may defecate or vomit in the course of sexual activity, or immediately after it. Such disturbances are not infrequently credited to the supposedly unsanitary nature of some sexual technique, but the true explanation of such activity is certainly psychologic.

A marked quiescence of the total body is the most widely recognized outcome of orgasm. The famous aphorism, *post coitum triste*—one is sad following coitus—, is not only a distortion of Galen's original statement, but an inadequate description of the usually quiescent state of

---

[53] The secretion of urine, or desire to urinate during sexual excitement, is also noted by: Van de Velde 1930:244. Havelock Ellis 1936(II,1):154. Hardenbergh 1949:227.

a person who has experienced orgasm.[54] There is neither regret nor conflict nor any tinge of sadness for most persons who have experienced orgasm. There is, on the contrary, a quiescence, a calm, a peace, a satisfaction with the world which, in the minds of many persons, is the most notable aspect of any type of sexual activity.

Sometimes, especially in youth, the post-orgasmic relaxation is hardly more than momentary. There are some individuals who, within a matter of seconds or minutes after the cessation of the orgasmic spasms, are ready for any type of vigorous exercise or mental activity. The average individual may require four or five minutes of repose before coming back to a normal state. Many persons promptly fall asleep after the termination of sexual activity, especially if it occurs in the evening when they are already fatigued.

It has been said that sexual excitement after orgasm recedes more slowly in the female than in the male. We do not know of data which warrant such a generalization, although it frequently recurs in the literature, sometimes with impressive but wholly imaginary charts to illustrate the concept.[55] The small body of actual measurements that are available (Figures 120–138) do not warrant any general distinction between females and males.

A few persons may be fatigued for some hours, and a rare individual may be exhausted for some days following orgasm, but such cases represent ill health which is overtaxed by the additional expenditure of energy required by sexual activity. The physical exhaustion which is usually ascribed to sexual excess is probably the outcome of the late hours, the lack of sleep, the alcoholic dissipation, or other excesses which may accompany the sexual activity. Sexual activity itself is limited by a self-regulating mechanism which controls the possible frequencies of orgasm. When sexual performance has reached the limits of one's nervous capacity, one is no longer interested erotically, and no longer responds to sexual stimuli. Once or twice in a lifetime, each male may deliberately attempt to set a record and force himself to

[54] Galen [ca. 130–200 A.D.] actually said: *Triste est omne animal post coitum, praeter mulierem gallumque* (every animal is sad after coitus, except the human female and the rooster).

[55] That excitement after orgasm diminishes more slowly in the female is often asserted in the literature, as in: Roubaud 1876:17. Rohleder 1907(1):313. Moll 1912:26. Talmey 1912:63; 1915:94. Urbach 1921:129 ff. Krafft-Ebing 1922:41. Bauer 1927(1):158. Hirschfeld 1928(2):230. Van de Velde 1930: 181, 248. Havelock Ellis 1936(II,1):168. Haire 1937:200, 211. Kahn 1939: fig.31. Faller in Hornstein and Faller 1950:239. Stone and Stone 1952:186. Few of the references give exact measurements. Klumbies and Kleinsorge 1950a:956; 1950b:61, found no difference between the female and male in this respect, but the case is complicated by the fact that the female had several consecutive orgasms. On the other hand, the Boas and Goldschmidt female case (1932:99) returned to a physiologic norm somewhat more rapidly than the male.

perform beyond the limits of his spontaneous erotic responses; but such performances are infrequent in the histories of most males. Similarly, females do not respond in sexual activities which go beyond their physiologic capacities, although they may, of course, be forced into physical relationships in which they do not respond.

Individual Variation. To summarize the data which have already been presented, we note that variations in the gross aspects of sexual response and orgasm may involve the following: (1) the amplitude of the muscular movements; (2) the particular parts of the body which are most prominently involved in the responses; (3) the speed of the muscular movements; (4) the number of pelvic thrusts or other movements which are made prior to orgasm; (5) the time which may elapse from the beginning of the activity to the peak at orgasm (and this may vary from ten seconds to an hour or two); (6) the magnitude and duration of the spasms which follow orgasm.

These variations offer endless possibilities for combination and recombination.[56] Consequently the responses of each individual may be quite unlike those of any other individual, although the basic physiologic patterns of sexual response and orgasm are remarkably uniform among all individuals, both female and male, and throughout all of the species of mammals.

Conscious Satisfactions. Even though sensory perception and intellectual activity may be at a minimum during sexual activity, there is usually enough conscious realization of satisfaction to provide a considerable stimulus for a continuation of the activity, usually to the point of orgasm. Although it is not entirely clear what the sources of those satisfactions may be, they appear to be influenced by the following:

¶ The nature and the intensity of the physical and psychologic stimuli which effect the sexual response.

¶ The innate physiologic capacities of the responding individual.

¶ The psychologic capacities of the individual, such as his or her ability to win sexual partners, to develop effective psychologic situations during the overt activity, and to respond sympathetically to the partner's performance.

¶ The physiologic level which is attained at orgasm. It is possible that orgasm which is accompanied by a pulse running at 150 may be

---

[56] It is commonly understood that there is individual variation in the conspectus of sexual response and orgasm, although it is only occasionally mentioned in the literature. But see: Roubaud 1876:16. Talmey 1915:95. Negri 1949:78, 82–83. Brown and Kempton 1950:207. Stone and Stone 1952:186.

more stimulating than orgasm which is reached with the pulse running at 100 or so; but the data are not conclusive. Similarly, other physiologic changes may be more important when they represent a maximum departure from the normal state.

¶ The previous sexual experience of the individual, and the manner in which he or she has been conditioned by such experience.

¶ The individual's previous experience with the particular sexual partner. The satisfactions to be obtained from relationships which are continued over long periods of years may steadily increase because of the increasing appreciation of the partner's psychologic and physiologic needs, and his or her preferences in sexual matters.

¶ The novelty of the sexual situation, which may stimulate when old situations have lost some of their former attraction.

¶ The directness and the speed with which orgasm is attained. Some individuals prefer direct and uninterrupted activity, others prefer a leisurely, long-drawn-out performance in which there are deliberate interruptions for the sake of delaying orgasm.

¶ The extent to which the sexual activities are accepted psychologically. The presence or absence of guilt feelings is, for many persons, the most important factor in determining the level of the satisfactions which may be obtained from a sexual relationship.

SUMMARY AND COMPARISONS OF FEMALE AND MALE

1. More than a score of the elements which have been recognized in the physiology of sexual response have been discussed in the present chapter. Females and males do not differ in regard to any of these basic elements. Because of differences in anatomy, the processes in the two sexes may differ in details. Tumescence, for instance, is most noticeable in the male in the erection of his penis, and most noticeable in the female in the erection of the nipples of her breasts and of the clitoris and labia minora; but the physiologic bases of these several events are essentially identical. The female and male are quite alike, as far as the data yet show, in regard to the changes in pulse rate, the changes in blood pressure, the peripheral circulation of blood, the tumescence, the increase in respiration, the possible development of an anoxia, the loss of sensory perception, and the development of rhythmic muscular movements, even including rhythmic pelvic thrusts.

2. Orgasm is a phenomenon which appears to be essentially the same in the human female and male. This is somewhat surprising, since orgasm appears to occur only infrequently among the females of the infra-human mammalian species.

3. Females appear to be capable of responding to the point of orgasm as quickly as males, and there are some females who respond more rapidly than any male. The usual statement that the female is slower in her capacity to reach orgasm is unsubstantiated by any data which we have been able to secure. But because females are less often stimulated by psychologic factors, they may not respond as quickly or as continuously as males in socio-sexual relationships.

4. In general, the after-effects of orgasm in the female do not differ in any essential way from those in the male. Ejaculation occurs in the adult male, and there is no such phenomenon in the female; but ejaculation depends upon a minor anatomic distinction between the female and male, and not upon any fundamental differences in the physiology of the two sexes.

In spite of the widespread and oft-repeated emphasis on the supposed differences between female and male sexuality, we fail to find any anatomic or physiologic basis for such differences. Although we shall subsequently find differences in the psychologic and hormonal factors which affect the responses of the two sexes (Chapters 16, 18), males would be better prepared to understand females, and females to understand males, if they realized that they are alike in their basic anatomy and physiology.

Chapter 16

PSYCHOLOGIC FACTORS IN SEXUAL RESPONSE

It might properly be contended that all functions of living matter are physiologic, but it is customary to distinguish certain aspects of animal behavior as psychologic functions. The distinctions can never be sharp, and they probably do not represent reality; but they are convenient distinctions to make, particularly in regard to human behavior.

Usually physiologists have been concerned with the functions of particular parts of the plant or animal, and with an attempt to discover the physical and chemical bases of such functions. Psychologists, on the other hand, have more often been concerned with the functioning—the behavior—of the organism as a whole. Many of the psychologic studies record—and properly record—the behavior of an animal without being able to explain the bases of that behavior in the known physics or chemistry of living matter. When psychologists try to explain behavior in physico-chemical terms, it is difficult to say, and quite pointless to try to say, whether such studies lie in the field of psychology or physiology.

It is important to understand how nebulous the distinctions are between the psychologic and physiologic aspects of behavior, for there are some who seem to believe that there are three universes: an animal's anatomy, its physiology, and its psychology. Such a misinterpretation formerly led biologists to think of a dualistic relationship between the physiologic capacities of an organism and its form and structure. The same sort of misinterpretation has led to the dualistic distinction of mind and body to which many persons have been inclined. But form and function are coordinate qualities of any living cell, and of any more complex assemblage of living cells.

Such specious distinctions between form and function have, unfortunately, lent encouragement to the opinion that the psychologic aspects of human sexual behavior are of a different order from, and perhaps more significant than, the anatomy or physiology of sexual response and orgasm. Such thinking easily becomes mystical, and quickly identifies any consideration of anatomic form and physiologic function as a scientific materialism which misses the "basic," the "human," and the "real" problems in behavior. This, however, seems an unnecessary judgment. Whatever we may learn of the anatomy and physiology and of the basic chemistry of an animal's responses, must contribute to our understanding of the totality which we call behavior. Those aspects of behavior which we identify as psychologic can be nothing but certain aspects of that same basic anatomy and physiology.

This, however, will not prevent us from recognizing the existence of many phenomena, such as the processes of learning and conditioning, the development of preferences in the choice of sexual objects, and the development of whole patterns of behavior, which cannot yet be explained in terms of the physics and chemistry which may be involved. Indeed, some of the aspects of sexual behavior which most critically affect the lives of human animals, and which are most often involved in the relations between human females and males, are among these still unexplained phenomena.

LEARNING AND CONDITIONING

One of the best known and distinctive qualities of living matter, although it is one which is still unexplainable in terms of physics and chemistry, is its capacity to be modified by its experience. The first time that an animal meets a given situation, its reactions may represent little more than direct responses to immediate stimuli; but in its subsequent contacts with similar stimuli, the organism may react differently from the way it did on the first occasion. In some fashion which no biologist or biochemist understands, living plant and animal cells, and groups of cells and tissues and organs in more complex animal bodies, are modified by their experience. The organism's later behavior represents a composite of its reactions to the stimuli which are immediately present, and its reactions to the memory of its previous experience. This depends on the processes which are known, in psychologic terminology, as learning and conditioning.

Learning and conditioning are, of course, familiar parts of the everyday experience of the human animal. Other things being equal, the first experiences, the most intense experiences, and the latest experiences may have the maximum effect on an individual's subsequent behavior. Freud and the psychiatrists, and psychologists in general, have correctly emphasized the importance of one's early experience, but it should not be forgotten that one may continue to learn and continue to be conditioned by new types of situations at any time during one's life. It is incorrect to minimize the importance of all except childhood experiences in the development of adult patterns of behavior.

Learning and conditioning in connection with human sexual behavior involve the same sorts of processes as learning and conditioning in other types of behavior. But man, because of his highly developed forebrain, may be more conditionable than any of the other mammals. The variations which exist in adult sexual behavior probably depend more upon conditioning than upon variations in the gross anatomy or physiology of the sexual mechanisms.

The sexual capacities which an individual inherits at birth appear to be nothing more than the necessary anatomy and the physiologic capacity to respond to a sufficient physical or psychologic stimulus. All human females and males who are not too greatly incapacitated physically appear to be born with such capacities. No one has to learn to become tumescent, to build up the neuromuscular tensions which lead to the rhythmic pelvic thrusts of coitus, or to develop any of the other responses which lead to orgasm.

But apart from these few inherent capacities, most other aspects of human sexual behavior appear to be the product of learning and conditioning. From the time it is born, and probably before it is born, the infant comes into contact with some of the elements that enter into its later sexual experience. From its first physical contacts with other objects, and particularly from its contacts with other human bodies, the child learns that there are satisfactions which may be obtained through tactile stimulation. In its early sexual experience with other individuals, the child begins to learn something of the rewards and penalties which may be attached to socio-sexual activities. From its parents, from other adults, from other children, and from the community at large, it begins to acquire its attitudes toward such things as nudity, the anatomic differences between males and females, and the reproductive functions; and these attitudes may have considerable significance in determining its subsequent acceptance or avoidance of particular types of overt sexual activity.

The type of person who first introduces an individual to particular types of socio-sexual activities may have a great deal to do with his or her subsequent attitudes, his or her interest in continuing such activity, and his or her dissatisfactions with other types of activity. Above all, experience develops a certain amount of technical facility, and an individual *learns* how to masturbate and *learns* how to utilize particular techniques in petting, in coitus, or in homosexual or other relations. As we shall subsequently see (Chapter 17), we may sharply distinguish the inherent sexual capacities with which an animal is born, from those aspects of its sexual behavior which are acquired by the processes of learning and conditioning.

DEVELOPMENT OF PREFERENCES

As a result of its experience, an animal acquires certain patterns of behavior which lead it to react positively to certain sorts of stimuli, and to react negatively to other sorts of stimuli. But there are also various degrees of response, and an animal learns to react toward or against certain stimuli more intensely than it does to others.[1] When there is any possibility of choosing, the animal may show strong preferences for one rather than another type of activity.

An individual may come to prefer particular types of individuals as sexual partners; may prefer tall persons or short persons; may prefer blondes or brunettes; may prefer sexual partners who are much younger or much older, or of his or her own age [2]; may develop an incapacity to respond to any except a single sexual partner, or a preference for variety in sexual experience; may prefer a heterosexual or a homosexual pattern of behavior; may prefer masturbation to the pursuit of socio-sexual contacts; may prefer a considerable amount of petting prior to actual coitus, or immediate coitus without preliminary play; may find satisfaction or be offended by the use of certain genital, oral, or anal techniques; may come to desire a variety of positions in coitus, or the more or less exclusive use of a single position; may choose a farm animal instead of a human partner for sexual relationships. All of these choices and reactions to particular stimuli may seem reasonable enough and more or less inevitable to the person who is involved, even though some of them may seem un-understandable, unnatural, and abnormal to the individual who has not been conditioned by the same sort of experience.

Even some of the most extremely variant types of human sexual behavior may need no more explanation than is provided by our understanding of the processes of learning and conditioning. Behavior which may appear bizarre, perverse, or unthinkably unacceptable to some persons, and even to most persons, may have significance for other individuals because of the way in which they have been conditioned.

---

[1] Students of animal behavior have noted that if a male's initial coital experience involves pain or fright, the male may become reluctant to mate thereafter. See, for example: Beach 1947b:265. Rice and Andrews 1951:180–181.

[2] The extreme variation which may be found among both men and women in the types of preferred sexual partners, is well illustrated in the listing in Hamilton 1929:502–505. That the males of infra-human species often develop strong preferences for particular females or types of females, is recorded in: Tinklepaugh 1928:296–300 (rhesus monkey). Yerkes and Elder 1936a:32–34, 38 (chimpanzee). Carpenter 1942:139 (monkey). Enders 1945 (fox). Hafez 1951 (ram). Ford and Beach 1951:91–93 (general discussion). Shadle (verbal communic.) records two porcupines, each of which developed strong preferences for a particular female. That estrual females of sub-primate species almost never show strong preferences, even in species in which preferences are strongly marked in the male, is noted in: Rowlands and Parkes 1935 (fox). Enders 1945.

Flagellation, masochism, transvestism, and the wide variety of fetishes appear to be products of conditioning, fortified sometimes by some other aspect of an individual's personality and by inherent or acquired anatomic and physiologic capacities. Sexual reactions to stockings, to underclothing, to other articles of clothing, to shoes, or to long hair may be no more difficult to explain than attractions to the body of a sexual partner, or to particular parts of that body, to the legs of females, to the breasts of females, to male genitalia, to buttocks, or to other portions of the human anatomy.

The male who reacts sexually and comes to erection upon seeing a streetcar, may merely reflect some early experience in which a streetcar was associated with a desirable sexual partner; and his behavior may be no more difficult to explain than the behavior of the male who reacts at the sight of his wife undressing for bed. There may be more social advantage in the one type of behavior than in the other. In rare instances some of the so-called aberrant types of behavior, meaning the less usual types of conditioned responses, may be definitely disadvantageous, but in most instances they are of no social concern. The prominence given to classifications of behavior as normal or abnormal, and the long list of special terms used for classifying such behavior, usually represent moralistic classifications rather than any scientific attempt to discover the origins of such behavior, or to determine their real social significance.

VICARIOUS SHARING OF EXPERIENCE

A fair amount of the conditioning which occurs in connection with human sexual behavior depends upon the fact that the human animal, with its extraordinary capacity for communication through verbal interchange, through the printed word and pictorial material, and through other modern devices, may vicariously share the sexual experience of many other persons. Learning of their satisfactions or difficulties in particular types of sexual activity may influence one's own decision to engage or not to engage in similar types of activity.

Many persons find considerable stimulation in listening to accounts of the sexual experience of other persons, in hearing fictional tales of sexual exploits, in reading of such experience, and in seeing photographs and drawings of sexual objects and activities. Many individuals become strongly conditioned toward or against having particular types of sexual activity, before they have ever had any actual experience of the sort.

An individual's pattern of sexual behavior usually depends to a great extent upon the longstanding and sometimes ancient social codes concerning the various types of sexual activity. The social attitudes may begin to condition the child at a very early age, and may force it to confine its attitudes, its responses, and its overt activities to sexual expressions which are acceptable to the particular culture.

REACTIONS TO ASSOCIATED OBJECTS

An animal may become conditioned to respond not only to particular stimuli, but to objects and other phenomena which were associated with the original experience. Pavlov's classic experiment with the dog which was so conditioned that it salivated upon hearing a dinner bell, as well as when it came in contact with the food with which the bell was originally associated, stands as the prototype of such associative conditioning.[3]

Sexual behavior, among all species of mammals, may involve a great deal of conditioning by phenomena which were associated with previous experience. Male cats and dogs and many other mammals respond sexually when they approach places in which they have had previous experience [4]; male rabbits, guinea pigs, skunks, raccoons, bulls, and horses may respond to odors left by female secretions.[5] They often respond to the odor of the urine of a female, especially if the female is in estrus. In the laboratory, male animals may respond to particular dishes, to particular boards, or to particular pieces of other furniture with which some female has had contact. They may respond more intensely to particular animals with which they have had previous sexual contact, they may respond less intensely to animals with which they have not had previous contact—although another phenomenon, psychologic fatigue, may lead to an exact reversal of this pattern of response.[6]

If satisfactory relations were previously had with an individual of the opposite sex, animals are more likely to respond to other individu-

---

[3] For descriptions of Pavlov's experiments with dogs, see: Woodworth and Marquis 1947:525–530. Andrews 1948:44. Freeman 1948:180.

[4] Male animals which showed sexual arousal on returning to a place where they had previously had coitus are described by: McKenzie and Berliner 1937:18 (ram). Zitrin and Beach in Hartman 1945:42–44 (cat). That a strange environment may prevent or handicap a male in having coitus although the female may not be thus affected, is recorded in: Marshall and Hammond 1944:12. Beach 1947b:264. Root and Bard 1947:81.

[5] One of the best illustrations of a male's interest in odors left by females is recorded for the porcupine by: Shadle, Smelzer, and Metz 1946:118, 120. Shadle 1946:159–160. See also: Beach and Gilmore 1949:391–392.

[6] That some male animals will copulate only with familiar females and ignore strange females is recorded in: Hartman 1945:39 (monkey). Shadle 1946:160–161 (porcupine). Beach 1947b:264–265 (general statement). That estrual females, on the contrary, will usually accept any male, familiar or strange, is recorded by: Bean, director of Brookfield Zoo (verbal communic.). Beach 1947b:264. Psychologic fatigue, the lessening of sexual response through prolonged living together, has been noted in animals by: Hamilton 1914:301–302. Miller 1931:397–398, 406.

als of that sex. If the previous experience was with an individual of their own sex, they are, because of the association with the previous experience, more likely to respond again to individuals of their own sex. If the laboratory investigator was present when the animal was previously involved in sexual activity, it is likely to react more intensely on later occasions if the investigator is again present. A dog which has been masturbated by its owner may subsequently come to full erection whenever it sees the human agent, move toward him, and try to renew the relationship. We have the record of one dog which did not go to its owner when it saw him, but ran to the place where it had been previously masturbated and there awaited a renewal of that experience.

We have already pointed out (p. 590) that sexual stimulation by things that one sees, hears, smells, or tastes often depends upon the associations which they evoke, rather than upon the direct physical stimulation of the sense organs through which those things are perceived. While this is true of all the higher mammals, it is particularly true of the human animal. From its earliest years the child comes to associate a considerable number of particular objects and phenomena with things that make him comfortable or in some other way prove satisfying. In the course of time, an adult comes to associate sexual activities with warmth, tactile satisfactions, particular types of food, alcoholic drinks, furniture, the clothing of the sexual partner, particular odors, particular intensities of light, particular sounds, certain musical compositions, particular sorts of voices, particular words which have been used to describe particular types of sexual performance, the sort of room or outdoor setting in which satisfactory sexual relations previously occurred, the use of particular techniques in a sexual relationship, and an endless list of other particular things.

Sometimes an individual may reach a point at which he reacts to these associated phenomena as intensely or more intensely than he reacts to the physical stimulation of a sexual contact. Not a few individuals find that they are more intensely aroused by the anticipation of an opportunity to engage in sexual activity than they are when they arrive at the activity itself.

SYMPATHETIC RESPONSES

Among most species of mammals, most males and some females become erotically aroused when they observe other individuals engaging in sexual activity. Animals which have not reacted to the mere presence of the other animals may become interested if the other animals begin sexual activity. Most males are likely to respond quite immediately to such stimuli, to come to erection, and to seek the opportunity for sexual activity of their own. This is as true of the human male as it is of the males of other species of mammals.[7]

These are, technically speaking, **sympathetic responses**. The one animal feels or reacts (*pathos*) with (*sym*) the other. Of all the situations to which an animal may become conditioned, none is as likely to evoke sexual responses as sexual activity itself. The restrictions which most human societies place upon the public performance of sexual acts probably did not arise out of any innate perception of what was shameful or wrong, but from an attempt to control the sympathetic responses of the bystanders and the social consequences of group sexual activity. Among laboratory animals and animals in the wild, vigorous competition and violent conflict are the usual outcome of group sexual activity. They are as likely to be the outcome of group activity in the human species, unless the individuals control their jealousies in a conscious attempt to obtain the especial stimulation which may be found in group activity.

In a socio-sexual relationship, the sexual partners may respond to each other and to the responses made by each other. For this reason, most persons find socio-sexual relationships more satisfactory than solitary sexual activities.

When there is physical contact, all of one's sense organs may aid in making one aware of the responses of the partner and of the movements of the partner's body, particularly when there are such extensive contacts as completely nude bodies may provide. Tensions developing in the body of the one partner may be reflected instantaneously in the reactions of the other partner's body. As the one partner approaches orgasm, his or her extreme reactions may stimulate the other partner into simultaneous orgasm. Such a simultaneity of response may occasionally originate in the fact that the two partners are so constituted that they respond in exactly the same period of time, but it usually depends upon some sympathetic interaction between the two.

SIGNIFICANCE OF CONDITIONING IN FEMALES AND IN MALES

In general, males are more often conditioned by their sexual experience, and by a greater variety of associated factors, than females. While there is great individual variation in this respect among both females and males, there is considerable evidence that the sexual re-

---

[7] Sympathetic responses in mammals are also noted in Ford and Beach 1951:71. Such responses are, of course, not invariable; for example, a herd of grazing ruminants may ignore a copulating couple in their midst, acc. Bean, director of Brookfield Zoo (verbal communic.). It should be noted that the sight of coitus is without effect upon inexperienced chimpanzees, but is sexually arousing to experienced animals of both sexes, acc. Nissen (verbal communic.).

sponses and behavior of the average male are, on the whole, more often determined by the male's previous experience, by his association with objects that were connected with his previous sexual experience, by his vicarious sharing of another individual's sexual experience, and by his sympathetic reactions to the sexual responses of other individuals. The average female is less often affected by such psychologic factors. It is highly significant to find that there are evidences of such differences between the females and males of infra-human mammalian species, as well as between human females and males.[8]

While we found no basic differences in the anatomy which is involved in the sexual responses of females and of males, and no differences in the physiologic phenomena which are involved when females and males respond sexually, we do find, in these responses to psychologic stimuli, an explanation of some of the differences that we have reported in the incidences and frequencies and the patterns of sexual behavior among females and males. We shall subsequently find (Chapter 18) that hormonal differences between the human female and male may account for certain other differences between the two sexes.

It cannot be too strongly emphasized that there is tremendous individual variation in the way in which different individuals may be affected by psychologic stimuli. We have already pointed out some of these differences. For instance, we have shown (p. 164) that there is a considerable proportion of the females who masturbate without associated fantasies, and a considerable proportion of our female sample who had never had specifically sexual dreams while they slept. In this respect, such a female differs considerably from the average male, for nearly all males do fantasy while masturbating, and nearly all of them have nocturnal sex dreams. On the other hand, we have also recorded (p. 164) that there are some females who invariably fantasy while they are masturbating, who have an abundance of sex dreams, and who have daytime fantasies which may so arouse them that they reach orgasm without any physical stimulation of any part of their bodies. It is only one male in a thousand or two who can fantasy to orgasm. In our sample, the range of variation in responses to psychologic stimuli is, therefore, much greater among females than it is among males. While we may emphasize the differences which exist between the average female and the average male, it should constantly be borne in mind that there are many individuals, and particularly many females, who widely depart from these averages.

---

[8] In summing up the situation among infra-human mammals, Ford and Beach 1951:241 state: "We are strongly impressed with the evidence for sexual learning and conditioning in the male and the relative absence of such processes in the female."

1. **Observing the Opposite Sex.** A third (32 per cent) of the males in the sample reported that they were considerably and regularly aroused by observing certain females (clothed or nude), including their wives, girl friends, and other females of the sort with whom they would like to have sexual relations. Another 40 per cent recorded some response. Only half as many of the females (17 per cent) in the sample reported that they were particularly aroused upon observing males, whether they were their husbands, boy friends, or other males, and another 41 per cent recorded some response. The specific data are as follows:

Observing the opposite sex

EROTIC RESPONSE	BY FEMALES	BY MALES
	%	%
Definite and/or frequent	17	32
Some response	41	40
Never	42	28
Number of cases	5772	4226

The responses of these males upon observing females were the physiologic responses characteristic of sexual arousal; they often included genital reactions, and often led the male to approach the female for physical contact. Females who had been aroused with similar intensities did occur in the sample, but most of the females who had been aroused had not responded with such marked physiologic reactions.

Responses upon observing potential sexual partners are also characteristic of the males of most of the infra-human species of mammals, but the females of most of the mammalian species less frequently show signs of erotic arousal before they have made physical contact with the sexual partner (p. 230). Psychologic arousal in the female occurs most frequently when she is in estrus. Female dogs, female chimpanzees, sometimes cows, female porcupines, and the females of some other species may become quite aroused while they are in estrus and become aggressive in making sexual approaches to the male; although among even these species the females are not aggressive as often as the males.[9]

[9] That aggressive sexual behavior is common among estrual females of infra-human mammalian species, is recorded by: Elliott acc. Miller 1931:382, 405 (fur seal; the male is so busy guarding his harem that the female must take the initiative). Zuckerman 1932:227–229, 243 (baboon). Carpenter 1942:131, 136, 154 (monkey; the female may repeatedly approach the male despite having been driven off and wounded). Yerkes and Elder 1936a:25–26 (chimpanzee; female almost always goes to the male, not the male to the female). Roark and Herman 1950:7 (cows sometimes pursue the bull). McKenzie (verbal communic., says between 10 and 30 per cent of mares may pursue stallions). Carpenter 1942:129 (in gibbon, both sexes are about equally aggressive in pugnacity and reproductive behavior). We have also observed such female aggressiveness in the dog and porcupine.

2. Observing One's Own Sex. The recognition of erotic arousal upon observing other individuals of one's own sex is, of course, a basically homosexual phenomenon. In our culture, with its strong condemnation of male homosexuality, most males who want to think of themselves as completely heterosexual are therefore afraid to admit that they see even esthetic merit in other males. On the other hand, females are allowed to find esthetic satisfactions in observing the nude female form, or in observing well dressed females, and our cultural traditions make it possible for a female to express her admiration of another female without being suspected of homosexual interests. Actually, the female's interest in other females is often a matter of identification with a person she admires, and lacks any erotic element.

In view of this lesser social acceptance of male interests in males, and of the readier acceptance of female interests in females, it is particularly interesting to find that males recognize and admit their erotic responses to other males as often or even more often than females recognize and admit their erotic responses to other females. The record is as follows:

Observing one's own sex

EROTIC RESPONSE	BY FEMALES	BY MALES
	%	%
Definite and/or frequent	3	7
Some response	9	9
Never	88	84
Number of cases	5754	4220

3. Observing Portrayals of Nude Figures. Something more than half (54 per cent) of the males in our sample had been erotically aroused by seeing photographs or drawings or paintings of nude females, just as they were aroused upon observing living females. Most homosexual males are similarly aroused by seeing portrayals of nude males. Fewer (12 per cent) of the females in the sample had ever been aroused by seeing photographs or drawings or paintings of either male or female nudes. The specific record is as follows:

Observing portrayals of nude figures

EROTIC RESPONSE	BY FEMALES	BY MALES
	%	%
Definite and/or frequent	3	18
Some response	9	36
Never	88	46
Number of cases	5698	4191

It is difficult for the average female to comprehend why males are aroused by seeing photographs or portrayals of nudes when they cannot possibly have overt sexual relations with them. Males on the other hand, cannot comprehend why females who have had satisfactory sexual relations should not be aroused by nude portrayals of the same person, or of the sort of person with whom they have had sexual relations. We have histories of males who have attempted to arouse their female partners by showing them nude photographs or drawings, and most of these males could not comprehend that their female partners were not in actuality being aroused by such material. When a male does realize that his wife or girl friend fails to respond to such stimuli, he may conclude that she no longer loves him and is no longer willing to allow herself to respond in his presence. He fails to comprehend that it is a characteristic of females in general, rather than the reaction of the specific female, which is involved in this lack of response.

Striking evidence of the differences in the reactions of females and males is to be found in the commercial distribution of portrayals of nude human figures. There is a tremendous business of this sort, including the sale of good nude art, of photographic prints, and of moving picture, physical culture, and nudist magazines. There are nude or near nude figures in the main pages and in the advertising sections of nearly all illustrated magazines. Much of this material is distributed without any deliberate intent to provide erotic stimulation, and much of it has artistic and other serious value; but all of it may provide erotic stimulation for many of the male consumers.[10]

Photographs of female nudes and magazines exhibiting nude or near nude females are produced primarily for the consumption of males. There are, however, photographs and magazines portraying nude and near nude males—but these are also produced for the consumption of males. There are almost no male or female nudes which are produced for the consumption of females. The failure of nearly all females to find erotic arousal in such portrayals is so well known to the distributors of nude photographs and nude magazines that they have considered that it would not be financially profitable to produce such material for a primarily female audience.

4. Erotic Fine Art. There may be a diversity of erotic elements in art, but the most obvious is the portrayal of the human body or portions of the human body in a fashion which gives evidence of the

---

[10] A survey and discussion of the popularity of near nudity in magazines and in advertising is in: A. Ellis 1951:104–107. He recognizes that females are not aroused by portrayals of male nudity, but explains it on the basis of the lack of taboo. For other recognition of the lack of female arousal from erotic pictures, see: Brettschneider in Wulffen et al. 1931:106. Wallace 1948:22. Friedeburg 1950:24 (47 per cent males, 11 per cent females erotically aroused by photographs or pictures in German survey).

artist's erotic interest in his or her subject matter, or provides erotic stimulation for the individual observing the work.[11]

An extensive study which we are making of the erotic element in art indicates that a very high proportion of the male artists who portray the human form, either female or male, do so in a fashion which indicates an erotic interest in that form. Even though there may be no portrayal of genitalia and no suggestion of sexual action, the nude body itself may be drawn or painted in a fashion which is erotic to the artist and to most males who subsequently observe the drawing or painting. Such artists as Michelangelo, Leonardo da Vinci, Raphael in his drawings, Rubens, Rodin, Renoir, and Maillol—to cite a few specific cases—rarely drew nudes which, in the judgment of the qualified artists whom we have consulted, did not show such an erotic element.

It is, of course, possible to portray the nude form, as was regularly done in Egyptian art, for instance, in a way which is not erotic; but among the males who have drawn or painted nude figures, it is rare to find any in European or American art who have done so without evident erotic interest. We have not found more than a half dozen male artists of moment who have regularly drawn nudes which have not shown an erotic content.

While the number of female artists has been much less than the number of males who have done painting or drawing, there are many hundreds of them in the history of European and American art. But in some years of searching, we have been able to find only eight instances of important female artists who have drawn the human figure in a fashion which qualified artists, female or male, judge to be erotic.[12] In conjunction with the data (p. 652) which indicate that relatively few females are aroused upon observing nude paintings or drawings, it is understandable that female artists themselves should not be erotically responsive to the nude subjects which they are drawing or painting; and this is evident in their finished work.

---

[11] Freud 1922:78–79 in his study of Leonardo da Vinci recognized this in the following words: "A kindly nature has bestowed upon the artist the capacity to express in artistic productions his most secret psychic feelings hidden even to himself, which powerfully affect outsiders who are strangers to the artist. . . ."

[12] For encyclopedic lists of artists, the standard works are: Thieme and Becker 1907–1947, 35 v. Mallett 1935; 1940. Bénézit 1948–1952, 5 v. For a convenient anthology, see: Sparrow 1905, Women painters of the world. Bulliet in McDermott and Taft 1932:233–252 lists 40 artists, all of whom are male, known for their female nudes. Havelock Ellis 1929:376–378 (cites Ferrero who explains the small part played by women in art as due to their less keen sexual emotions. Ellis, however, interprets male power of creation in fine arts as compensation for the male's lesser role in producing and moulding the race). Wallace 1948:21 recognizes that female painters and sculptors do not use nude subjects as frequently as male artists.

It is to be noted that seven out of the eight female artists whose work seems erotic had confined themselves to portraying the nude female form.

5. Observing Genitalia. Most heterosexual males are aroused by observing female breasts or legs, or some other part of the female body. They are usually aroused when they see female genitalia. A smaller percentage of the females in the sample (of 617 to whom the question was put) reported erotic arousal as a product of their observation of male genitalia, and more than half (52 per cent) reported that they had never been aroused by observing male genitalia. The record is as follows:

Observing genitalia of opposite sex

EROTIC RESPONSE	BY FEMALES	BY MALES
	%	%
Definite and/or frequent	21	many
Some response	27	many
Never	52	few
Number of cases	617	

Many females are surprised to learn that there is anyone who finds the observation of male genitalia erotically stimulating. Many females consider that male genitalia are ugly and repulsive in appearance, and the observation of male genitalia may actually inhibit their erotic responses. It may be true, as psychoanalysts suggest, that the negative reactions of females to male genitalia may originate in unpleasant sexual experiences with males; but there seems no doubt that these reactions largely depend upon the fact that most females are not psychologically stimulated, as males are, by objects which are associated with sex.

Among the infra-human species of mammals there seem to be something of the same differences between the reactions of females and males to the genitalia of the opposite sex. For instance, a female monkey or ape grooming the body of a male who may become aroused erotically, may pay no attention to the male's erect genitalia. On the other hand, when male apes and monkeys groom females, they usually show considerable interest in the female genitalia, and explore around and within the genital cavity. Male rats, guinea pigs, dogs, raccoons, skunks, porcupines, and many other male animals may similarly explore at considerable length about the genitalia of the female, but the females of these species less often explore about the genitalia of the male.[13] Any interpretation of the human female's lack of interest in

[13] We base this statement on our own observations and on discussions with several who have been on the staff of the Yerkes Laboratories at Orange Park,

male genitalia must take into account the similar situation among these infra-human species.

Most human males with homosexual interests are aroused, and in most instances strongly aroused, by seeing male genitalia. Genital exposures and genital exhibitions are frequently employed to interest other males in homosexual contacts. In the course of a homosexual relationship among males, considerable attention may be given to the genital anatomy and genital reactions. Moreover, many males who are not conscious of homosexual reactions are interested in their own genitalia and in the genitalia of other males. But only a small percentage of the homosexual females is ever aroused erotically by seeing the genitalia of other females.

6. Observing Own Genitalia. A great many of the males in the sample (56 per cent) had been aroused by observing their own genitalia as they masturbated, or by viewing their genitalia in a mirror. Few (9 per cent) of the females in the sample had found any erotic stimulation in looking at their own genitalia. The specific data are as follows:

Observing own genitalia

EROTIC RESPONSE	BY FEMALES	BY MALES
	%	%
Definite and/or frequent	1	25
Some response	8	31
Never	91	44
Number of cases	5725	3332

There were more males (56 per cent) who were aroused by observing their own genitalia than there were females (48 per cent) who were aroused by observing male genitalia. The male's arousal may have a homosexual element in it, but many of the males who have never consciously recognized any other homosexual interests and have never had homosexual contacts may be aroused at seeing their own genitalia or the genitalia of other males.

7. Exhibitionism. Because of their interest in their own genitalia and their arousal upon seeing the genitalia of other persons, males quite generally believe that other persons would be aroused by seeing their genitalia. This seems to be the prime factor which leads many

Florida. Dr. Nissen of that staff sums up their data by observing that in the chimpanzee the head, back, limbs, and anus of the male receive the most grooming attention by the grooming partner (whether female or male), and that the ventral body surface and genitalia receive the least. When the male grooms the female he concentrates considerable attention upon the sexual skin and anal area, and may make anal or vaginal insertions occasionally. Shadle (verbal communic.) is the source of the data on the skunk and porcupine.

males to exhibit their genitalia to their wives, to other female partners, and to male partners in homosexual relationships.[14]

It is difficult for most males to comprehend that females are not aroused by seeing male genitalia. Some males never come to comprehend this. Many a male is greatly disappointed when his wife fails to react to such a display, and concludes that she is no longer in love with him. On the contrary, many females feel that their husbands are vulgar, or perverted, or mentally disturbed, because they want to display their genitalia. We have seen difficulties develop in marital histories because of this failure of females to understand male psychology, and of males to understand female psychology. Divorces had grown out of some of these misunderstandings.

The male who exposes himself in a public place similarly secures erotic satisfaction primarily because he believes that the females who observe him are going to be aroused as he would be at seeing a genital exhibition. Sometimes the exhibitionist is aroused by the evident fright or confusion or other emotional reactions of the females who see him and, responding sympathetically, he may be stimulated by such an emotional display. But a considerable portion of the erotic arousal which the exhibitionist finds is a product of his anticipation that the female will be aroused, and this is evidenced by the fact that he is usually in erection before any passerby sees him. His reactions, therefore, may not depend entirely or even primarily upon the responses of the passing female (p. 655).

There are some females who will show their genitalia to the male partner because they intellectually realize that this may mean something to him. But only an occasional female among those who exhibit receives any erotic arousal from this anticipation of the male's responses. There are no cases in our sample, and practically none in the literature, of females publicly exhibiting their genitalia because they derived erotic satisfaction from such an exhibition.[15]

Stage, night club, burlesque and other commercial exhibitions of female nudity almost never, as far as our sample indicates, provide erotic

---

[14] The male chimpanzee frequently solicits the female by coming to erection, spontaneously or by masturbation, and exhibiting the erection to the female, according to: Yerkes and Elder 1936a:9. Nissen, verbal communic. Our observation.

[15] That exhibitionism is infrequent among females in comparison to males, has also been recognized by: [Jacolliot] Jacobus X 1900:347. Hirschfeld 1920(3):319; 1948:504. Kronfeld in Marcuse 1923:121. Bilder-Lexikon 1930(1):241. Brown 1940:383. Guyon 1948:319. Allen 1949:108. Exhibitionism is considered non-existent among females by: Walker and Strauss 1939:177–178. Fenichel 1945:346. Rickles 1950:49 (considers it due to the fact that women "have nothing to expose"). In a few pre-literate cultures, women may solicit men by deliberately exposing their own genitalia; see: Ford and Beach 1951:93.

stimulation for the exhibiting females. Our specific data provide no physiologic evidence (Chapter 15) of arousal among the females staging such exhibitions, although some of them may acquire considerable facility in making body movements which are taken by many of the males in the audience to indicate that the exhibiting females are tremendously aroused. Most of the females in our histories who had been involved in such stage exhibitions, were highly distainful of males who could so easily be misled into believing that there was any real eroticism in such a performance.

8. Interest in Genital Techniques. While the genitalia may be the chief focus of a considerable amount of sexual activity, this does not depend wholly on the fact that these organs are well supplied with end organs of touch. There are many other parts of the body which are similarly supplied with end organs, and the importance attached to the genitalia in a sexual relationship must partly depend upon the fact that most males and some females are psychologically conditioned to consider the genitalia as *the* structures which are primarily associated with sexual response.

This interpretation is favored by the fact that males attach much more importance to the genitalia than females do in a sexual relationship. But there is no reason for believing that the genitalia of the male are more richly supplied with end organs than the genitalia of the female. While genital erection may draw the male's attention to his own genitalia, this does not suffice to interest most females in his genitalia.

Most males, whether heterosexual or homosexual, are inclined to initiate a sexual relationship through some genital exposure or genital manipulation. Most females prefer to be stimulated tactilely in various other parts of the body before the activity is concentrated on the genitalia. It is the constant complaint of married females that their husbands are interested in "nothing but the intercourse," and by that they mean that he is primarily concerned with genital stimulation and an immediate genital union. On the other hand, it is the constant complaint of the married male that his wife "will do nothing to him," which means, in most instances, that she does not tactilely stimulate his genitalia.

These same differences in the significance of genital activities are to be found in the homosexual activities of females and males. A high proportion of the homosexual contacts among males is initiated through some genital exposure or some sort of genital manipulation (groping). During the actual relationships most homosexual males are likely to prefer more genital than non-genital stimulation. But in female homosexual relationships, the stimulation of all parts of the body may proceed for some period of time before there is any concentration of attention on the genitalia. We have histories of exclusively homosexual females who had had overt relationships for ten or fifteen years before they attempted any sort of genital stimulation.

Homosexual females frequently criticize homosexual males because they are interested in nothing but genitalia; homosexual males, in turn, may criticize homosexual females because "they do nothing" in a homosexual relationship. The idea that homosexuality is a sexual inversion is dispelled when one hears homosexual females criticizing homosexual males for exactly the same reasons which lead many wives to criticize their husbands, and when one hears homosexual males criticize homosexual females for exactly the things which husbands criticize in their wives. In fact, homosexual males, in their intensified interest in male genitalia and genital activity, often exhibit the most extreme examples of a typically male type of conditioning.

9. Observing Commercial Moving Pictures. Portrayals of more or less erotic situations are so common in present-day commercial moving pictures that their significance as sources of erotic arousal, for either females or males, is probably less than it was in an earlier day and certainly much less than most official and unofficial censors would believe. It is not impossible that many males would more often respond erotically to love scenes, to close-ups of petting and kissing, and to exhibitionistic displays of semi-nude bodies if they were to observe such pictures in the privacy of their homes or in conjunction with some sexual partner. In the average public theatre, however, the openly expressed reactions of the audience suggest that they are more amused than aroused by the sort of eroticism which is usually presented. However, their vocal responses, cat calls, and whistling may indicate that they are reacting emotionally and trying to deny it by way of a contrary response.

The males and females in our sample recorded their reactions to commercial moving pictures as follows:

Observing moving pictures

EROTIC RESPONSE	BY FEMALES	BY MALES
	%	%
Definite and/or frequent	9	6
Some response	39	30
Never	52	64
Number of cases	5411	3231

This means that the females found the moving pictures erotically stimulating somewhat more often than the males. This is one of the few sources of psychologic stimulation which seem to have been more significant for the females in the sample.

Some of the stimulation provided by a moving picture may depend on the romantic action which it portrays, and some of it may depend on the portrayal of some particular person. In a larger number of instances, the erotic stimulation may depend on the emotional atmosphere created by the picture as a whole, just as viewing a landscape, reading a book, or sitting with another person before an open fire may lead to emotional responses which then become erotic. Sometimes the erotic element in the picture may have no obvious sexual meaning except to the individual who has been conditioned by the particular element. Sometimes the erotic arousal may depend upon the presence of the companion with whom one is attending the performance.

10. Observing Burlesque and Floor Shows. Burlesque shows more or less openly attempt to provide erotic stimulation for the attending audience, and a considerable proportion of those who go to such shows do so with the anticipation that they are going to be aroused erotically. More skilled versions of the burlesque routines are the chief elements of the average night club's floor show.

Most males are aroused by the advertisements at the entrance to a burlesque show, and considerably aroused by anticipating what they are going to see. Most of the males (62 per cent) in our sample had been aroused by the show itself upon their first visit or two, but most of them had not found the shows particularly stimulating after that. Since some of these males had continued to attend such shows, it may be that they did receive some generalized erotic satisfaction from them even though it was not as specific as on the first occasion. Some of them may have gone because they were attracted by the humorous elements in such performances; but in many instances they continued to go because they hoped that they would again be stimulated as they had been on their first few visits.

The erotic reactions of those females and males in our sample who had ever seen a burlesque or night club floor show were reported as follows:

Observing burlesque and floor shows

EROTIC RESPONSE	BY FEMALES	BY MALES
	%	%
Definite and/or frequent	4	28
Some response	10	34
Never	86	38
Number of cases	2550	3377

A decade or two ago the burlesque audiences were almost exclusively male; today the audiences may include a more equal number of females and males. It is difficult, however, to explain this attendance by females in view of the fact that so few of them (14 per cent in our sample) are aroused erotically by such shows. Apparently most females attend burlesque shows because they are social functions about which they are curious, and which they may share with their male companions. They may find some pleasure in the humorous elements in such a show. Only a very few of them are seeking homosexual stimulation from observing the females in the show.

11. Observing Sexual Action. A considerable proportion of those males in our sample who had had the opportunity to observe other persons in sexual activity had responded sympathetically during their observation. The females in the sample who had had the opportunity to observe sexual activity rarely reported such sympathetic responses. Most of them had been indifferent in their responses, if they had not been offended by the social impropriety of such an exhibition.

It is, therefore, no accident, and not merely the product of the cultural tradition, that commercialized exhibitions of sexual activity, since the days of ancient Rome, have been provided for male but almost never for female audiences. There are many males who would not accept an opportunity to attend such exhibitions because they consider them morally objectionable, but even they usually recognize that they would be aroused if they were to observe them.

There is an inclination to explain these differences in the responses of females and of males as products of the cultural tradition, and there is a widespread opinion that females are more inclined to accept the social proprieties because they are basically more moral than males. On the other hand, the same sorts of differences between the sympathetic responses of females and of males may be observed in other species of mammals. The males of practically all infra-human species may become aroused when they observe other animals in sexual activity.[16] Of this fact farmers, animal breeders, scientists experimenting with laboratory animals, and many persons who have kept household pets are abundantly aware. The females of the infra-human species less often show such sympathetic responses when they observe other animals in sexual activity. These data suggest that human females are

---

[16] The sight of coitus between animals of the same species arouses such diverse animals as the bull (our observation, and McKenzie, verbal communic.), and chimpanzee (Nissen, verbal communic.). Zitrin and Beach in Hartman 1945:43–44 report that male cats become aroused by seeing female cats reacting to being masturbated by a human experimenter.

more often inclined to accept the social proprieties because they are stimulated psychologically and respond sympathetically less often than most males do.

12. Observing Portrayals of Sexual Action. In spite of state and federal laws, and in spite of the considerable effort which law enforcement officers periodically make to prevent the distribution of photographs, drawings, moving pictures, and other portrayals of sexual action, such materials exist in considerable abundance in this country and probably in greater abundance in most other countries. Graphic portrayals of sexual action have existed in most cultures, throughout history. This is a measure of the considerable significance which such materials may have for the consuming public which, however, is largely male.

Practically all of the males in the sample had had the opportunity to observe portrayals of sexual action, and had taken the opportunity to observe them. Most of the males (77 per cent) who had seen such material indicated that they had been aroused erotically by seeing it. A smaller proportion of the females in the sample had had the opportunity to see, or had taken the opportunity to see such portrayals of sexual action.[17] Only a third of them (32 per cent) had found any erotic arousal in observing such material. The specific record is as follows:

Observing portrayals of sexual action

EROTIC RESPONSE	BY FEMALES	BY MALES
	%	%
Definite and/or frequent	14	42
Some response	18	35
Never	68	23
Number of cases	2242	3868

Many females, of course, report that they are offended by portrayals of sexual action, and denounce them on moral, social, and aesthetic grounds. This is ordinarily taken as evidence of the female's greater sense of propriety; but in the light of our other data on the relative significance of psychologic stimulation for females and for males, it seems more likely that most females are indifferent or antagonistic to the existence of such material because it means nothing to them erotically.

---

[17] Clark and Treichler 1950 report an increase in acid phosphatase (an enzyme found in the urine) in four males aroused by erotic pictures, a decrease in a male who was repelled by them, and no change in two females who saw the pictures.

Most males find it difficult to comprehend why females are not aroused by such graphic representations of sexual action, and not infrequently males essay to show such materials to their wives or other female partners, thinking thereby to arouse them prior to their sexual contacts.[18] The wives, on the other hand, are often at a loss to understand why a male who is having satisfactory sexual relations at home should seek additional stimulation in portrayals of sexual action. They are hurt to find that their husbands desire any stimulation in addition to what they, the wives, can provide, and not a few of the wives think of it as a kind of infidelity which offends them. We have seen considerable disturbance in some of the married histories because of such disagreements over the husband's use of erotic objects, and there are cases of wives who instituted divorce proceedings because they had discovered that their husbands possessed photographs or drawings of sexual action.

Local drives against so-called obscene materials, and state, federal, and international moves against the distribution of such materials, are not infrequently instituted by females who not only find the material morally and socially objectionable, but probably fail to comprehend the significance that it may have for most males and for some females.

13. Observing Animals in Coitus. Many human males and some females respond sympathetically upon observing animals of other species in coitus. The specific data show that 32 per cent of the males in our sample had so responded. Watching dogs or cattle in coitus had been the inspiration for the involvement of some of the farm boys in sexual relationships with the animals themselves. There were some females but fewer (16 per cent) who were aroused by observing the sexual activities of other animals. The data are as follows:

Observing animals in coitus

EROTIC RESPONSE	BY FEMALES	BY MALES
	%	%
Definite and/or frequent	5	11
Some response	11	21
Never	84	68
Number of cases	5250	4082

14. Peeping and Voyeurism. There are probably very few heterosexual males who would not take advantage of the opportunity to ob-

---

[18] This masculine misconception is not new. Brantome (16th century, "Lives of Fair and Gallant Ladies," First Discourse, ch. 5 (1901:55), tells of a prince who served wine to women in a cup covered with copulatory figures. He would then ask, "Now feel ye not a something that doth prick you in the mid part of the body, ladies, at the sight?" and the women would reply, "Nay!, never a one of all these droll images hath had power enow to stir me!"

serve a nude female, or to observe heterosexual activity, particularly if it were possible to do so surreptitiously so they would not suffer the social disgrace that the discovery of their behavior might bring. To many males, the observation of a female who is undressing may be erotically more stimulating than observing her when she is fully nude, for the undressing suggests, in fantasy, what they may ultimately be able to observe. Consequently, we have the peeper who gets into difficulty with the law, the peep show which was formerly common in this country and which is still available in many other countries, and the more surreptitious and unpublicized peeping in which most males engage, at some time in their lives, from the windows of their homes, from hotel windows, and from wherever they find the opportunity to observe. Our data are insufficient for determining what percentage of the male population is ever involved, but Hamilton found some 65 per cent of the males in his study admitting that they had done some peeping.[19] The percentages for the population as a whole are probably higher.

The erotic significance of what the peeper observes obviously depends on his capacity to be stimulated psychologically. But there are few instances in our own study, or in other studies, or in the medical and psychiatric literature, of females as peepers. Out of curiosity some females are undoubtedly sometimes involved, and a few of them may find erotic stimulation in such peeping; but such behavior is certainly rare among females.

15. Preferences for Light or Dark. We have previously (1948:581–582) pointed out that many (40 per cent) of the males in our sample preferred to have their coitus or other sexual activities where there was at least some light. Fewer (19 per cent) of the females in the sample preferred sexual relations in the light. This, again, is ordinarily taken to represent the greater modesty of the female, but it seems to depend upon the fact that the male is stimulated by seeing the sexual partner, by seeing the genitalia or other parts of the body of the sexual partner, by getting some chance to observe, as a voyeur, something of his own sexual action, and by the opportunity to observe various objects with which he comes to associate sexual action. Females, as we have already shown, are much less often attracted by observing the male partner, his genitalia, or other objects associated with the sexual performance. The specific data are as follows:

---

[19] Hamilton 1929:456 found 16 per cent of females and 83 per cent of males had had desire to peep as adults; 20 per cent of females and 65 per cent of males had done actual peeping.

Preferences for Light or Dark

Preferences	Report on own preferences		Subject's report on	
	BY FEMALES	BY MALES	FEMALE SPOUSE	MALE SPOUSE
	Percent		*Percent*	
Definitely prefer light	8	21	11	21
Prefer some light	11	19	13	10
Prefer dark	55	35	58	34
No preference	26	25	18	35
Number of cases	2042	798	662	1633

Anthropologic data indicate that there are different customs in regard to having coitus in the light or dark in various cultural groups,[20] and the cultural tradition may be a factor in determining the practice in our own culture; but this does not explain why, within our own single culture, males are more likely than females to prefer some light during their sexual activities. The differences provide another illustration of the greater capacity of the male to be conditioned by experience.

16. Fantasies Concerning Opposite Sex. Practically all males who are not exclusively homosexual may be erotically aroused by thinking of certain females, or of females in general. Fewer males of the lower educational levels are aroused by such fantasies, and older males sometimes lose their capacity to be stimulated by fantasies, and males who are exclusively homosexual may not fantasy concerning females. But most of the males in our sample (84 per cent) indicated that they were at least sometimes, and in most instances often aroused by thinking of sexual relations with females—by thinking of the sexual relations that they had previously had, or by thinking of the sexual relations that they anticipated they might have or would like to have. Such erotic stimulation probably occurs more often than any other single type of psychologic stimulation among males.

A smaller percentage (69 per cent) of the females in the sample reported that they had ever had erotic fantasies about males, and nearly a third (31 per cent) insisted that they had never been aroused by thinking about males or of sexual relations with them. They had not even been aroused by thinking of their husbands or of their boy friends. Most of the females who were not aroused by the contemplation of males were heterosexual, and most of them had had sexual relations with males in which they had regularly responded to the point

---

[20] For varying practices in various cultures in regard to day or night preferences for coitus, see Ford and Beach 1951:73.

of orgasm; but even some of the females who were most responsive in physical relationships had never been aroused by fantasies about males.

The specific data showing these differences between the females and males in the sample are as follows:

Fantasies concerning opposite sex

EROTIC RESPONSE	BY FEMALES	BY MALES
	%	%
Definite and/or frequent	22	37
Some response	47	47
Never	31	16
Number of cases	5772	4214

These differences between females and males have a great deal to do with the fact that more males search for overt sexual experience, and fewer females search for such experience. These differences provide one explanation of the fact that males are usually aroused and often intensely aroused before the beginning of a sexual relationship and before they have made any physical contact with the female partner. These differences account for the male's desire for frequent sexual contact, his difficulty in getting along without regular sexual contact, and his disturbance when he fails to secure the contact which he has sought. The differences often account for the female's inability to comprehend why her husband finds it difficult to get along with less frequent sexual contacts, or to abandon his plans for coitus when household duties or social activities interfere.

Too many husbands, on the other hand, fail to comprehend that their wives are not aroused as they are in the anticipation of a sexual relationship, and fail to comprehend that their wives may need general physical stimulation before they are sufficiently aroused to want a genital union or completed coitus. Too often the male considers the wife's lesser interest at the beginning of a sexual relationship as evidence that she has lost her affection for him. Sexual adjustments between husbands and wives could be worked out more often if males more often understood that the reactions of their particular wives represent characteristics which are typical of females in general, and if females more often understood that the sexual interests shown by their particular husbands represent qualities which are typical of most males.

17. Fantasies Concerning Own Sex. Sexual arousal from fantasies about other males, or of sexual relations with other males, is as frequent among homosexual males as heterosexual fantasies are among heterosexual males. Such erotic fantasies are less frequent among homosexual females, but they do occur in as high or a higher percentage of the homosexual females (74 per cent) as heterosexual fantasies occur among heterosexual females. The specific data are as follows:

Fantasies concerning own sex

EROTIC RESPONSE	BY FEMALES
	%
Definite and/or frequent	28
Some response	46
Never	26
Cases with homosexual history	194

18. Fantasies During Masturbation. Some 89 per cent of the males in our sample had utilized erotic fantasies as one of their sources of stimulation during masturbation. Some 72 per cent had more or less always fantasied while masturbating. Such fantasies usually turn around memories of previous sexual experience, around sexual experience that the male hopes to have in the future, or around sexual experience which he may never allow himself to have but which he anticipates might bring erotic satisfaction if the law and the social custom made it possible for him to engage in such activity. In not a few instances males develop rather elaborate fictional situations which they regularly review as they masturbate. Quite a few males, particularly among the better educated groups, may, at least on occasion, utilize erotic photographs or drawings, or make their own erotic drawings, or read erotic literature, or write their own erotic stories which they use as sources of stimulation during masturbation. Some 56 per cent of the males in the sample indicated that they observed their own genitalia at least on occasion during masturbation, and while this is more likely to be true of males with homosexual histories, it is also true of many others who give no other evidence of homosexual interests. But in any event, these males find the observation of their own genitalia an additional source of erotic stimulation. So dependent are many males on psychologic stimuli in connection with masturbation that it is probable that many of them, especially middle-aged and older males, would have difficulty in reaching orgasm if they did not fantasy while masturbating.

Fantasies during masturbation

FANTASIES PRESENT	AMONG FEMALES	AMONG MALES
	%	%
Almost always	50	72
Sometimes	14	17
Never	36	11
Number of cases	2475	2815

The record shows (p. 164) that only 64 per cent of the females in our sample who had ever masturbated, had fantasied while masturbating.[21] Only 50 per cent of the females who had masturbated, had regularly fantasied for any period of their lives. We have nearly no cases of females utilizing erotic books or pictures as sources of stimulation during masturbation.

19. Nocturnal Sex Dreams. Nearly all males have nocturnal sex dreams which are erotically stimulating to them. Ultimately some 75 per cent of the females (p. 196) may have such nocturnal dreams.

To judge from our sample, approximately 83 per cent of the males (the accumulative incidence figure) ultimately have sex dreams which are erotically stimulating enough to bring them to orgasm during sleep. The corresponding figure for the females in the sample was 37 per cent (p. 196).[22]

The frequencies of sex dreams show similar differences between males and females. Among those males in the sample who were having any sex dreams (the active sample), the median frequencies averaged about 10 times per year in the younger age groups, and about 5 times per year in the older age groups. The median female in the sample had had sex dreams which were sufficiently erotic to bring her to orgasm with frequencies of 3 to 4 per year (p. 197). For perhaps 25 per cent of the females who had had any dreams which had resulted in orgasm, the experience had not occurred more often than 1 to 6 times in their lives. These differences depend, again, on differences in the significance of psychologic stimulation for the average female and the average male.

20. Diversion During Coitus. We have already pointed out (p. 384) that effective female responses during coitus may depend, in many cases, upon the continuity of physical stimulation. If that stimulation is interrupted, orgasm is delayed, primarily because the female may return to normal physiologic levels in such periods of inactivity. This appears to be due to the fact that she is not sufficiently aroused by psychologic stimuli to maintain her arousal when there is no physical stimulation. We have pointed out that the male, on the contrary, may go through a period in which physical activity is interrupted without losing erection or the other evidences of his erotic arousal, primarily because he continues to be stimulated psychologically during those periods.

---

[21] Hamilton 1929:429 reports 36 per cent of females, 69 per cent of males record fantasies during masturbation.
[22] Hamilton 1929:318–319 reports 66 per cent of females, 42 per cent of males without nocturnal sex dreams.

Similarly, because the male is more strongly stimulated by psychologic factors during sexual activities, he cannot be distracted from his performance as easily as the female. Many females are easily diverted, and may turn from coitus when a baby cries, when children enter the house, when the doorbell rings, when they recall household duties which they intended to take care of before they retired for the night, and when music, conversation, food, a desire to smoke, or other non-sexual activities present themselves. The male himself is sometimes responsible for the introduction of the conversation, cigarettes, music, and other diversions, and he, unwittingly, may be responsible for the female's distraction because he does not understand that the sources of her responses may be different from his.

It is a standard complaint of males that their female partners in coitus "do not put their minds to it." This is an incorrect appraisal of the situation, for what is involved is the female's lack of stimulation by the sorts of psychologic stimuli which are of importance to the male. Such differences between females and males have been known for centuries, and are pointed out in the classic and Oriental literature. From the most ancient to the most modern erotic art, the female has been portrayed on occasion as reading a book, eating, or engaging in other activities while she is in coitus; but no artist seems to have portrayed males engaged in such extraneous activities while in coitus.

Various interpretations may be offered of these differences between females and males. Many persons would, again, be inclined to look for cultural influences which might be responsible. But some sort of basic biologic factor must be involved, for at least some of the infra-human species of mammals show these same differences. Cheese crumbs spread in front of a copulating pair of rats may distract the female, but not the male. A mouse running in front of a copulating pair of cats may distract the female, but not the male. When cattle are interrupted during coitus, it is the cow that is more likely to be disturbed while the bull may try to continue with coitus.[23] It explains nothing to suggest that this is due to differences in levels of "sex drive" in the two sexes. There are probably more basic neurologic explanations of these differences between females and males (p. 712).

21. Stimulation by Literary Materials. Erotic responses while reading novels, essays, poetry, or other literary materials may depend upon the general emotional content of the work, upon specifically romantic

---

[23] As examples of the fact that the female is more easily distracted, see: Beach 1947b:264 (bitches will eat during coitus, most male dogs refuse food in this situation; female cats may investigate mouse holes during coitus). Robert Bean, director of Brookfield Zoo, reports (verbal communic.) females of various species eating during coitus.

material in it, upon its sexual vocabulary (particularly if it is a vernacular vocabulary), or upon its more specific descriptions of sexual activity. The reader may thus, vicariously, share the experience of the characters portrayed in the book, and reactions to such literary material are some measure of the reader's capacity to be aroused psychologically.

The reactions of the females and males in our sample were as follows:

Reading literary materials

EROTIC RESPONSE	BY FEMALES	BY MALES
	%	%
Definite and/or frequent	16	21
Some response	44	38
Never	40	41
Number of cases	5699	3952

It will be noted that the females and males in the sample had responded erotically in nearly the same numbers while reading literary materials.[24] Twice as many of the females in the sample had responded to literary materials as had ever responded to the observation of portrayals of sexual action (p. 662), and five times as many as had responded to photographs or other portrayals of nude human figures (p. 652). At this point we do not clearly understand why this should be so. There are possible psychoanalytic interpretations, but in view of all the evidence that there may be basic neurophysiologic differences between females and males, we hesitate to offer any explanation of the present data.

22. Stimulation by Erotic Stories. Practically all of the males in the sample, even including the youngest adolescent boys, had heard stories that were deliberately intended to be erotically stimulating, usually through their descriptions of sexual action. Nearly half (47 per cent) of the males in the sample reported that they had been aroused, at least on occasion, by such stories. There had been differences in the responses among males of the various educational levels. Most of the better educated males had responded, while fewer of the males of the lower educational levels were aroused by such stimuli. Some 53 per cent of the males in the total sample said that they had never been aroused by such stories.

Some 95 per cent of the females in the sample had heard or read stories that were deliberately intended to bring erotic response, but

[24] That the erotic stimulation which females derive from reading romantic stories or seeing moving pictures equals or exceeds that which is derived from those sources by males, is also recognized by: Friedeburg 1950:24. See also Dickinson and Beam 1934:111, 427.

only 14 per cent recalled that they had ever been aroused by such stories. The specific record is as follows:

Stimulation by erotic stories

EROTIC RESPONSE	BY FEMALES	BY MALES
	%	%
Definite and/or frequent	2	16
Some response	12	31
Never	86	53
Number of cases	5523	4202

Note that some 86 per cent of the females who had heard obscene stories had never received any erotic arousal from them. Some of the females had been offended by the stories, and it is not impossible that their failure to be aroused represented a perverse attitude which they had developed in consequence of the general opinion that such stories are indecent and immoral. On the other hand, a surprising proportion of the females in the sample indicated that they enjoyed such stories, usually because of their intrinsic humor. Sometimes their interest in the stories represented a defiance of the social convention. There is some indication, although we do not have the data to establish it, that there is an increasing acceptance of such stories among females in this country today. The older tradition which restricted the telling of such stories in the presence of a female has largely broken down within the last decade or two, and since there is this freer acceptance of such stories by many females, it is all the more surprising to find how few ever find erotic stimulation in them.

23. Erotic Writing and Drawing. What is commonly identified as pornography is literature or drawing which has the erotic arousal of the reader or observer as its deliberate and primary or sole objective. Erotic elements may be involved in the production of other literary material and in the fine arts; but in the opinion of most students, and in various court decisions, such literary and fine art materials are distinguished by the fact that they have literary or artistic merit as their prime objective, and depend only secondarily on erotic elements to accomplish those ends.[25]

---

[25] For examples of court decisions holding that despite certain obscene or indecent passages the books themselves were not obscene; see: In re Worthington Co. 1894:30 N.Y. Supp. 361. Halsey v. N.Y.Soc. for the Suppression of Vice 1922:136 N.E.(N.Y.) 219 ("It contains many paragraphs, however, which taken by themselves are undoubtedly vulgar and indecent. . . . Printed by themselves they might, as a matter of law, come within the prohibition of the statute. So might a similar selection from Aristophanes or Chaucer or Boccaccio or even from the Bible. The book, however, must be considered broadly as a whole"). U.S. v. One Book Entitled Ulysses 1934:72F(2d)705. Com. v. Gordon 1949:66 D. and C. (Pa.) 101 (Judge Bok's scholarly discussion of the changing concepts of obscenity, citing our male volume on 116). But cf. Com. v. Isenstadt 1945:62 N.E.2d (Mass.)840. Detailed discussions are found in: Alpert 1938, and Jenkins 1944.

In every modern language, the amount of deliberately pornographic material that has been produced is beyond ready calculation. Some thousands of such documents have been printed in European languages alone, and the literature of the Orient and other parts of the world is replete with such material. Similarly, there is an unlimited amount of pornographic drawing and painting which has been produced by artists of some ability in every part of the world, and there is no end to the amateur portrayals of sexual action.

But in all this quantity of pornographic production, it is exceedingly difficult to find any material that has been produced by females. In the published material, there are probably not more than two or three documents that were actually written by females. It is true that there is a considerable portion of the pornographic material which pretends to be written by females who are recounting their personal experience, but in many instances it is known that the authors were male, and in nearly every instance the internal content of the material indicates a male author. A great deal of the pornographic literature turns around detailed descriptions of genital activity, and descriptions of male genital performance. These are elements in which females, according to our data, are not ordinarily interested. The females in such literature extol the male's genital and copulatory capacity, and there is considerable emphasis on the intensity of the female's response and the insatiability of her sexual desires. All of these represent the kind of female which most males wish all females to be. They represent typically masculine misinterpretations of the average female's capacity to respond to psychologic stimuli. Such elements are introduced because they are of erotic significance to the male writers, and because they are of erotic significance to the consuming public, which is almost exclusively male.

Among the hundreds and probably thousands of unpublished, amateur documents which we have seen during the past fifteen years, we have been able to find only three manuscripts written by females which contain erotic elements of the sort ordinarily found in documents written by males. Similarly, out of the thousands of erotic drawings which we have seen, some of them by artists of note and some of them by lesser artists and amateurs, we have been able to find less than a half dozen series done by females.

Females produce another, more extensive literature which is called erotic, and do drawings which are called erotic; but most of these deal with more general emotional situations, affectional relationships, and love. These things do not bring specifically erotic responses from males, and we cannot discover that they bring more than minimal responses from females.

24. Wall Inscriptions. Making inscriptions (graffiti—literally, *writings*) of various sorts on walls of buildings, walls lining country lanes, walls in public toilets, and walls in still other public places, is a custom of long and ancient standing. Among the inscriptions made by males, an exceedingly high proportion is sexual and obviously intended to provide erotic stimulation for the inscribers as well as for the persons who may subsequently observe them.

Relatively few females, on the contrary, ever make wall inscriptions.[26] When they do, fewer of the inscriptions are sexual, and only a small proportion of the sexual material seems to be intended to provide erotic stimulation for the inscribers or for the persons who observe the inscriptions.

With the collaboration of a number of other persons, we have accumulated some hundreds of wall inscriptions from public toilets, making sure that the record in each case covered all of the inscriptions, sexual or non-sexual, heterosexual or homosexual, which were on the walls in the place (p. 87). The record is as follows:

Incidences of sexual inscriptions

	FEMALE	MALE
	%	%
Places with any sexual inscription	50	58
% of inscriptions which were erotic	25	86
Number of places surveyed	94	259
Number of sexual inscriptions	331	1048

A high proportion (86 per cent) of the inscriptions on the walls of the male toilets were sexual. The sexual materials were drawings, lone words, phrases, and sometimes more extended writing. There were three chief subjects in these inscriptions: genitalia (either female or male), genital, oral, or anal action (either heterosexual or homosexual), and vernacular vocabularies which, by association, are erotically significant for most males.

On the contrary, not more than 25 per cent of the toilet wall inscriptions made by the females dealt with any of these matters. Most of the female inscriptions referred to love, or associated names ("John and Mary," "Helen and Don"), or were lipstick impressions, or drawings of hearts; but very few of them were genital or dealt with genital action or sexual vernaculars. A brief summary of the material which we have accumulated shows the following:

---

[26] That other collections of graffiti clearly reflect this scarcity of female authors of such material, can be seen by surveying the articles on graffiti in Anthropophyteia 1907(4):316–328; 1908(5):265–275; 1909(6):432–439; 1910(7): 399–406; 1912(9):493–500.

Content of sexual inscriptions

SUBJECT OF INSCRIPTIONS	MADE BY FEMALES	MADE BY MALES
	%	%
Heterosexual	17	21
Genitalia of opposite sex	5	3
Coital contacts	7	8
Oral contacts	2	11
Anal contacts	1	—
Other erotic items	2	3
References to dating	0	5
Homosexual	11	75
Genitalia of own sex	7	15
Oral contacts	1	30
Anal contacts	0	18
Other erotic items	2	8
References to dating	1	21
Erotic, not classifiable as heterosexual or homo-sexual	5	6
Non-erotic references to love		
With own sex	12	3
With opposite sex	35	3
Sex not specified	9	0
Hearts	6	0
Lips	69	0
Number of inscriptions	331	1048

Again it will be suggested that females are less inclined to make wall inscriptions of any sort, and less inclined to make erotic wall inscriptions, because of their greater regard for the moral codes and the social conventions. In view of our data showing that most females are not erotically aroused by the psychologic stimuli that are of significance to the male, and in view of the data showing that most females are not erotically aroused by observing sexual action, by portrayals of sexual action, or by fantasies about sexual action, there seems little doubt that the average female's lack of interest in making wall inscriptions must depend primarily upon the fact that they mean little or nothing to her erotically. The male usually derives erotic satisfaction from making them, and he may derive even greater satisfaction in anticipating that the inscriptions he makes will arouse other males, amounting sometimes to hundreds and thousands of other males who may subsequently see them.

It is notable that the wall inscriptions in male toilets are concerned with male genitalia and male functions more often than they are concerned with female genitalia or functions. This, at first glance, makes them appear homosexual, but we are not yet ready to accept this interpretation. It is possible that homosexual males are actually more inclined, while heterosexual males are less inclined to make wall inscriptions. It is possible that homosexual males are more inclined be-

674

cause they may be more aroused in making such inscriptions, and because they anticipate how other males will react upon seeing them. The heterosexual male has no such incentive, since he knows that no females will see his writing. But we are inclined to believe that many of the inscriptions that deal with male anatomy and male function are made by males who are not conscious of homosexual reactions and who may not have had overt homosexual experience, but who, nevertheless, may be interested in male anatomy and male functions as elements which enter into heterosexual activities.

But whatever the conscious intent of the inscriber, the wall inscriptions provide information on the extent and the nature of the suppressed sexual desires of females and males. The inscriptions most frequently deal with activities which occur less frequently in the actual histories. This means that the males who make the inscriptions, and the males who read them, are exposing their unsatisfied desires. The inscriptions portray what they would like to experience in real life. Usually the inscriptions are anonymous. They are usually located in restricted, hidden, or remote places. Most of the males who make them would not so openly express their erotic interests in places where they could be identified.

Comparisons of the female and male inscriptions epitomize, therefore, some of the most basic sexual differences between females and males.

25. Discussions of Sex. Males are much more inclined, and females are less inclined to discuss sexual matters with other persons. Striking evidence of this has already been presented in discussing the sources of information which start females masturbating. The data are as follows:

First information on masturbation

SOURCES	FEMALE [*]	MALE [*]
	%	%
Self-discovery	57	28
Verbal and printed sources	43	75
Petting experience	12	Very few
Observation	11	40
Homosexual experience	3	9
Number of cases	2675	3999

[*] The totals amount to more than one hundred per cent because some individuals were simultaneously affected by two or more sources of information.

Note that 57 per cent of the females in the sample had first masturbated as a result of their own discovery that such a process was possible, and that relatively few had started because they had heard of masturbation through verbal sources. Even some of the females who had not started until they were in their thirties or forties had not known that masturbation was possible for a female until they discovered it through their own exploration. On the contrary, most of the males had heard about masturbation, or observed other persons masturbating, before they themselves ever began; and only 28 per cent had learned how to masturbate from their own self-discovery. This is a measure of the extent to which sex is discussed among pre-adolescent and early adolescent males. Such discussions of sex also occur among older males, from adolescence into old age.

For most males, discussions of sex often provide some sort of erotic stimulation. They do not provide anything like the same sort of stimulation for the average female, and in consequence she does not have the same inspiration for engaging in such conversations. Moreover, many of the females in the sample who had overheard discussions of reproductive and sexual functions when they were children, or even when they were adults, had not tried to understand what was being discussed, primarily, as they asserted, because they were "not interested" in sex. We have already recorded (p. 140) that there is frequently a lapse of some years between the time that the female first hears of masturbation, and the time that she attempts to masturbate herself.

On the contrary, from an early age the average male is interested in all that he can learn about sex and searches for sexual information, in part because it may mean something to him erotically. Practically all males who have reached adolescence attempt masturbation almost immediately upon hearing of it. This is further evidence of the importance of psychologic stimulation for the male.

26. Arousal From Sado-Masochistic Stories. Some persons are aroused sexually when they think of situations that involve cruelty, whipping, flagellation, torture, or other means deliberately adopted for the infliction of pain. More individuals are emotionally disturbed when they contemplate such sado-masochistic situations, and they may not recognize such a disturbance as sexual; but at this stage in our knowledge, it is difficult to say how much of the emotional disturbance, or even the more specifically sado-masochistic reactions, may involve sexual elements (p. 88).

A distinctly higher percentage of the males in the sample had responded to sado-masochistic situations in a way which they recognized as sexual.[27] The specific data are as follows:

---

[27] That sado-masochism is less frequent among women is also noted by: [Jacolliot] Jacobus X 1900:347–348. Talmey 1910:136. Forel 1922:236. Krafft-Ebing

Arousal from sado-masochistic stories

EROTIC RESPONSE	BY FEMALES	BY MALES
	%	%
Definite and/or frequent	3	10
Some response	9	12
Never	88	78
Number of cases	2880	1016

That fewer of the females and more of the males had responded, appears again to have depended on the fact that reactions to sado-masochistic stories rely on fantasy. As many females as males seem to react erotically when they are bitten (see below) or when they engage in more specifically sado-masochistic contacts, and this further emphasizes the differences in the psychologic reactions of the two sexes.

It is quite probable that many more males and some more females would respond to such sado-masochistic stimuli if they were to find themselves in sexual situations which were associated with sadism. The development of sado-masochistic responses in a number of our histories had begun in that way.

27. Responses to Being Bitten. It is difficult to know how much of the response of an individual who is being hurt is the product of the physical stimulation, and how much is the product of the stimulation provided by psychologic conditioning, the association of sexual and sado-masochistic phenomena, and the psychologic satisfactions which are to be found in submitting to a sexual partner. It is also very difficult to determine how many of the physical and emotional responses which are manifest in a sado-masochistic situation are sexual and how many are more properly identified as some other sort of emotional response.

During heterosexual petting and coitus, and in homosexual relations, the most frequent manifestation of sado-masochistic responses is to be found in the nibbling and biting which many persons inflict on various parts of the body of a sexual partner. Such behavior is widespread among all of the mammals, and much more widespread in human sexual patterns than most persons comprehend.[28] Definitely

1922:129. Kronfeld in Marcuse 1923:314. Bilder-Lexikon 1930(2):538. Negri 1949:206. That sadism is more typical of males and masochism of females is asserted by: Wexberg 1931:182–183. Rosanoff 1938:156. Brown 1940:383. T. Reik 1941:216. Scheinfeld 1944:243. Thorpe and Katz 1948:326–327. Hamilton 1929:458, 461 (an equal number, 28–29 per cent, of males and females reported pleasure from pain being inflicted on them, but one-third of the females as against one-half of the males reported pleasant thrills at some time from inflicting pain on a person or animal).

[28] We have records of biting (usually by the male) as a part of sexual activity in a number of mammalian species, including the baboon, various monkeys,

sexual responses consequent on such biting were recognized by about equal numbers of the males and females in our sample. The specific data are as follows:

Responses to being bitten

EROTIC RESPONSE	BY FEMALES	BY MALES
	%	%
Definite and/or frequent	26	26
Some response	29	24
Never	45	50
Number of cases	2200	567

Twice as many males had responded erotically to being bitten as had responded to sado-masochistic stories. There were more than four times as many females who had responded erotically when they were bitten as had ever responded to sado-masochistic stories. This provides one more body of data to show that males may be aroused by both physical and psychologic stimuli, while a larger number of the females, although not all of them, may be aroused only by physical stimuli.

28. Fetishism. Practically all heterosexual males, as we have already noted, are aroused erotically when they observe the female body or particular parts of it. When the part of the partner's body which brings the erotic response is farther removed from the genital area, as the hair of the head, the feet, and the fingers are, such responses have commonly been identified as fetishes. But the definitions are obviously nebulous, for all of these reactions depend on nothing more than associative conditioning, and it is difficult to draw the line between the sexual responses of the average male when he sees the genitalia or the breasts or some other portion of the partner's body, and his responses to objects which are more remote but still associated with his previous sexual experience.

When an individual responds to objects which are entirely removed from the partner's body, as clothing (especially underclothing), stockings, garters, shoes, furniture, particular types of drapery, or objects which are still more remote from the particular female with whom the sexual relations were originally had, the fetishistic nature of the response seems more pronounced. But in any event, it still depends upon the sort of psychologic conditioning which is involved in most erotic responses.

Persons who respond only or primarily to objects which are remote from the sexual partner, or remote from the overt sexual activities with

---

mink, marten, sable, ferret, skunk, horse, zebra, pig, sheep, rat, dog, guinea pig, chimpanzee, lion, cat, tiger, leopard, rabbit, raccoon, sea lion, shrew, opossum, and bat. Such biting is particularly violent, and even savage, among the Mustelidae (ferret, mink, sable, skunk).

a partner, are not rare in the population. This is particularly true of individuals who are erotically aroused by high heels, by boots, by corsets, by tight clothing, by long gloves, by whips, or by other objects which suggest sado-masochistic relationships, and which may have been associated with the individual's previous sexual activity.

It has been known for some time, and our own data confirm it, that fetishism is an almost exclusively male phenomenon.[29] We have seen only two or three cases of females who were regularly and distinctly aroused by objects that were not directly connected with sexual activity. Our data on the limited number of females who respond to seeing male genitalia or any other portion of the nude or clothed male body, would lead one to expect that females would not be aroused by objects which are still more remote from the sexual partner himself. There seems no question that the differences in the incidences of fetishes among males and females depend upon the fact that the male is more easily conditioned by his sexual experience and by objects that were associated with those experiences.

29. Transvestism. An individual who prefers to wear the clothing of the opposite sex, and who desires to be accepted in the social organization as an individual of the opposite sex, is a transvestite (from *trans,* a transference, and *vesta,* the clothing). But it should be emphasized that transvestism involves not only a change of clothing. The occasional adoption at a masked ball, or in a stage production, of clothing characteristic of the opposite sex is not transvestism in any strict sense, for true transvestism also involves a desire to assume the role of the opposite sex in the social organization.

True transvestism is a phenomenon which involves many different situations and has many different origins.[30] There are persons who are permanent transvestites, who try to identify with the opposite sex in their work as well as in their homes, at all times of the day and through

---

[29] The infrequent occurrence of fetishism among females has also been recognized by: Talmey 1910:136. Hirschfeld 1920(3):1–79 (many case histories, including a few female cases). Krafft-Ebing 1922:24. Forel 1922:240. Hamilton 1929:463 (13 per cent of females, 33 per cent of males). Stekel 1930(2):341. Walker and Strauss 1939:175. Brown 1940:381. Scheinfeld 1944:243. Fenichel 1945:344.

It is notable that fetish magazines currently found on the newsstands are all slanted toward male purchasers.

[30] Transvestism in various pre-literate societies is recorded, for example, for the Navajo, Kwakiutl, Crow, Eskimo (North America); Tanala, Lango, Mbundu (Africa); Uripev, Dyak (Oceania); Chukchee, Yakut, Yukaghir (Siberia); Lushais (India). In many instances the transvestites are respected and thought to possess magical powers; in other instances they are merely tolerated. The great majority of transvestites are anatomic males. Ford 1945:32 points out that "Cases of women adopting the dress and habits of men are much more rare." For further anthropologic data, see: Parsons 1916:521–528 (Zuni). Hill 1935:273–279 (Navajo). Devereux 1937:498–527 (Mohave). Dragoo 1950 ms. (in more than twenty societies in North America). Ford and Beach 1951:130–131 (general, brief discussion).

all of the days of the year. There are persons who are partial transvestites, who adopt their changed roles only on occasion, as at home in the evening, or occasionally on week ends, or on other special occasions.

Psychologically the phenomenon sometimes depends upon an individual's erotic attraction for the opposite sex. A male, for instance, may be so attracted to females that he wishes to be permanently identified with them. He wants to have sexual relationships with them, and he wishes to live permanently with them, as another female might live with them. The neighbors may believe it to be two females who are living together, although it is sexually a heterosexual relationship which is involved.

Sometimes transvestism depends upon an individual's violent reactions against his or her own sex. In such a case, he may or may not be erotically attracted to the opposite sex. If he is attracted, he may have heterosexual relationships. But he may so idealize females that he is offended by the idea of having sexual relationships with them, and then he may be left without any opportunity for socio-sexual contacts, because his dislike for individuals of his own sex will prevent him from having sexual relationships with them.

There are some psychiatrists who consider all transvestism homosexual, but this is incorrect.[31] Transvestism and homosexuality are totally independent phenomena, and it is only a small portion of the transvestites who are homosexual in their physical relationships. A misinterpretation on this point may generate tragedy when psychiatrists insist, as we have known them to do in several cases, that all transvestites, including those who are basically opposed to everything connected with their own sex, must frankly accept their "homosexuality" and accept overt homosexual relationships if they wish to resolve their psychologic conflicts.

On the other hand, some males are transvestites and wish to be identified with the opposite sex because they are homosexual and because they hope to attract the type of male who would hesitate to engage in homosexual relationships if the other individual were not identifiable, to at least some degree, with femininity.

In not a few instances transvestism develops out of a fetishistic interest in the clothing or some part of the clothing of the opposite sex. The adoption of the clothing of the opposite sex may not modify the original sexual history of the individual, whether it was heterosexual or homosexual.

---

[31] The assumption that transvestism is always associated with or an expression of homosexuality may be noted, for example, in: Krafft-Ebing 1922:398. Forel 1922:251. Thorpe and Katz 1948:314. Allen 1949:146–147.

There are many cases of transvestism which are associated with sado-masochism. Then the masochistic male wishes to be identified as a female in order to be subjugated as males might, conceivably, subjugate a female.

It is clear that transvestism depends very largely upon the individual's capacity to be conditioned psychologically. There are few phenomena which more strikingly illustrate the force of psychologic conditioning. It is, therefore, highly significant to find that an exceedingly large proportion of the transvestites are anatomically males who wish to assume the role of the female in the social organization. At this point we cannot give percentages, although we are attempting to secure a sample which will ultimately allow us to estimate the number of transvestites in the United States; but it is our present understanding that there may be a hundred anatomic males who wish to be identified as females, for every two or three or half dozen anatomic females who wish to be identified as males.[32] This last is particularly interesting because females often assume some of the clothing of males in working around their homes, on farms, in factories, and elsewhere; but we find no evidence that such females are interested in being identified socially with the opposite sex, and such an adoption of male attire has little or nothing to do with transvestism. Males, of course, do not usually wear any part of the female costume unless they are true transvestites.

Transvestism provides one of the striking illustrations of the fact that males are more liable to be conditioned by psychologic stimuli, and females less liable to be so conditioned. The males who wish to be identified as females are in reality very masculine in their psychologic capacities to be conditioned.

30. Discontinuity in Sexual Activity. We have already pointed out that the sexual activities of females are often very discontinuous. Between periods of activity there may be weeks or months and sometimes years in which there is no activity of any sort. This is true of masturbation in the female (p. 148), of nocturnal dreams to the point of orgasm (p. 197), of pre-marital petting (p. 236), of pre-marital coitus (p. 289), of extra-marital coitus (p. 419), and of homosexual experience (p. 456). It is most strikingly true of the female's total sexual outlet. Some females who at times have high rates of outlet, may go for weeks or months or even years with very little outlet, or none at all. But then after such a period of inactivity the high rates of outlet may

---

[32] An example of the sort of unwarranted statement that gets into and is perpetuated in the professional literature, although it is unsubstantiated by any specific data, is the estimate in Allen 1949:145 that transvestism is as common among females as among males.

develop again. Discontinuities in total outlet are practically unknown in the histories of males.

These differences in the continuity of sexual activities may depend upon a variety of factors. They certainly depend in part upon the differences in the way in which females and males respond to psychologic stimuli. Because males are so readily stimulated by thinking of past sexual experiences, by anticipating the opportunity to renew that experience, and by the abundant associations that they make between everyday objects and their sexual experience, the average younger male is constantly being aroused. The average female is not so often aroused. In some instances the male's arousal may be mild, but in many instances the arousal may involve genital erection and considerable physiologic reaction. Nearly all (but not all) younger males are aroused to the point of erection many times per week, and many of them may respond to the point of erection several times per day. Many females may go for days and weeks and months without ever being stimulated unless they have actual physical contact with a sexual partner. Because of this constant arousal, most males, particularly younger males, may be nervously disturbed unless they can regularly carry their responses through to the point of orgasm. Most females are not seriously disturbed if they do not have a regular sexual outlet, although some of them may be as disturbed as most males are without a regular outlet. The failure to recognize these differences in the needs of the two sexes for a regular sexual outlet may be the source of a considerable amount of difficulty in marriage. It is the source of many social disturbances over questions of sex. In establishing sex laws, in considering the sexual needs of females and males in penal and other institutions, in considering the need among females and among males for non-marital sources of sexual outlet, and in various other social problems, we cannot reach final solutions unless we comprehend these considerable differences between the sexual needs of the average female and the average male.

31. Promiscuity. Among all peoples, everywhere in the world, it is understood that the male is more likely than the female to desire sexual relations with a variety of partners. It is pointed out that the female has a greater capacity for being faithful to a single partner, that she is more likely to consider that she has a greater responsibility than the male has in maintaining a home and in caring for the offspring of any sexual relationship, and that she is generally more inclined to consider the moral implications of her sexual behavior. But it seems probable that these characteristics depend upon the fact that the female is less often aroused, as the average male is aroused, by the idea of promiscuity.

An attempt to analyze the reason for the greater promiscuity of the male suggests that it depends upon a variety of psychologic capacities which are not so often found in the female. Several of these we have already discussed. The male is aroused at observing his potential sexual partner, as most females are not. The male is aroused because he has been conditioned by his previous experience, as most females have not. The male is aroused by anticipating new types of experience, new types of sexual partners, new levels of satisfaction that may be attained in the new relationships, new opportunities to experiment with new techniques, new opportunities to secure higher levels of satisfaction than he has ever before attained. In both heterosexual and homosexual relationships, promiscuity may depend, in many instances, upon the male's anticipation of variation in the genital anatomy of the partner, in the techniques which may be used during the contacts, and in the physical responses of the new partner. None of these factors have such significance for the average female.

Male promiscuity often depends upon the satisfactions that may be secured from the pursuit and successful attainment of a new partner. There are some heterosexual males, and a larger proportion of the homosexual males, who may limit themselves to a single contact with any single partner. Once having demonstrated their capacities to effect sexual relations with the particular individual, they prefer to turn to the pursuit of the next partner.

The male's greater inclination to be promiscuous shows up in the record of his petting experience, his experience in pre-marital coitus, in extra-marital coitus, and in homosexual relations. In all of these types of relationships, few females have anywhere near the number of partners that many a promiscuous male may have. The specific data are as follows:

Number of Partners

NUMBER OF PARTNERS	PRE-MARITAL PETTING		PRE-MARITAL COITUS		HOMOSEXUAL CONTACTS	
	Female	Male	Female	Male	Female	Male
	Percent		*Percent*		*Percent*	
1	10	6	53	27	51	35
2–5	32	20	34	33	38	35
6–10	23	16	7	17	7	8
11–20	16	21	4	11	3	6
21–30	8	10	1	4	—	2
31–50	6	11	—	3	—	3
51–100	4	8	—	4	—	3
101+	1	8	—	1		8
Number of cases	2415	1237	1220	906	591	1402

It has sometimes been suggested that the male's capacity to be erotically aroused by *any* female, and even by a physically, mentally, and aesthetically unattractive, lower level prostitute, is a demonstration of the fact that he is not as dependent as females are upon psychologic factors for the achievement of satisfactory sexual relationships. On the contrary, the capacity of many males to respond to *any* type of female is actually a demonstration of the fact that psychologic conditioning, rather than the physical or the psychologic stimuli that are immediately present, is a chief source of his erotic response. As far as his psychologic responses are concerned, the male in many instances may not be having coitus with the immediate sexual partner, but with all of the other girls with whom he has ever had coitus, and with the entire genus Female with which he would like to have coitus.

32. Significance of Sexual Element in Marriage. Our data indicate that the average female marries to establish a home, to establish a long-time affectional relationship with a single spouse, and to have children whose welfare may become the prime business of her life. Most males would admit that all of these are desirable aspects of a marriage, but it is probable that few males would marry if they did not anticipate that they would have an opportunity to have coitus regularly with their wives. This is the one aspect of marriage which few males would forego, although they might be willing to accept a marriage that did not include some of the goals which the average female considers paramount.

Conversely, when a marriage fails to satisfy his sexual need, the male is more inclined to consider that it is unsatisfactory, and he is more ready than the female to dissolve the relationship. We have no statistical tabulation to substantiate these generalizations, but we have discussed the reasons for their marriages, and the reasons for maintaining their marriages, with some thousands of the females and males who have contributed to the present study.

It is too simple to dismiss these differences in female and male attitudes toward marriage as the product of innate moral differences between the sexes. Neither does it suffice to consider that these differences are a product of the female's greater importance in childbearing and in the preservation of the species. Whatever truth there may be in either of these assertions, it seems certain that these differences between female and male approaches to marriage depend primarily upon the fact that the average male is so conditionable that he has a greater need than most females have for a regular and frequent sexual outlet.

33. Social Factors Affecting Sexual Patterns. For males, we found (1948) that social factors were of considerable significance in determining patterns of sexual behavior. In the present volume we have found that social factors are of more minor significance in determining the patterns of sexual behavior among females.

For instance, we found that the educational level which the male ultimately attained showed a marked correlation with his patterns of sexual behavior (Table 175). Thus, the males who had ultimately

Table 175. Correlations Between Social Factors and Patterns of Sexual Behavior

Sexual activity	Educ. ♀	Educ. ♂	Decade of birth ♀	Decade of birth ♂	Age at onset of adol. ♀	Age at onset of adol. ♂	Relig. ♀	Relig. ♂
Masturbation								
Accum. incid. (orgasm)	√	−	±	−	x	±	√	
Act. incid. (orgasm)	±	√	x	±	x	√	√	±
Freq., act. med. (orgasm)	x	√	x		x	±	−	√
Percent total outlet	±	√	−		x	x	±	x
Nocturnal dreams								
Accum. incid. (orgasm)	x	√	−		x	x	±	
Act. incid. (orgasm)	x	√	x	x	x	x	±	−
Freq., act. med. (orgasm)	x	±	x	x	x	x	x	x
Percent total outlet	x	±	x		x	x	x	±
Petting								
Accum. incid. (exper.)	x	x	√	±	x	x	−	
Accum. incid. (orgasm)	x	√	√	±	x	x	√	
Act. incid. (orgasm)	x	√	√	±	−	x	±	±
Freq., act. med. (orgasm)	x	−	x		x	x	x	x
Percent total outlet	x	√	±		x	x	x	−
Pre-marital coitus								
Accum. incid. (exper.)	x	√	√	−	−	√	√	
Accum. incid. (orgasm)	−	√	√	−	−	√	√	
Act. incid. (exper.)	±	√	√	−	−	±	√	√
Act. incid. (orgasm)	±	√	√	−	−	±	√	√
Freq., act. med. (exper.)	x	√	x		x	x	±	√
Freq., act. med. (orgasm)	x	√			x	x	±	√
Percent total outlet	±	√	√		−	−	±	√
Marital coitus								
Act. incid. (orgasm)	±	x	√	x	x	x	x	x
Freq., act. med. (exper.)	x	x	√			±	x	√
Freq., act. med. (orgasm)		x	±			±	x	√
Percent total outlet	−	±	±		x		±	−
Extra-marital coitus								
Act. incid. (exper.)	±	±	√		x		√	√
Freq., act. med. (exper.)	x	√	x		x		x	√
Percent total outlet	±	±	−		x		−	√
Homosexual								
Accum. incid. (orgasm)	±	−	x	x	x	√	√	±
Act. incid. (orgasm)	√	√	x	−	x	±	√	√
Freq., act. med. (orgasm)	±	−				x	−	−
Percent total outlet	±	±	x		x	±	−	−
Total outlet								
Accum. incid.	±	x	±	x	x	x	√	x
Active incid.	−	x	±	x	−	x	√	−
Freq., act. med.	−	−	−		x	√	√	√

√ = marked correlation.   ± = some correlation.   − = little correlation.   x = no correlation.   Blanks = no data available.   ♀ = female.   ♂ = male.

gone on into college depended primarily on masturbation and much less frequently on coitus for their pre-marital outlet. On the other hand, the males who had not gone beyond grade school or early high school had drawn only half as much of their pre-marital outlet from masturbation, but they had drawn five times as much of their pre-marital outlet as the upper level males had from coitus. Similarly, kissing habits, breast manipulations, genital manipulations, mouth-genital contacts, positions in coitus, nudity during coitus, the acceptance of nudity or near-nudity during non-sexual activities, and many of the other items in the sexual behavior of a male, are usually in line with the pattern of behavior found among most of the other males in his social group. We have emphasized that such differences do not depend upon anything that is learned in school, for both lower level and upper level males may be together in the same grade school and high school, and the patterns are, for the most part, set soon after the mid-teens and before the average male ever goes on into college. We have emphasized that these differences in patterns of sexual behavior depend upon differences in the sexual attitudes of the different social levels in which the male is raised or into which he may move. This means that he is psychologically conditioned by the attitudes of the social group in which he is raised or toward which his educational attainments will lead him.

In contrast, in connection with most types of sexual activity we have found that patterns of sexual behavior among females show little or no correlation with the educational levels which the females ultimately attain (Table 175). In her pre-marital petting, pre-marital coitus, and extra-marital coitus, and in her total sexual outlet, there are some differences in the incidences and/or frequencies which appear to be correlated with the educational levels of the females, but the apparent differences prove to depend on the fact that marriage occurs at different ages in the different educational groups; and when the pre-marital activities are compared for the females who marry at about the same age, the average incidences and frequencies of these various types of sexual activity prove to be essentially the same in the several educational levels. This appears to mean, again, that females are not conditioned to the extent that males are conditioned by the attitudes of the social groups in which they live.

We have also shown that the age at onset of adolescence and the rural or urban backgrounds do not show as marked a correlation with the patterns of behavior among females as they do among males.

For both the females and males in our sample, degrees of religious devotion did correlate with the incidences of the various types of sexual activity, and devoutly religious backgrounds had prevented some of the females and males from ever engaging in certain types of sexual activity. The incidences of nearly all types of sexual activity except marital coitus were, in consequence, lower among the religiously more devout females and males, and higher among the religiously less devout (Table 175).

The degree of religious devotion, however, had continued to affect those males who finally did become involved in the morally disapproved types of activity, and the median frequencies of such activities were lower among the more devout males and higher among the less devout males (Table 175); but among those devout females who had become involved in morally disapproved types of activity, the average rates of activity were, on the whole, the same as those of the less devout females. This was true, for instance, of masturbation, of nocturnal dreams to orgasm, of pre-marital petting, of pre-marital coitus, and of homosexual contacts among females (Table 175). While religious restraints had prevented many of the females as well as the males from ever engaging in certain types of sexual activity, or had delayed the time at which they became involved, the religious backgrounds had had a minimum effect upon the females after they had once begun such activities.

SUMMARY AND COMPARISONS OF FEMALE AND MALE

We have, then, thirty-three bodies of data which agree in showing that the male is conditioned by sexual experience more frequently than the female. The male more often shares, vicariously, the sexual experiences of other persons, he more frequently responds sympathetically when he observes other individuals engaged in sexual activities, he may develop stronger preferences for particular types of sexual activity, and he may react to a great variety of objects which have been associated with his sexual activities. The data indicate that in all of these respects, fewer of the females have their sexual behavior affected by such psychologic factors.

It was in regard to only three of these items (moving pictures, reading romantic literature, and being bitten) that as many females as males, or more females than males, seem to have been affected. Fewer females than males were affected in regard to twenty-nine of the thirty-three items. There are instances in which the percentages of females who were affected were only slightly below the percentages of males who were affected; but in regard to twelve of these items, the number of females who were erotically aroused was less than half the number of males who were aroused.

There is tremendous individual variation in this regard, and there may be a third of the females in the population who are as frequently affected by psychologic stimuli as the average of the males. At the extreme of individual variation, there were, however, 2 to 3 per cent of the females who were psychologically stimulated by a greater variety of factors, and more intensely stimulated than any of the males in the sample. Their responses had been more immediate, they had responded more frequently, and they had responded to the point of orgasm with frequencies that had far exceeded those known for any male. A few of the females were regularly being stimulated by psychologic factors to the point of orgasm, and this almost never happens among any of the males.

Many of these differences between the sexual responses of females and males have been recognized for many centuries, and there have been various attempts to explain them. It has been suggested that they depend upon differences in the abundance or distribution of the sensory structures in the female and male body. It has been suggested that they depend upon differences in the roles which females and males take in coitus. It has been suggested that they are in some way associated with the different roles that females and males play in connection with reproduction. It has been suggested that there are differences in the levels of "sex drive" or "libido" or innate moral capacities of the two sexes. It has been suggested that the differences depend upon basic differences in the physiology of orgasm in females and males.

But we have already observed that the anatomy and physiology of sexual response and orgasm (Chapters 14 and 15) do not show differences between the sexes that might account for the differences in their sexual responses. Females appear to be as capable as males of being aroused by tactile stimuli; they appear as capable as males of responding to the point of orgasm. Their responses are not slower than those of the average male if there is any sufficiently continuous tactile stimulation. We find no reason for believing that the physiologic nature of orgasm in the female or the physical or physiologic or psychologic satisfactions derived from orgasm by the average female are different from those of the average male. But in their capacities to respond to psychosexual stimuli, the average female and the average male do differ.

The possibility of reconciling the different sexual interests and capacities of females and males, the possibility of working out sexual adjustments in marriage, and the possibility of adjusting social concepts to allow for these differences between females and males, will depend upon our willingness to accept the realities which the available data seem to indicate.

What physicochemical bases there may be for the similarities and differences between the psychosexual capacities of females and males is a matter that we shall undertake to explore in the chapters that follow (Chapters 17 and 18).

Chapter 17

NEURAL MECHANISMS OF SEXUAL RESPONSE

The data which we have now accumulated on the gross physiology and psychology of sexual response and orgasm make it possible to recognize some of the internal mechanisms which may be involved.

Since there are no essential differences between the responses of females and males to tactile and other sensory stimulation (Chapters 14, 15), such responses must depend upon internal mechanisms which are essentially the same in the two sexes. On the other hand, since there are marked differences between females and males in their responses to psychologic stimuli, it seems apparent that those responses must depend upon some mechanism which functions differently in the two sexes.

It is the function of the exploring scientist to describe what he finds, whether or no the observed phenomena are explainable in terms of the known anatomy and known physiologic processes. We have described, as far as we have been able to obtain the data, what happens to the mammalian body when it responds sexually. While it now seems possible to identify some of the internal mechanisms which may account for that behavior, at points we shall find that there is nothing yet known in neurologic or physiologic science which explains what we have found. These are the areas in which, it may be hoped, the neurologist and physiologist may do further research.

It has been important to understand the gross behavior of the sexually responding animal, for too much of the physiologic and experimental work has, so far, been concentrated on explaining the nature of genital responses, and has ignored the fact that all parts of the body may be involved whenever there is sexual response and orgasm. There have been studies of the effects of the stimulation of end organs of touch in genital and perineal areas, studies of the nerves that connect those end organs with lower portions of the spinal cord, studies of the sexual function of that end of the cord, and studies of the nerves that transmit impulses from the cord to the genital and pelvic structures which are involved in a sexual response. In addition, there have been studies of the possible function of certain portions of the brain, particularly of the cerebrum, in connection with sexual response; but, once again, they have been studies of cerebral function in connection with genital response.[1] The neurologic and experimental studies should now be extended in directions which may explain why the whole animal body is involved whenever there is any sort of sexual response.

EVIDENCE OF NERVOUS FUNCTION

There is nothing which needs to be emphasized more than this fact that the entire body of the animal is involved whenever there is any sexual response (Chapters 14, 15). We have pointed out that the tactile or other sensory stimulation of *any* part of the body, and not of the genitalia alone, may initiate these responses. We have pointed out that psychologic stimulation may, in many instances, bring responses that are quite identical with those effected by tactile stimulation. We have pointed out that sexual responses may involve changes in the function of the circulatory system, of the respiratory system, of the sensory capacities of the animal, of all of the glands of the body, and of the muscular activities in every part of the body. We have shown that orgasm is similarly a function of the whole animal body. The internal mechanisms which are responsible for such activity must be mechanisms which can affect all parts of the body. There are three such mechanisms which, conceivably, might accomplish that end.

1. Chain Reactions. The action of any part of an animal's body may be directly responsible for a chain of activities—a series of successive steps in which each act initiates the succeeding act in the chain. Such a chain of responses may be involved, for instance, when the driver of an automobile steers his course along a particular path, and is able to pass other cars without consciously planning the movements of his steering gear. His sensory perception of the objects which lie ahead, his adjustments for the distances which are involved, for the speed of his own car, for the speed of the approaching car, and for the movements of his hands or arms which may steer him safely past the approaching car, may represent such a chain of responses. The initial action is responsible for the next action and that in turn determines the following act, until the ultimate end is achieved.

It has been suggested that sexual activities similarly represent chains of responses. It has been suggested that the stimulation of end organs,

---

[1] For neurologic studies of the genital area see, for instance: Eberth 1904. Marshall 1922:264–272. Stone 1923b:88–90, 104. Semans and Langworthy 1938. Bard 1940:556 ff. Kuntz 1945:304–323. Hooker in Howell (Fulton edit.) 1949:1202–1205. Kuntz 1951:101–106. For studies of the brain in relation to sexual response, see for example: Bard 1934, 1936, 1939, 1940, 1942. Brooks 1937. Rioch 1938. Stier 1938. Klüver and Bucy 1938, 1939. Davis 1939. Maes 1939. Dempsey and Rioch 1939. Beach 1940, 1942a, 1942b, 1943, 1944, 1947a, 1947b. Brookhart, Dey, and Ranson 1941. Brookhart and Dey 1941. Dey, Leininger, and Ranson 1942. Clark 1942. Langworthy 1944.

and the consequent stimulation of the nerves which connect those end organs with the lower end of the spinal cord, and of the nerves which go from the cord to the muscle fibers in the walls of the circulatory system, may directly account for the rises in pulse rate and blood pressure and for the increased peripheral circulation which is responsible for the tumescence of various body structures during sexual response. It has been proposed that these circulatory disturbances are then responsible for the changes which are to be observed in the breathing rate, and that these physiologic disturbances are responsible for the spectacular muscular activities which characterize sexual response. But the present data do not show that these phenomena appear in sequence, one after the other.[2] On the contrary, there is an instantaneous and simultaneous appearance of all of these physiologic changes as soon as the animal is stimulated and begins to respond.

It is not impossible that there are some aspects of the later developments in sexual response—like the ultimate build-up of neuromuscular tensions, the increasing loss of sensory perception, and the ultimate disturbance of the breathing rate as an individual approaches orgasm —which may be products of physiologic developments which appear earlier in the course of the sexual activity. But this cannot be true of most of the phenomena which appear during sexual response and orgasm. A rise in pulse rate, a rise in blood pressure, a rise in breathing rate, a diminution of the capacity for sensory perception, glandular secretions, and a development of neuromuscular tensions over the whole body, appear to develop simultaneously, sometimes within a fraction of a second, as soon as the animal is stimulated and begins to respond.

2. Blood-Distributed Agents. The coordination of the functions of separated parts of the body is sometimes effected through blood-circulated agents, such as hormones (Chapter 18). For instance, many of the physiologic aspects of certain emotional reactions, such as anger, may be duplicated by the injection of adrenaline into the blood stream, and it is generally understood that the appearance of raised pulse rates, raised blood pressures, and still other aspects of an angry animal may be a direct product of the secretion of adrenaline from the adrenal glands into the blood stream. But the speed with which sexual responses may occur is far greater than that which could be effected by any blood-circulated substance such as adrenaline. While it may not take more than a few seconds for adrenaline to be carried by the blood over short distances in the animal's body, it takes a longer time for it to circulate to all parts of the body. Sexual responses, how-

---

[2] Beach 1947b:241 also points out the inadequacy of a chain reaction hypothesis in sexual response.

ever, may be initiated, carried through their complex course, and completed within a matter of seconds (p. 605).

It is possible that there is adrenaline secretion during the more advanced stages of any protracted sexual activity, and this in the late stages of sexual response may reenforce some of the physiologic changes which the nervous system has initiated; but adrenaline cannot be responsible for most of the changes which appear as soon as there is sexual stimulation.

3. Nervous Mechanisms. Our best reasons for believing that sexual responses must depend primarily upon nervous mechanisms are the speed with which all parts of an animal's body may become involved, the steady and convulsive build-up of neuromuscular tensions as the action develops, the abrupt build-up of tensions at the approach of orgasm, the remarkable rigidity which may develop just before orgasm, the explosive discharge of neuromuscular tensions at orgasm, and the abrupt cessation of tension after orgasm. There are no other means of intercommunication which act as quickly as nervous mechanisms, and no other mechanism that can bring so nearly simultaneous reactions from all parts of the body. The gradual and steady accumulation of neuromuscular tensions during sexual arousal is a known characteristic of some other nervously controlled responses. The explosive discharge which characterizes orgasm is the sort of phenomenon that cannot be ascribed to anything except a neural mechanism.

Finally, the electroencephalograms which are now available (Figure 140, p. 630) show that there are remarkable changes in brain potentials in the course of sexual response, and that it is the development of these and their sudden release at orgasm which provide the most characteristic aspects of sexual response and orgasm.[3] For these several reasons, the search for the mechanisms of sexual response may be concentrated primarily upon the structure and function of the vertebrate nervous system.

TACTILE STIMULATION AND REFLEX ARCS

It is obvious from the record which we have already given, as well as from everyday experience, that physical contacts, touch, and pressure may bring sexual responses only because there are sensory structures in the external surface of the animal's body which respond to such stimulation. These sensory structures are the end organs of touch, and it is on these that many sexual responses depend.

---

[3] Dr. Abraham Mosovich has been kind enough to communicate to us the as yet unpublished results of his electroencephalographic research and has sent us some sample electroencephalograms. See page 630.

We have already pointed out that end organs which are sensitive to such other physical stimuli as light, heat, and sound may also be involved, but we have emphasized that there are no experimental data which show exactly how these other end organs function in connection with sexual response (p. 590).

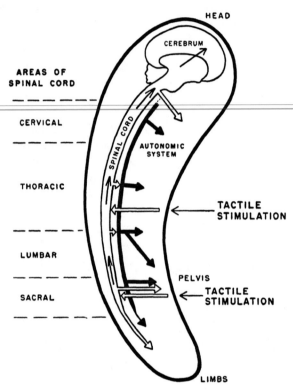

Figure 141. Diagram of neural mechanisms in mammalian sexual response to tactile stimulation

The stimulus received by any such organ is transmitted by afferent (in-going) nerves to the spinal cord. The cord in turn may transmit the impulses by efferent (out-going) nerves to those muscles with which they are directly connected (Figure 141). This may lead to a spontaneous and quite involuntary action of muscles which can be and are ordinarily voluntarily controlled. The prominent muscular action of the pelvis and lower limbs may be accounted for as the direct product of such a simple reflex arc. Because the stimulation of any part of the cord may be extended to all parts of the cord, the reflex arcs may involve not only the particular part of the body which was originally stimulated, but other parts of the body as well. Genital stimulation, for instance, brings not only pelvic responses but responses of the muscles in every other part of the body (see pp. 586, 618 for a further discussion).[4]

A major portion of the experimental work that has been done on the physiology of sexual response has been concerned with these relatively simple aspects of the problem, and the location of those areas of the spinal cord which are concerned with a major portion of the pelvic and lower limb responses is specifically known.

FUNCTION OF THE SACRO-LUMBAR AREA

The experimental work has shown that lower portions of the cord, namely, the sacral and lumbar areas (Figure 141), are the mediating centers upon which genital reflexes and pelvic responses depend.[5] As long as the end organs and the afferent nerves are intact, and as long as the sacro-lumbar areas of the cord and the efferent nerves are intact, it is possible to secure pelvic responses and responses of the lower limbs even though other portions of the cord and considerable portions of the brain of the animal may be damaged. This is demonstrated when the cord is actually cut above the sacro-lumbar area in laboratory experiments with animals. It is also demonstrated by human paraplegics. These are individuals who have had the spinal cord injured or cut at some point above the sacro-lumbar area as the result of an accident, some surgical operation, or some other damage.

An animal that has the cord damaged above the sacro-lumbar area may respond to tactile stimulation of its genitalia, perineum, or other pelvic areas, and may still become tumescent in those areas, develop a genital erection, and reach orgasm.[6] Unfortunately, because the full

---

[4] Impulses originating in tactile stimulation of the pelvic area are transmitted to the sacral area of the cord chiefly by way of the internal pudendal nerve. The efferent impulses include spinal impulses which travel via the internal pudendal nerve to the muscles about the genitalia, causing them to tense. Parasympathetic impulses travel via the erigens nerves to the blood vessels of the genital area, causing tumescence. The striped muscles of the pelvic area and lower limbs are thrown into spasm by spinal impulses from the sacral area; the smooth muscle is thrown into spasm by sympathetic impulses originating in the lumbar area, traveling to the genitalia mainly via the hypogastric nerves. See, for instance: Stone 1923b:89–90, 94, 104. Semans and Langworthy 1938. Kuntz 1945:308–312. Munro et al. 1948:903–910. Hooker in Howell (Fulton edit.) 1949:1203–1204. Whitelaw and Smithwick 1951: 121–130.

[5] That a sacral center mediates tumescence and a lumbar center mediates ejaculation and corresponding muscular contractions in the female is generally accepted in the literature. See: Kuntz 1945:309, 322–323. Talbot 1949:266. However, sacral nerves may play a part in ejaculation according to Munro et al. 1948:910, and lumbar nerves may also be involved in tumescence according to: Kuntz 1945:308–309. Root and Bard 1947:87–89.

[6] Erection and ejaculation in an animal with a severed spinal cord is noted, for example, in: Marshall 1922:264–265. Bard 1940:556. Beach 1942a:213, 215. Munro et al. 1948. Fulton 1949:140. Talbot 1949:266–267 (a concise discussion of the effects of cord damage to sexual function).

extent of the physiologic changes which occur in sexual response has not hitherto been comprehended, there are no good data on the responses which probably occur in the lower limbs of such an experimental animal.

When the human paraplegic receives genital or pelvic stimulation, he does not feel the stimulation or the consequent physiologic changes

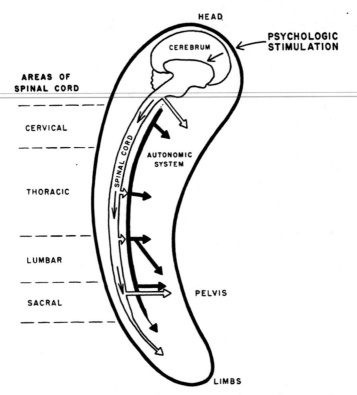

Figure 142. Diagram of neural mechanisms in mammalian sexual response to psychologic stimulation

in those areas, although he may observe them; and he is not conscious of sexual satisfaction when he has coitus or engages in other sexual activities, even though he may come to erection and reach orgasm. This indicates that a brain, and a cord which is intact between the level of stimulation and the brain, are necessary for the attainment of that change of physiologic state which the human animal recognizes as sexual satisfaction. A paraplegic may, however, be conscious of tactile stimulation of portions of the body innervated by nerves which connect with those parts of the cord that are located above the point of damage, and may receive erotic satisfaction from the stimulation of an upper portion of his or her body, even when the pelvis and lower limbs are unresponsive (p. 700).[7]

Psychologic stimulation, depending primarily on the function of the cerebrum (p. 710), may also effect pelvic and lower-limb responses. This also depends, to a considerable degree, on the sacro-lumbar areas of the cord (Figure 142). If those portions of the cord are completely removed in an experimental animal, or seriously damaged in a human subject, or if those areas are separated from the brain by a complete cutting (transection) of the cord above the sacro-lumbar areas, then psychologic stimulation no longer brings pelvic or genital or lower-limb responses.[8]

The cat is one of the experimental animals in which the sacral area of the cord and all of the lumbar area except the uppermost (the first lumbar) segment may be removed without eliminating the animal's responses to the tactile stimulation of genital or other pelvic areas, or its responses to psychologic stimulation. But in the case of this animal it has been demonstrated that such responses are largely eliminated if the first lumbar segment is also removed. This upper end of the lumbar area is connected with the genital area by the hypogastric nerve which is one of the branches of the so-called autonomic system, and this appears to be at least one of the paths by which nervous impulses which are the product of psychologic stimulation are transmitted to the genital area (Figure 142).[9] On the other hand, in man and in some of the other mammals in which there are other connections (additional to the hypogastric nerve) between portions of the spinal cord which are anterior to the sacro-lumbar areas, and other parts of the autonomic nervous system, it is not improbable that there may be some residue of pelvic response when the entire sacro-lumbar area is removed. There is need for a more exact examination of this matter in further work with experimental animals and with clinical cases in the human animal. However, since the human paraplegics who have had the connection between the brain and the sacro-lumbar area completely severed, report failures to achieve erection through

---

[7] The inability of the true paraplegic to perceive orgasm is noted in Munro et al. 1948:905, 909–910, implied in Talbot 1949:269, and recorded by Dr. L. W. Freeman (verbal communic.). Hyndman and Wolkin 1943:144 add "For . . . orgasm to be sensually appreciated, the spinal connections with the brain must, of course, be intact."

[8] The inability of psychologic stimulation to effect genital response after damage to the cord above the lumbar area is discussed in: Hyndman and Wolkin 1943:143–144. Kuntz 1945:309. Root and Bard 1947:87–88. Talbot 1949: 266–267. Hooker in Howell (Fulton edit.) 1949:1203. Even compression of the cord may produce erectile impotence in humans. Cf. Elsberg 1925:66, 132, 308.

[9] By-passing of a transection of the cord via the hypogastric nerve is described for the cat by Root and Bard 1947:86–89.

psychologic stimulation, it is probable that the major portion of the responses of the pelvic areas and of the lower limbs even in the human species depend upon the integrity of at least the upper segments of the lumbar area, and upon the existence of intact connections between the brain and that portion of the cord.

The importance of the sacro-lumbar area of the cord was one of the first things known about the neurology of sexual response. As early as 1863 it was known that the direct electrical stimulation of certain sacral areas would bring genital erection and other sexual responses from laboratory mammals, and this subsequently was demonstrated for human subjects.[10]

It has also been known for half a century that ejaculation may be effected by the direct stimulation of certain sacro-lumbar areas.[11] Unfortunately, however, most of the neurologists and laboratory experimentalists have taken orgasm and ejaculation to be synonymous phenomena. Since we now understand that this is not so (pp. 634–636), it seems possible that the prostate and other glands which are necessary to effect ejaculation may be brought into action by the direct stimulation of sacral centers without the development of responses that are sexual in any strict sense. Consequently all of this work needs to be reviewed for evidence that a true orgasm may be effected as a result of the direct stimulation of these sacro-lumbar centers. It should be recalled that the mechanical stimulation of the prostate gland via the rectum, as in a prostatic massage, may bring an exudation of semen from the penis and sometimes a propulsive exudation which appears to be an ejaculation, even when there is no genital erection or other evidence of sexual response.

On the other hand, it is also to be noted that the stimulation of the interior of the rectum, of the sort that is involved in a prostatic massage, may also exert pressure upon the perineal nerve masses (page 585), and it is known that this may bring specifically sexual responses with orgasm. This is frequently realized by persons who have been the recipients in anal coitus, and apparently may sometimes be involved in the ejaculatory responses which occur during a prostatic massage.

Both the laboratory experimental work and the work with human subjects have further demonstrated that genital erection and responses

---

[10] In 1863 Eckhardt found that stimulation of the erigens nerves produced erection in the dog. Also see footnote 4.

[11] For a detailed description of the production of erection and ejaculation through electrical stimulation of the sacral segments, see: Durfee, Lerner, and Kaplan 1940. Also see footnote 4. As early as 1863 Eckhardt noted that tactile stimulation of the penis could not cause erection in a dog whose pudendal nerve was severed, and in 1879 Nikolsky reported contraction of penile blood vessels on cutting the erigens nerves. See Kuntz 1945:308.

to the point of orgasm usually depend upon the integrity of the pudendal and erigens nerves, which are efferent nerves connecting the sacral area of the cord with the muscles of the genital and pelvic areas. If these nerves are accidentally damaged or surgically cut, erection in the male is no longer effected. The effect of damage to the pudendal nerve of the female has not been clearly recorded in the experimental or medical literature.

In the course of mammalian evolution the forepart of the brain has been the portion of the central nervous system which has evolved most elaborately. There are fewer basic differences between the sacro-lumbar areas of the cord in the human animal and the sacro-lumbar areas in the lower mammals. The importance of the sacro-lumbar areas in sexual response explains why we find human sexual behavior much closer to the sexual behavior of the infra-human species, and even of the lower mammalian species, than most persons might have anticipated and than some persons would like to believe.

ROLE OF UPPER PORTIONS OF CORD

In regard to sexual response there is very little precise information concerning the role of those portions of the cord which lie above (anterior) to the sacro-lumbar area. As long as genital and pelvic responses were considered the major portion or even the whole of sexual response, and as long as the demonstrated function of the sacro-lumbar area seemed to account for all genital and pelvic responses, there was a tacit assumption that the thoracic and cervical areas of the cord had little or no direct connection with sexual behavior. The integrity of these upper portions of the cord was obviously necessary for the transmission of psychologic stimulation from the cerebrum to the genital and pelvic areas, but, apart from that, there has been little experimental work to show what role this upper end of the cord may have in sexual response.[12]

If, however, it is understood that tactile stimulation of other portions of the mammalian body quite apart from the genital and pelvic areas may bring some sort of sexual response, and even full and complete sexual response to the point of orgasm, it becomes difficult to believe that the stimulation of the upper end of the trunk of the animal or of the mouth or other portions of the head of the animal is transmitted

---

[12] Bilateral anterior chordotomy, the partial transection of the cord to relieve intractable pain, usually interferes with or prevents tumescence and orgasm in both sexes. The transection is usually done above the lumbar area. Hyndman and Wolkin 1943:143 conclude that the operation ". . . is almost certain to abolish erection and orgasm in the male and orgasm in the female . . . the desire for intercourse is not abolished." Stookey 1943:430 adds, "In the male, ejaculation becomes impossible; erection is not interfered with but orgasm is lost. In the female also there may be a loss of orgasm."

to the sacro-lumbar areas of the cord before it brings response. It is more reasonable to believe that the stimulation of end organs in the upper end of the mammalian body is mediated through thoracic or cervical spinal centers, and that these directly effect sexual response not only of the upper end of the body itself, but of the whole of the animal body which is involved during the sexual response. Because of the complexity of the neural anatomy at this point, and because of the lack of specific experimental work in this area, we can do no more than draw attention to the possibility and hope that additional data may be gathered on the point.

We have already drawn attention (p. 694) to the evidence which suggests that the stimulation of any area of the cord spreads throughout the length of the cord, and that this effects the movement of muscles in the upper portions of the body by way of the spinal nerves which lead out of corresponding portions of the cord. These nerves connect directly with the muscles in the upper portions of the body and are apparently responsible for their action during sexual response and orgasm.

One body of data which shows something of the function of the upper end of the cord comes from a single case of a human female paraplegic in whom an operation had severed the damaged cord above the sacro-lumbar area. While psychologic stimulation did not bring any pelvic reactions from the subject, she claimed that tactile stimulation of the upper end of her body (apparently centered about the breasts) did bring sexual response which led to complete orgasm. There appeared to be a build-up of neuromuscular tensions and the explosive release which characterizes orgasm, although the orgasm involved only the upper parts of the body and did not involve the pelvic area. Because the portion of the body which was involved in orgasm was connected with the brain, the subject was conscious of the sexual arousal and found satisfaction in the orgasm.[13]

There are some other cases in which an awareness of tactile stimulation remains in the intact portion of the body even though no erotic satisfaction is derived from such tactile stimulation. It would be very important to accumulate data on these points from additional cases of true paraplegics.

Unfortunately, the medical literature is badly confused by the fact that many of these supposedly paraplegic cases may have had some injury which did not sever the cord. Some of the reported cases of paraplegics suddenly becoming capable of full sexual function when

---

[13] This case was brought to our attention by Dr. L. W. Freeman of the Indiana University Medical Faculty.

there is a sufficient stimulus, probably depend upon the fact that the cord, although it may have been considerably damaged, was never completely severed.[14] It is doubtful whether any true paraplegic ever maintained voluntary control over the lower portion of his or her body. Consequently it is important to sharply distinguish the records on the sexual performances of supposed paraplegics from the records of individuals in whom the cord is definitely known to have been severed.

ROLE OF AUTONOMIC NERVOUS SYSTEM

While the central nervous system, meaning the brain and the spinal cord, is directly connected by afferent nerves with the peripheral end organs which receive sensory stimuli, and by efferent nerves which transmit impulses to other organs of the body and consequently effect muscular action, the autonomic nervous system (Figures 141, 142) is not directly connected with any peripheral end organs and cannot be brought into direct action by the stimulation of end organs of touch or other types of sense organs. The activities which animals voluntarily undertake are controlled by the central nervous system. There is ordinarily no voluntary control of the action of the autonomic nervous system.

The main trunks of the sympathetic division of the autonomic nervous system lie parallel and close to the spinal cord. Throughout the length of the cord there are nerve connections between it and the autonomic system, and consequently the autonomic system may be brought into action by the stimulation of the cord at various levels. It may also be brought into action by the direct stimulation of blood-circulated substances such as adrenaline. Most of the autonomic nervous system consists of fine and highly branched nerves that spread throughout the body in a fashion that sometimes makes it very difficult for the neurologist to determine with any exactitude what nerves are connected with a given organ.

It is customary to think of the autonomic nervous system as consisting of two main parts, the sympathetic and parasympathetic groups of nerves. Although physiologists and neurologists have become increasingly aware that these two parts do not function independently, and that there is no sharp differentiation of the action of the two at many points, the classification has provided a useful basis for thinking in regard to certain points, and there are some of the body functions

---

[14] Without additional operations or autopsy it is difficult to determine if transection is complete, and information on sexual function is confusing. For example, Talbot, J. Nerv. Ment. Dis. 1952:360–361, states that 20 per cent of a group of paraplegics and quadriplegics experienced erection from psychologic stimulation. Note that Talbot 1949:266–268 states that complete transection precludes all cortical influence but found 8 presumably complete transection cases experiencing erection from psychologic stimuli.

which may depend primarily upon one or the other of the branches of the autonomic system. Anatomically the two portions of the autonomic system are not wholly discrete, for they lie closely parallel at many points and sympathetic nerves may actually fuse at some points with parasympathetic nerves.

Moreover, the stimulation of the sympathetic branch of the system produces a chemical substance (sympathin) which acts as a hormone, producing many of the same results which follow the secretion of adrenaline from the adrenal glands or the injection of the drug adrenaline. The most specific action of sympathin is its effect upon the parasympathetic system, and this is brought into action whenever the sympathetic system is stimulated. Similarly, however, stimulation of the parasympathetic system leads to the secretion of acetylcholine, which has a specific action on sympathetic nerves and brings that portion of the system into play. This accounts for the fact that in many emotional situations the responses of an animal may involve physiologic changes which may be attributed to both sympathetic and parasympathetic controls. This is exactly what we find in regard to the physiology of sexual response; and in the present volume we shall not attempt to suggest which aspects of sexual response may depend upon one or the other portion of the autonomic system.

Actually both portions of the autonomic system are usually involved when any portion of the system comes into action in the course of a sexual response, for the parts of the body which are primarily under autonomic control (such as the heart, the glands, and the smooth muscles) are supplied with both sympathetic and parasympathetic nerve fibers. These two types of nerve fibers act antagonistically on the organ and they may thus exert a slight but constant (tonic) effect on it. A given organ, therefore, will be stimulated or inhibited almost instantaneously depending on (1) the strength of the impulse, and (2) whether the impulse reaches it through the sympathetic or parasympathetic system. Because of the complexity of the interrelationships and the delicate balance which exists here, it is not surprising that some organs may be dominated by the parasympathetic system at the same time that other organs are being controlled by the sympathetic system. It is, moreover, to be remembered that most of the organs which are reached by the autonomic nervous system also have sensory nerves which connect with the central nervous system, and may be affected by reflexes which develop in that part of the nervous system.

Many of the physiologic activities which we have found characteristic of sexual response probably do not represent any direct reaction via a reflex arc. The efferent spinal nerves leading from the sacrolumbar areas can account for only some of the genital, pelvic, and lower-limb responses during sexual activity. Most of the activity must depend primarily upon the action of the autonomic nervous system which is brought into play through nerves which originate in the sacro-lumbar areas and in other parts of the cord. The instantaneous involvement during sexual response of those portions of the body, such as the heart, which are for the most part beyond voluntary control, constitutes strong evidence that the autonomic nervous system is involved.[15]

The following aspects of the gross physiology of sexual response are among those that apparently depend upon the action of the autonomic system:

Increase in pulse rate
Increase in blood pressure
Vasodilation
Increase in peripheral circulation of blood
Tumescence of distensible parts of body
Reduced rate of bleeding
Hyperventilation
Irregular breathing
Increase in genital secretions
Increase in salivary secretion
Increase in lacrimal secretion
Gastro-intestinal activity inhibited
Hair raised
Eye pupil dilated
Ejaculation

Reference to Table 176 will show that more of the elements in sexual response appear to depend on parasympathetic than on sympathetic function.

The autonomic system, like the sacro-lumbar area, is ancient in mammalian heritage, and the similarities between human and infra-human sexual responses probably depend, for the most part, upon the fact that these two are the portions of the nervous system which are chiefly responsible for human sexual responses. The physiology of human sexual response depends only to a minimum degree upon the more highly evolved human brain (p. 708).

THE SEXUAL SYNDROME

We have already pointed out that the most distinctive aspect of any sexual response is the fact that it is a group, a cluster, a syndrome of physiologic elements, all of which usually appear whenever there is

---

[15] Kuntz 1945:313 is one of the few others who have concluded that the entire autonomic system is involved in sexual response.

any sort of sexual response. But nearly all of the elements of sexual response are found in other situations, particularly in other emotional responses and most particularly in anger (Table 176).[16] Actually the sexual elements which are not found in anger are to be found in epilepsy, but there is no other sort of behavioral response which in-

Table 176. Physiologic Elements in the Sexual Syndrome, Anger, Fear, Epilepsy, and Pain

Physiologic element	Neural mech.	Sexual resp.	Anger	Fear	Epilepsy	Pain
Increase in pulse rate	symp.	√	√	√	may	√
Increase in blood pressure	symp.?	√	√	√	may	√
Vasodilation	para.	√	may	no	√	
Increased periph. circ. of blood	para.	√	may	no	may	
Tumescence	para.	√	rare	rare	rare	rare
Reduced rate of bleeding	symp.	√	√	√		√
Hyperventilation	symp.	√	√	√		√
Breathing irregularity		√	√	√	may	√
Anoxia		?			√	
Diminished sensory perception		√	√	√	√	√
Increase in genital secretions		√	no	no		
Increase in salivary secretion	para.	√	√	no	√	no
Increase in lacrimal secretion		√				√
Increase in perspiration	para. + symp.	occas.		√		√
Adrenaline secretion	symp.	√?	√	√		√
Increase in muscular tensions	spinal symp.	√	√	√	√	√
Increase in muscular capacity		√	√	√		
Involuntary muscular activity	spinal	√	√	√	√	√
Rhythmic muscular movements	spinal	√	no?	no	may	no
Gastro-intest. activ. inhibited	symp.	√?	√	√		√
Hair raised	symp.	may	√	√	.	may
Eye pupil dilated	symp.	√	√	√	√	√
Ejaculation	spinal symp.	√	rare	rare	may	
Involuntary vocalization		√	√	√	√	√

√ = physiologic element is present. No = element is absent. May = element sometimes but not always present. Symp. = sympathetic division of autonomic system. Para. = parasympathetic portion of autonomic system. Blanks = data not available.

volves all of the elements which may be found in sexual response. One might hypothesize that if certain of the physiologic elements were prevented from developing in a sexual response, or taken away from a sexual response, the individual might be left in a state of anger or fear, or in some other emotional state. The fact that frustrated sexual responses so readily turn into anger and rage might thus be explained. On the other hand, it not infrequently happens, both in the lower mammals and in man, that anger, fighting, and quarrels suddenly turn into sexual responses.

The close relationship of sexual responses and these other emotional states can best be seen by a more detailed examination of the physiology of anger, fear, and epilepsy.

---

[16] The physiology of the emotions is summarized in such sources as: Prince 1914. Cannon 1927. Bard 1934, 1939. Shock in Reymert 1950.

Anger. The closest parallel to the picture of sexual response is found in the known physiology of anger. Table 176 shows fourteen items which usually appear in both situations. There is, of course, some variation under differing conditions, but the items which are usually identical or closely parallel in anger and the sexual syndrome are the following:

Increase in pulse rate
Increase in blood pressure
Vasodilation (sometimes)
Increase in peripheral circulation of blood
Reduced rate of bleeding
Hyperventilation
Anoxia (probably)
Diminution of sensory perception
Adrenaline secretion (probably)
Increase in muscular tensions
Reduction of fatigue and/or increased muscular capacity
Gastro-intestinal activity inhibited (in sex?)
Hair raised (at least in other mammals)
Involuntary vocalization

There are four respects in which the physiology of anger does not fit the known physiology of sex.

1. In sexual response there is an invariable increase in surface temperatures, color, and tumescence during sexual arousal; but in anger there is sometimes (but not always) a vasoconstriction which makes the face of the angry person white, although this may alternate with a flushing of the face which indicates a peripheral flow of blood. Genital tumescence is the more or less inevitable outcome of sexual response in the uninhibited adult animal, short of old age, and while penile erections may sometimes appear in anger, particularly among the infra-human mammals and among pre-adolescent human males, they are not usual.

2. Genital (Cowper's, Bartholin, and cervical) secretions appear with sexual arousal, but apparently not in anger.

3. The most distinctive aspect of sexual physiology is the rhythmic muscular movement which develops when an individual is sexually aroused. These movements include the pelvic thrusts which constitute the copulatory movements among mammals. Such muscular movements do not appear when one is simply angry.

4. Orgasm is a phenomenon which is unique to sexual response. On those rare occasions in which it develops in anger, it is possible that it represents the development of a true sexual response.

Fear. The physiology of fear involves a number of the elements which appear in sexual activity, but fear is not as close as anger is to the sexual syndrome (Table 176). There are nine items in the physiology of fear that are identical with those in a sexual response:

Increase in pulse rate
Increase in blood pressure
Hyperventilation
Diminution of sensory perception
Adrenaline secretion (probably)
Increase in muscular tensions
Reduction of fatigue and/or increased muscular capacity
Hair raised
Involuntary vocalization

On the other hand there are five aspects of the sexual syndrome which are opposites of those found in fear: a peripheral circulation of blood, a vasodilation, genital secretions, salivary secretions, and involuntary and rhythmic muscular movements. The most distinctive aspect of fear which is lacking in a sexual response, is the vasoconstriction which causes the blanching of the face. The rhythmic muscular flow which is so characteristic of a sexual response is lacking in fear. Genital erections only occasionally appear when an individual is afraid, and orgasm is still more rare.[17] In such a case, it is possible that the fear has given way to a sexual response.

Epilepsy. The physiology of epilepsy includes eight or more elements of the sexual syndrome (Table 176).[18] In both epilepsy and sexual response there are:

Increase in pulse rate (sometimes)
Increase in blood pressure
Vasodilation
Anoxia
Diminution of sensory perception
Increase in salivary secretion
Increase in muscular tensions
Rhythmic muscular movements
Involuntary vocalization

The most remarkable parallel lies in the similarity of the spasmodic muscular movements in epilepsy and the tensions and muscular move-

---

[17] Ejaculation from fright was noted as early as Aristotle [384–322 B.C.]: Problems, Bk. IV:877a; Bk. XXVII:949a. We have a few instances in our own case histories.
[18] For the symptoms of epilepsy, see such a standard work as Penfield and Erickson 1941.

ments which are part of the build-up toward orgasm. The extreme rigidity which develops in the build-up to orgasm, and particularly just before orgasm, provides an especially close parallel to the states of tension in epilepsy. On at least some (rare?) occasions orgasm may occur during epilepsy. An even more striking similarity to epileptic movements is to be seen in the more extreme types of spasm which may follow orgasm. Persons who have seen both epileptic seizures and the more extreme types of orgasm have invariably been impressed by the similarities between the two.[19] The electroencephalograms which are now available show a striking resemblance between sexual response and epileptoid reactions (Figure 140). This is true in the period of the build-up toward orgasm, in the rigidity which precedes orgasm, and in the often violent, convulsive movements which follow orgasm.

If the cerebrum is considerably damaged in an accident or by some operation, the body of the individual may become continuously rigid in what is known as a tonic decerebrate rigidity. Such a rigid state is described as a true release phenomenon which results from an interference with the function that the brain ordinarily exercises in controlling muscular tension. The situation in which muscular tensions alternate with movements of the sort which characterize an epileptic seizure has been interpreted to be a "transient decerebrate rigidity."[20] Persons in a state of tonic decerebrate rigidity present the most striking parallel we have seen with certain moments in sexual activity, and particularly in the tensions which immediately precede orgasm. In decerebrate rigidity the body is stretched out to its maximum, the gluteal muscles are tensed and the buttocks tightly appressed, the legs and the arms are usually extended and stiffly held under tension, the feet may be pointed in line with the legs, the toes may be curled or spread, the fingers are flexed and strongly tensed, the back may be arched, and the neck may be so tensed that the head is held at a stiff angle. In all of these respects, tonic decerebrate rigidity matches what may be seen, in a transient state, in both epilepsy and in sexual activity.[21]

---

[19] Some of the earliest writers to draw attention to the similarities between epilepsy and sexual response included Democritus (ca. 420 B.C.), who was quoted as saying that orgasm is like a small epileptic seizure (see: Clement of Alexandria; Paedagogus Bk. II, ch. X, and K. Freeman 1949:306). Aretaeus (2nd–3rd cent. A.D.) cites the similarity in "On the Cure of Chronic Diseases" Bk. I, ch. IV. The latter also noted erection and ejaculation in epilepsy in his "On the Causes and Symptoms of Acute Diseases" Bk. I, ch. V. Erection and ejaculation are sometimes listed as epileptic sequelae in modern medical literature, as in: Hyman 1945:1515.

[20] The concept of the spasms in epilepsy depending on a transient decerebrate rigidity is introduced, for example, in Penfield and Erickson 1941:87.

[21] The postural and muscular similarities existing between decerebrate rigidity and the body just before orgasm are exemplified by the photograph in Penfield and Erickson 1941:fig. 21, p. 86. The rotation of the arm, so characteristic of decerebration, is not, however, a part of the sexual syndrome.

ROLE OF THE BRAIN IN SEXUAL RESPONSE

In its embryonic origin the brain has three main areas: the forebrain, the midbrain, and the hindbrain. In the evolution of the vertebrates, the forebrain is the portion which has become most highly evolved, reaching its acme of complexity in the primates and some other higher mammals and, of course, particularly in the human animal. The chief contributions which this developed forebrain makes to human sexual behavior are: (1) an increase in the capacity to be psychologically stimulated by a diversity of erotic situations; (2) an increase in the possibilities of conditioning; and (3) an increase in the capacity to develop inhibitions which interfere with the spinal and autonomic controls of sexual behavior. This curious mixture is what many persons identify as "an intelligent control of the sexual instincts."

The cerebrum of the mammal is derived from the embryonic forebrain. Sexual and emotional functions have been ascribed to certain areas of the cerebrum, including the frontal lobes, the occipital lobes, the parietal lobes, and the temporal lobes. These four areas constitute the great bulk of the human brain.

Frontal Lobes. Although the frontal lobes occupy a considerable space in the human brain, their exact function has constituted a mystery. Accidental damage to the frontal lobes and surgical operations on the frontal lobes have variously been reported as having no effect, or a variety of diverse and often contradictory effects. This has been true, for instance, of the reports on the effects of frontal lobe operations on the sexual function. There were early reports to the effect that frontal lobe operations considerably reduced the "sex drive" of the individual, and optimistic clinical claims that the patterns of sexual behavior might be modified by such operations.[22] The possibilities of such operations have, once again, been seized upon by those who are interested in controlling persons whose sexual activities they consider socially undesirable, and some clinical reports have encouraged the idea that homosexual could be changed into heterosexual patterns of behavior, that exhibitionists would lose their compulsions to exhibit, that highly responsive persons might become mild and relatively unresponsive. On the other hand, there have been reports of persons whose anxieties and inhibitions were supposed to be released and their sexual responses increased as a result of frontal lobe operations.[23]

---

[22] For a report of reduced sexual response following pre-frontal lobotomy, see, for example: Banay and Davidoff 1942.

[23] For reports of increased sexual response following pre-frontal lobotomy, see: Hemphill 1944. McKenzie and Procter 1946. Kolb 1949 (an excellent discussion). The Columbia Greystone Project is reported in: F. A. Mettler 1949 (Selective partial ablation of the frontal cortex. Hoeber publ.) and Mettler 1952 (Psychosurgical problems. Blakiston publ.). Our data from this project will be published by W. B. Pomeroy in the third volume of the Greystone Series.

We have had the opportunity to make a long-range study of 95 patients who had been subjects for frontal lobe operations. From these patients we secured histories before operation, obtaining a record of their sexual activities for some time prior to the operation. We similarly obtained records from these same patients some time (a median of 3.7 years) after operation. The detailed report is being presented by one of us in connection with the total report on the Columbia-Greystone Brain Research Project. In summary, it may be pointed out that among the females in the sample, the intensities of sexual arousal, the number of items that brought sexual arousal, the frequencies of sexual activities of particular sorts, and the frequencies of total sexual outlet did not show any significant change between a period antedating the institutionalization, and the period when the histories were retaken some years after the operation. The median lapse of time between the pre-institutional and post-institutional histories was 7.0 years.

On the other hand, among the males who had the frontal lobe operations, the frequencies of response to various stimuli and the frequencies of overt sexual activity had dropped in the course of the seven years; but the decline in responsiveness and in frequencies was not significantly different from the decline that may occur in seven years in that portion of the male population that has not had frontal lobe operations. In other words, the decline in male activity after such an operation appears to be the product of aging, rather than a direct effect of the operation. There is no decline in female responsiveness or activity because the female, unlike the male, does not show an aging effect at the ages which were involved in the experimental sample.

The critical review of the previously reported effects of frontal lobe operations, which is included in our detailed report on this material, further substantiates our conclusion that there is no demonstrated relation between the function of the frontal lobes of the human brain and any of the investigated aspects of sexual behavior. There are, inevitably, considerable shock effects from any brain operation as serious as a frontal lobe operation. Immediately following such an operation, a subject's responses and behavior may be seriously modified. Many of the reported effects of frontal lobe operations which are to be found in the clinical literature are in actuality such immediate effects. But in most of the experimental subjects with whom we had the opportunity of working, most of these effects had disappeared within a matter of six months or so; and psychologic tests, psychiatric examinations, and a variety of other tests showed little or no permanent change in the operated sample. Some of the research group working on the project felt that there had been a lessening of emotional tensions in a statistically significant portion of the group, but this did not seem certain to some of the others working on the project, and seems not to have been reflected in any of the data which we have on the sexual behavior of the sample. The samples, however, in both our own and in most of the other studies have been small. Most of the investigators have not secured anything like precise data on the sexual behavior of the patients.

Occipital, Parietal, and Temporal Lobes. Neither is there any clearly demonstrated relation between the functions of the occipital, parietal, or temporal lobes of the mammalian brain, and any aspect of sexual behavior. There is one body of work that reports that operations on the temporal lobes have a depressing effect upon emotional responses in general, but a stimulating effect on sexual responses; but no confirmation of these results by other investigators has yet been published.[24]

Cerebrum. On the other hand, the cerebrum as a whole seems to be significant in the sexual behavior of the mammal. Memory, various aspects of learning, various aspects of motor control, and other behavioral functions are considerably disturbed when there is accidental damage to, or operation on various parts of the cerebrum, and these may considerably affect the animal's sexual function.

Within the last thirty years some dozen different investigators, working with a total of six different species of mammals (rats, cats, dogs, monkeys, rabbits, and guinea pigs), have performed operations on the cortex (the outer layers) of the cerebrum and noted the effects of such operations on the sexual performances of these animals.[25]

Any sort of damage to the cortex may seriously affect an animal's motor coordination, and thus affect its physical capacity to perform effectively as a sexual partner. In addition, the cortical damage may reduce the animal's capacity to react to psychosexual stimuli. The degree of interference is more or less directly proportional to the extent of the damage to the cortex. There are differences in the serious-

---

[24] The relative unimportance, sexually, of the parietal lobes is noted in Beach 1950:263. Lashley reports (verbal communic.) the same finding. Occipital lobe ablations in the cat produce no sexual effect other than that ascribable to blindness, according to Beach 1944:129; 1950:266. Dr. C. C. Turbes (verbal communic.) reports occipital ablations without sexual effects in dogs and monkeys. However, Klüver and Bucy 1939 describe increased sexual activity in male monkeys with bilateral temporal lobectomies; we understand that similar results were obtained with monkeys in research done under Dr. Berry Campbell at the University of Minnesota. On the other hand, Beach 1950:263–264 reports no sexual changes following bilateral temporal ablations in the cat, and Poirier 1952:234 likewise failed to find such sexual effects.

[25] For experiments on the relation of the cerebrum to sexual response, see: Bard 1934, 1936, 1939, 1940, 1942 (cat, dog). Brooks 1937 (rabbit). Rioch 1938 (cat). Klüver and Bucy 1938, 1939 (monkey). Davis 1939 (rat). Maes 1939 (cat). Dempsey and Rioch 1939 (guinea pig). Stone in Allen and Doisy 1939 (rabbit, rat). Beach 1940, 1942a, 1942b, 1943, 1944 (rat, cat). Langworthy 1944 (cat).

ness of the effects on different species of mammals.[26] In general, cortical operations reduce the animal's capacity to recognize (be stimulated by) sexual objects, and very much reduce its aggressiveness in approaching a sexual partner.[27] Since sexual relationships so largely depend on the aggressiveness of the male, and only to a lesser extent on the aggressiveness of the female, damage to the cortex of the male more seriously interferes with his effectiveness as a sexual partner. Although the female with cortical damage may similarly have her aggressiveness reduced, so that she no longer attempts to mount other females or males, it is still possible for an intact male to mount her (if she is in estrus) in effective copulation.[28]

There are a few cases of human males with cerebral damage (the exact nature of which is usually not determined) who similarly have had their responses to psychologic stimuli materially reduced by the injury.[29] They are not aroused by any memory of previous sexual experience, and they find it difficult or impossible to explain why the previous experience was stimulating or satisfying, although they may retain some intellectual realization that the experience was formerly pleasurable. They are not aroused by discussions of sexual activities, by seeing possible sexual partners, or by seeing other sexual objects, and they show no interest in any renewal of sexual experience. However, if such a male still has his sacro-lumbar centers intact, he may still be capable of responding to direct, tactile stimulation, and he may still come to erection, copulate, and reach orgasm. There is one case of a human female who similarly had her psychologic responsiveness reduced by a cerebral injury.[30] Unfortunately few persons with cerebral damage have had their sex histories reported.

---

[26] That the extent of any interference with sexual response is correlated with the extent of cortical lesion is carefully described for the rat by: Beach 1940: 204–205, who reports diminished copulatory activity in male rats with lesions of over 20 per cent of the cortex, and a complete loss of copulatory activity when the lesion exceeds 60 per cent. But Brooks 1937:549–550 reports that male rabbits can copulate effectively with all of the cortex removed, providing the olfactory bulbs are spared.

[27] Localized cortical damage has been reported in a few instances to intensify sexual response, as in: Klüver and Bucy 1939. Langworthy 1944. The first authors stress the "psychic blindness" (inability to visually recognize objects) of the operated animals, but this does not harmonize with the reported "hypersexuality" which, it may be noted, seems to have been chiefly autoerotic. Likewise the frequent and protracted copulation suggests not "hypersexuality," but an inability to achieve orgasm. The seemingly intensified responses reported by Langworthy were complicated by motor defects which interfered with effective copulation.

[28] This is well demonstrated by the work of Beach. See particularly: Beach and Rasquin 1942. Beach 1943. Ford and Beach 1951:240–241.

[29] Stier 1938 describes the sexual effects of brain injury in some 33 human males, and reports that deleterious effects were more pronounced in older males. Goldstein and Steinfeld 1942 give a detailed discussion of a single case. Both papers note a marked reduction in response to psychologic stimuli.

[30] This female case was originally reported by Symonds and is cited in other literature, e.g., Beach 1942a:216.

Although the data on the relation of the cortex to sexual behavior are limited, they do show that this is the part of the nervous system through which psychosexual stimuli are mediated. Since we have shown (Chapter 16) that there are considerable differences in the effectiveness of such psychologic stimuli between females and males, we may believe that this, the most striking disparity which exists between the sexuality of the human female and male, must depend on cerebral differences between the sexes. What the nature of such cerebral differences may be, we do not know. There have been one or two studies which report differences in the biochemistry of the cerebral cortex in female and male animals. The studies are important and highly suggestive, but further investigation is needed before we are warranted in making any generalization.[31]

Since there are differences in the capacities of females and males to be conditioned by their sexual experience, we might expect similar differences in the capacities of females and males to be conditioned by other, non-sexual types of experience. On this point, however, we do not yet have information.

Hypothalamus. While we may explain the similarities of female and male responses to tactile stimulation on the basis of the similarities of spinal and autonomic mechanisms, and while we are inclined to believe that differences between female and male responses to psychologic stimuli may depend on cerebral differences, we still have one of the most significant aspects of sexual response to explain. This is the fact that sexual responses constitute a syndrome of elements most of which are found in other emotional responses, although the combination in which they appear during sexual activity is not duplicated in any other type of behavior. It is still difficult to understand why touching an animal at one time should bring responses which we recognize as sexual, while touching it on some other occasion may make the animal angry or afraid.

It is inevitable that one should assume that such a cluster of responses must depend upon some mediating mechanism, a master switchboard which controls all of the individual elements but brings them together as a unit during sexual response.[32] Perhaps it is not reasonable or necessary to believe that there should be such a mediating

---

[31] For studies of chemical differences that appear to distinguish female from male brains, in both human and infra-human species, see: Weil 1943. Weil and Liebert 1943. Weil 1944.

[32] Such "sex centers" in the cerebrum have been postulated by various students, including: Loewenfeld 1908:597–598. Rohleder 1923:4, 19. Von Bechterew acc. Beach 1940:194. A summary is in Stone 1923b. However, recent and thorough research has discovered no such centers, and Penfield and Rasmussen 1950:26 stress that they were never able to elicit erotic sensations by stimulating the exposed human brain.

mechanism, but the possibility is sufficient to warrant continued search for a central control of the sexual syndrome.

The concept of a master switchboard is encouraged by experimental work which shows that damage to the hypothalamus, a small structure in the brain which lies below the cerebrum, may considerably modify the animal's capacity to be aroused in anger, in fear, and in still other emotional responses. This it seems to do through some control which it exerts on the autonomic nervous system. Nevertheless, those who have done the most extensive research on this particular portion of the brain conclude that there is no evidence—either for or against—that the hypothalamus in any way controls sexual responses. The statement was originally made some twenty years ago, but we are advised that it still represents our present state of knowledge or lack of knowledge on this matter.[33]

We have made considerable progress in understanding the anatomy and the gross physiology of sexual response. After three decades of research done by a score of students of human and infra-human sexual behavior, we are able to identify some of the internal mechanisms which account for the similarities between female and male sexual responses, and have located the portion of the brain which seems responsible for the differences which we have found in the capacities of females and males to respond to psychologic stimuli. With this much of the story pieced together, it should now be possible for the observer of gross behavior, the anatomist, the neurophysiologist, and the student experimenting with mammalian sexual behavior to recognize the areas in which we most need additional research.

---

[33] The role of the hypothalamus in mediating autonomic elements and in the expression of emotion is noted by: Fulton 1949:237, 243–245. The sexual role of the hypothalamus is emphasized by some, as in: Brookhart and Dey 1941. Dey, Leninger, and Ranson 1942. Ford and Beach 1951:240. But Clark 1942 and Bard 1940 report inconclusive results and feel that the evidence is still insufficient for final judgment. Bard 1940:574, 576, states "There are not yet available sufficient experimental facts to warrant any general statement about the relation of the hypothalamus to the excitation and execution of estrual behavior. . . . Further work must be done before any precise statement can be made concerning the relation of the hypothalamus to the central management of sexual behavior." This author recently (verbal communic.) considers the above statements still valid.

Chapter 18

HORMONAL FACTORS IN SEXUAL RESPONSE

We have seen that sexual responses depend upon a basic anatomy which is essentially the same in the female and the male (Chapter 14), and involve physiologic processes which, again, are essentially the same in the two sexes (Chapter 15). Throughout the present volume we have found, however, that there are differences in the sexual behavior of females and males, and we have presented data which suggest that some of these may depend upon differences in capacities to be affected by psychosexual stimuli.

Some of the most striking differences between the sexual patterns of the human female and male are not, however, explainable by any of

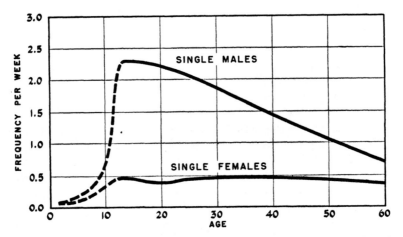

Figure 143. Comparison of aging patterns of total outlet in human female and male

Showing active median frequency of orgasm in total sexual outlet. Data estimated for pre-adolescence are shown by broken line. Data from Table 154 and our 1948:226.

the data which we have yet presented. Throughout the present volume we have emphasized, for instance, the later development of sexual responsiveness in the female and its earlier development in the male. We have pointed out that the male's capacity to be stimulated sexually shows a marked increase with the approach of adolescence, and that the incidences of responding males, and the frequencies of response to the point of orgasm, reach their peak within three or four years after the onset of adolescence (Figure 143). On the other hand, we have pointed out that the maximum incidences of sexually responding females are not approached until some time in the late twenties and in the thirties (Figures 99, 150), although some individuals become fully responsive at an earlier age.

We have pointed out that the frequencies of sexual response in the male begin to decline after the late teens or early twenties, and drop steadily into old age (Figure 143). On the other hand, we have shown that among females the median frequencies of those sexual activities which are not dependent upon the male's initiation of socio-sexual contacts, remain more or less constant from the late teens into the fifties and sixties (Figures 143–145). Nothing that we know about the anatomy or physiology of sexual response, or about the relative significance of psychologic stimuli in females and males, would account for these differences in the development of sexual responsiveness, and for these differences in the aging patterns of the two sexes.

ROLE OF THE HORMONES

In attempting to identify other factors which might affect sexual capacities, it should, again, be emphasized that sexual response is primarily a function of the nervous system. Muscles and blood vessels and other anatomic structures become involved only as a result of the stimulation of the nerves which control those organs. Factors which affect the level of an individual's capacity to respond sexually must be factors which in some way determine the capacities of the nervous system, or some portion of it, to be affected by sexual stimuli.

There is usually considerable variation in an animal's capacity to respond sexually at different periods in its life, and even on different occasions within a short span of time. The newly-born animal's capacity to be sexually aroused may be less than the capacity of the somewhat older animal. Individuals who have reached old age are no longer as capable of responding as they were at an earlier age. The capacity of an animal to respond in a particular sexual situation may be considerably reduced or may totally disappear if the stimulation is continued without interruption for a protracted period of time. Individuals who are physically exhausted, starved, or in ill health are not easily aroused sexually; or if they are aroused, they may not be capable of effective action and may fail to reach orgasm. Such data suggest that anything that modifies the physiologic level at which an animal functions may, through its effect upon the nervous system, modify the general nature of its sexual behavior.

Among the internal factors which may affect the way in which the animal body functions, the best understood are the hormones. These are chemical substances which are produced chiefly in endocrine organs, from which they are ultimately carried by the blood stream to

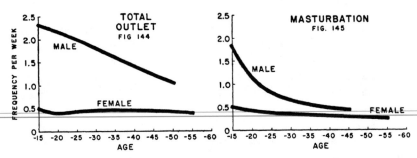

Figures 144–145. Comparison of aging patterns among single females and males

Showing contrasts in active median frequencies of orgasm in female activities which are not primarily dependent on the male. Data from Tables 154 and 23, and our 1948:226, 240.

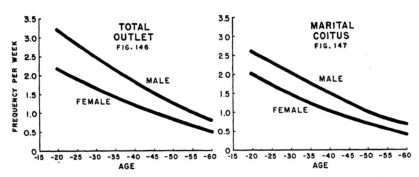

Figures 146–147. Comparison of aging patterns among married females and males

Showing how active median frequencies of orgasm attained in socio-sexual contacts in the female are affected by the patterns of male activity. Data from Tables 154 and 93, and our 1948:226, 241.

every part of the vertebrate body. Because of their accessibility to all parts of the body, hormones may have more effect on bodily functions than any other mechanism except the nervous system.

Discovery of the Hormones. Among all of the hormones that are now known to exist in the vertebrate body, the so-called sex hormones were practically the first to be discovered.

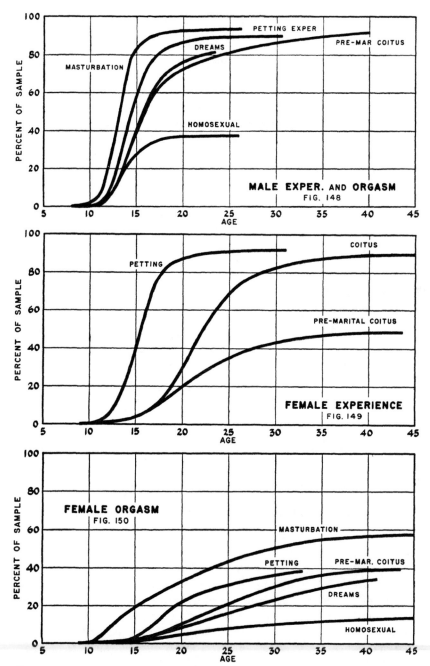

Figures 148–150. Comparisons of Female and Male Experience and Orgasm

Accumulative incidences. For most outlets, male experience and orgasm curves are nearly identical. Female curves showing experience in heterosexual activities are closer to male curves, because the male determines the pattern. Female curves showing orgasm, rise more slowly and do not reach their peak until the mid-twenties or still later. Data from Tables 25, 42, 56, 75, 131, and our 1948:500, 520, 534, 550, 624.

It is probable that most races of men, including even the most primitive, have recognized that the testes served two roles—one in connection with reproduction, and one in connection with the growth and function of the body as a whole.[1] Long before it was known that the ovaries and testes produce specific reproductive cells, the eggs and the sperm, it was known that some substance produced by the male had to be transferred to the female tract before she could reproduce; and the inability of a male to contribute to such reproduction after his

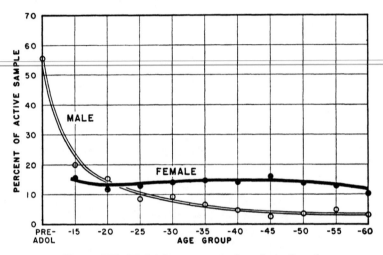

Figure 151. Multiple orgasm in female and male

Active incidences in coitus. For the male the curves show an aging effect, and for the female a plateau extending from the mid-teens into the late fifties. The differences between the two curves more or less parallel the differences between the curves for the total outlet of females and males (Figure 143). Data from Table 176 and our 1948:232.

testes had been removed provided early evidence that they were the source of an essential part of this fertilizing substance.

Although there was, of course, no understanding that a chemical mechanism was involved, both primitive and ancient peoples also recognized that the early physical development of the male animal and its capacity to engage in sexual activity also depended upon its possession of testes. Human male castrates as well as castrates among farm animals were known to the earliest peoples in all parts of the world, and consequently the effects of castration were well understood at an

---

[1] The Trobriand Islanders and certain Australian tribes are said not to be aware of the relationship between coitus and pregnancy, but even they consider that coitus paves the way for subsequent supernatural impregnation. The Trobrianders castrate boars "to improve their condition" and note that such boars cease copulation. This renders suspect Malinowski's statement that the testes are considered to be only ornamental appendages. See: Malinowski 1929:168, 180–181, 190–191. Ashley Montagu 1937:199–202.

Table 177. Multiple Orgasm

Average Number of Orgasms in Each Coital Experience

AGE	TOTAL WITH MULTIPLE ORGASM	NUMBER OF ORGASMS								CASES WITH ANY ORGASM
		1	1-2	2	2-3	3	3-4	4	Over 4	
	%	%			*Percent of females*					
Adol.–15	16	84	2	4	7	1	1	1	0	77
16–20	12	88	4	3	2	1	1	1	—	841
21–25	13	87	4	4	2	1	1	—	1	1770
26–30	14	86	5	3	2	2	1	—	1	1733
31–35	15	85	5	4	2	1	1	1	1	1366
36–40	14	86	5	3	2	2	1	—	1	966
41–45	16	84	6	3	2	1	1	1	2	570
46–50	14	86	5	4	1	1	1	1	1	303
51–55	13	87	6	3	0	1	0	2	1	127
56–60	*10*	*90*	*6*	*4*	*0*	*0*	*0*	*0*	*0*	49

Table based on all females who had had coitus at least 25 times, irrespective of marital status.

Italic figures throughout the series of tables indicate that the calculations are based on less than 50 cases. The dash (—) indicates a percentage smaller than 0.5.

Table 178. Accumulative Incidence: Onset of Menopause

By Educational Level

AGE	TOTAL SAMPLE	EDUCATIONAL LEVEL				TOTAL SAMPLE	EDUCATIONAL LEVEL			
		0–8	9–12	13–16	17+		0–8	9–12	13–16	17+
	%	*Percent*				*Cases*	*Cases*			
40	1	5	1	—	1	930	177	170	252	331
42	4	8	4	2	3	734	146	137	191	260
44	7	15	9	3	5	578	115	110	145	208
46	19	36	25	10	11	461	95	97	109	160
48	31	45	36	21	26	371	82	86	78	125
50	51	72	48	38	48	276	60	71	55	90
52	73	90	63		78	206	50	57		58
54	88					164				
56	97					123				
58	99					102				
60	100					68				

early date.[2] Because the ovaries are within the body cavity, female castrations are more difficult to perform, and were rarely done before

[2] The understanding and practice of castration may have begun as early as 7000 B.C. in the early Neolithic when animals were first domesticated. See: Steinach 1940:3, 25. Turner 1948:324. Many primitives who were at a Neolithic cultural level when first visited by Europeans practiced castration. See, for instance: Malinowski 1929:191 (Trobriand Islanders). Evans-Pritchard 1940:33 (Nuer of Africa). It is interesting to note that in cases of self-castration, as on the islands of Ponapé and Tonga (Westermarck 1922(1):561) and among the Hottentot of Africa (Bryk 1934:123), the males were careful to remove only one of the testes. In addition, there are references to castration in some of the oldest myths of Europe and Egypt. See: Möbius 1903:12. Pritchard 1950:181 (Middle Assyrian laws, 15th century B.C., tablet A, laws 15 and 20, provide castration as a punishment).

the days of modern medicine. Consequently the significance of the ovaries in regulating body functions was not understood at as early a date.

The ancients knew that the effects of castration depended upon the age at which the human or other male was castrated. They knew that when the testes of the human male were removed before the onset of adolescence, the effects were more marked than they were when the castration was performed on a male who was a fully grown adult.[3] The basic biology of these matters did not find an explanation, however, until the middle of the nineteenth century.

In 1835, Graves recognized the relation between thyroid pathologies and the physiologic disturbances which accompany the disease which now bears his name. In 1849, Berthold, studying castrations and testicular implants in fowl, concluded that the testes secreted one or more blood-borne substances which were responsible for the modifications which his experimental castrations had produced. Within the next decade, Addison had noted the deterioration of the adrenal cortex in victims of the disease which bears his name. By the mid-seventies, Gull had identified the role of the thyroid in certain pathologic conditions; by 1887 Minkowski associated acromegaly with pituitary hyperfunction, and two years later (in 1889) von Mering and Minkowski had removed the pancreas and experimentally produced diabetes. The work of Brown-Séquard in 1889 revived interest in the utilization of testicular extracts in clinical practice, and since then there has been a tremendous development of experimental work on the significance of both testicular and ovarian hormones.[4]

The list of endocrine glands which are now recognized in the vertebrate body include the ovaries and testes (or the *gonads,* as the two sets of organs are generically called), the pituitary gland which is located at the base of the brain, the thyroid and parathyroid glands which are located in the throat, and the adrenal glands which are located at the top of the kidneys near the small of the back. The thymus is a gland which reaches its maximum development in the early life of the animal but degenerates considerably after that.[5] The pineal

---

[3] Aristotle (384–322 B.C.), in the Historia Animalium, Bk. IX:631b–632b, has a lengthy discussion of the effects of castration in man and in other male animals, differentiates between pre-adolescent and adult castration, and refers to the removal of the ovaries of sows to lessen their sexual responsiveness.

[4] For brief histories of endocrinology, see: Corner 1942:228–233. C. D. Turner 1948:5–9.

[5] The removal of the thymus and/or injection of thymus extracts are usually without effect; but the development of the gonads at the onset of adolescence is associated with an atrophy of the thymus, and castrates retain the large thymus of childhood. See: Hoffman 1944:275. C. D. Turner 1948:21–22. Selye 1949:679, 683.

gland, in the brain, may be an endocrine organ.[6] The liver and the pancreas, in addition to secreting substances which directly affect digestion, also produce important hormones which influence the development and the maintenance of activities of other organs of the body.[7]

The testes are the chief source of the several hormones known as androgens (the so-called male hormones) in the body of the male. One of the best known androgens is testosterone. The ovaries are the chief source of the so-called female hormones in the female. The most prominent of the female hormones are the estrogens and progesterone.[8] As a group, hormones from the ovaries and testes may be referred to as gonadal hormones.

In recent years, the hormones produced by most of the endocrine glands have been isolated and identified as specific chemical substances. Many of them are closely related compounds of carbon, hydrogen, and oxygen. For instance, the androgens and estrogens, the 17-ketosteroids which are produced by the adrenal glands and by some other structures in the body, and the steroids which are chemicals characteristically found in all animal tissues, are all closely related chemical compounds, even though each may have a different and unique effect on the physiology of the body. Some of the other hormones, such as those produced by the thyroid and the pituitary glands, are totally different in their chemical composition.

Nature of the Hormones. A general knowledge of the hormones has become widespread in the population as a whole, but in regard to certain critical matters this knowledge is quite incorrect. Journalistic accounts of scientific research, over-enthusiastic advertising by some of the drug companies, over-optimistic reports from clinicians who have found a lucrative business in the administration of sex hormones, and some of the discussions among state legislators and public administrators who hope that hormone injections will provide one-package cure-alls for various social ills, have led the public to believe that endocrine organs are *the* glands of personality, and that there is such an exact knowledge of the way in which they control human behavior that properly qualified technicians should, at least in the near future, be able to control any and all aspects of human sexual behavior. It is,

[6] The functions of the pineal gland are poorly known and the data are conflicting, but see: Hoffman 1944:282–284. C. D. Turner 1948:19–21. Selye 1949:595.
[7] It is not fully demonstrated that the liver secretes a hormone, but secretions of the liver inactivate gonadal and probably other hormones. See: C. D. Turner 1948:25. Williams 1950:7.
[8] The estrogenic hormones found in human tissues and fluids are estradiol, estrone, and estriol. Estradiol is thought to be the true hormone and the others to be products derived from it, according to: Smith in Williams 1950:351. Talbot et al. 1952:296.

therefore, important that the general reader understand the nature of the hormones, and understand some of the difficulties that are involved in the accumulation and interpretation of data in the field of endocrinology.

Hormones are products of the physiologic processes that go on in certain of the gland cells that are to be found in the plant or animal body. Any cell which secretes a liquid content which becomes a significant part of the total volume of the cell, or which secretes materials which work their way out of the cell through a permeable cell wall or through some rupture of the wall, may be identified as a gland cell. Many of the cells of the body, and particularly those that line various body cavities, may be considered gland cells even though they are not part of a specific organ which is identifiable as a gland. Consequently, it is not always possible to identify all of the sources of the hormones in an animal's body, even including the androgens and estrogens and the 17-ketosteroids, for part of the hormone may come from cells or groups of cells which lie outside of the specific organs which are known to be the chief sources of these hormones. For instance, the removal of the ovaries or testes (as in a complete castration) may not eliminate all the sources of the sex hormones, and this is one reason that it is difficult to interpret some of the experimental data.[9]

In more complex glands, the secreting cells may pour their products into internal cavities from which ducts may carry them away. This is true, for instance, of the salivary glands. On the other hand, the glandular structures which give rise to the best known of the hormones do not have either internal cavities or ducts. Their secretions are picked up by the blood vessels which enter or surround the glands, and are thus carried away by the blood stream to other parts of the body. The structures are therefore known as ductless glands, or glands of internal secretion, or endocrine (meaning *internally secreting*) organs. There are, however, hormones produced by gland cells in other types of structures, such as the placenta and the duodenal mucosa; and there is some reason for believing that most of the organs in the mammalian body may produce, in actuality, substances which, when circulated through the blood stream, may influence the activities of at least some of the other organs.[10]

The hormones produced by any endocrine gland may affect other endocrine glands as well as organs which are not glandular. For in-

[9] For instance, after removal of the testes, the cortex of the adrenal gland enlarges and secretes an additional amount of androgenic (male) hormone in mice and guinea pigs. See: Hartman and Brownell 1949:331. Selye 1949:130–131.
[10] The placenta secretes estrogen, progesterone, and a gonadotropin. The duodenal mucosa secretes several hormones concerned with digestion. See: C. D. Turner 1948:13–16.

stance, the secretions of the testes and ovaries have a direct effect on the anterior lobe of the pituitary and on the adrenal glands, and each of these has a direct effect on the testes and ovaries. Consequently an increase or decrease in the secretory capacity of any one of these glands may be reflected in the activities of the other glands.[11] Some of the other endocrine organs, such as the thyroid, may similarly affect the secretory capacities of the ovaries and testes, and of the pituitary and adrenal glands.

Although the effectiveness of any hormone is usually proportionate to the amount which is available, there is usually a point of optimum effectiveness, and an increase in the amount of hormone beyond that point may have negative effects which, in certain respects, may be as extreme as those obtained when there is an under-supply or complete removal of the source of supply of the hormone.[12]

Usually the amount of hormone produced in an endocrine organ such as a testis or ovary is very small, and the amount that is to be found in the blood or the urine or at any other point in the body is so exceedingly minute that its recovery and chemical identification may be very difficult. Consequently, most of the reports of female and male hormone levels do not rest upon physical or chemical measurements, but upon such indirect evidence as can be obtained by injecting urine or blood extracts into experimental animals, and upon measurements of the changes which are thus effected in the growth or degeneration of some structure (like a rooster's comb) in the experimental animal. Difficulties in measurement have been the source of considerable error in much of the reported work, including the studies which have attempted to analyze the relationships of the sex hormones and sexual behavior.[13]

Moreover, when the amount of hormone in an animal's body is determined by measuring the hormone in its urine, it is questionable what relation the amount of excreted material may have to the amount of the hormone that the body is actually utilizing. The hormone in the

---

[11] For the interrelationships between the secretions of the anterior lobe of the pituitary, the ovaries, testes, and adrenal cortex, see: Heller and Nelson 1948: 229–243. C. D. Turner 1948:241–244, 462–468. Hartman and Brownell 1949: ch. 26. Selye 1949:22–23. Burrows 1949:329–343. Williams 1950:21–32. Talbot et al. 1952:385–392.

[12] For instance, both a thyroid deficiency and an excess of thyroid retard sexual maturation. See: Lisser 1942:29–32. Hoffman 1944:239–240. Williams 1950: 122, 175. Talbot et al. 1952:21, 31. See footnotes 37, 66.

[13] As an example of the difficulties encountered in analyses: testicular androgen is metabolized and excreted partly in the form of ketosteroids in the urine, and hence the injection of an androgen such as testosterone propionate may be followed by an increase in urinary ketosteroids. But another androgen, methyl testosterone, may have no effect on the levels of the urinary ketosteroids, or even decrease them. See: Selye 1949:628. Howard and Scott in Williams 1950:337.

urine may merely represent that portion which the body has been unable to utilize. Recent endocrinologic research indicates that this latter may be the correct interpretation, especially when an animal is receiving an over-supply of hormones. In an effort to allow for this, hormonal measures are often made on materials recovered from the blood; but it is not clear that this eliminates the difficulty, for it is still not certain how much of the hormone carried in the blood stream may ultimately be utilized by an animal. Consequently reports on hormonal levels in females and males, or in heterosexual and homosexual males, are acceptable only when allowances are made for possible errors in interpreting the measurements, when a sufficient allowance is made for variation in the same individual on different occasions, when the series of reported cases is of some size, and when the report is based on a statistically adequate experimental group which can be compared with an adequate control group.

GONADAL HORMONES AND PHYSICAL CHARACTERS

Development of Physical Characters in Young Mammals. The most certain effects of the gonadal hormones are their effects on the development of physical characters in the young animal. This includes the development in size and function of many parts of the body, including some of the characters (the secondary sexual characters) which most clearly differentiate adult females from males. Many of these characters do not fully differentiate until the onset of adolescence. The development of these secondary sexual characters, as well as the development of the adult anatomy as a whole, depends upon the animal's possession of intact gonads. This is true of the human female and male, and of the females and males of the lower species of mammals.

If the gonads fail to develop normally—as not infrequently happens when testes of the human male, for instance, are retained in the body cavity and fail to descend into the scrotum—or if the gonads of the pre-adolescent human female or male become diseased, or if they are removed by castration any time before the onset of adolescence, the normal development of adult characters is usually slowed up or completely stopped. And when a castrated animal does reach its full size, its body proportions are not typical of those usually found in the normal adult, *i.e.*, the animal becomes a typical capon, a gelding, or a eunuch in form and structure.[14]

------

[14] The failure of pre-adolescent castrates, either surgical or functional, to develop the secondary sexual characters typical of the adult, is described in such standard works as: Laughlin 1922:435. Lipschütz 1924:6 ff. Pratt in Allen et al. 1939:1267–1268, 1282. Hamilton 1941:1904. Greenblatt 1947:182, 254–257. Mazer and Israel 1951:231. Ford and Beach 1951:170.

In the normal human adolescent, female or male, the genitalia are among the first structures to acquire adult size.[15] When the gonads have been damaged before adolescence, the genitalia may develop even more slowly than the rest of the body and may remain more or less infantile even into later years.

In the human female and male, hair normally begins to develop in the armpits during adolescence. In normal development, pubic hair appears in both sexes, although the pubic hair of the female is ordinarily confined to a more limited triangle while it may spread over a more extensive area in the male. Ultimately, but often not until late in the twenties or thirties, the pubic hair of the male may develop upward along a midline (the *linea alba*) on the abdomen. The normal male also develops hair on his face, on his chest, on his legs, and elsewhere on his body, while such hair is usually absent or scant in the female. But if the testes of the pre-adolescent male are damaged or eliminated, the hair in these several parts of the body may fail to appear at the usual age. If it does subsequently develop in the castrated male, it may appear in a pattern which is in many respects more typical of the very young adolescent. The face and chest and other parts of the body of such a male may remain more or less hairless.[16] While the early castration of a female does not have as marked an effect on the development of her body hair, it may contribute to the appearance of facial hair and other hair developments which are not typical of her sex.[17]

The Adam's apple is characteristic of the adult human male. Associated with this, his voice is rougher and usually at a pitch which is lower than that characteristic of the female. The Adam's apple is ordinarily not developed in the female. The male who is castrated before adolescence fails to develop an Adam's apple and retains a high voice.

In the course of adolescent development the shoulders of a normal male widen more than they do in the normal female. In adult females the hips become characteristically larger and wider than they are in the male. The buttocks in an adult female usually become larger and more elongated, while the buttocks of the adult male remain smaller and are more often rounded. The male who is castrated before adolescence retains body proportions which are closer to those of the juvenile.

The breast of the female normally enlarges in size, and the colored, corrugated (areolar) area surrounding the nipple becomes consider-

---

[15] For data on the growth of the human male genitalia before the rest of the body is fully grown, see: Schonfeld and Beebe 1942:771.

[16] Absence of pubic hair and abnormal hair distribution in males with gonadal insufficiencies are well illustrated in Selye 1949:651–660.

[17] A diminution of axillary and pubic hair, and sometimes hirsutism in females with

ably expanded. The female who has diseased or damaged ovaries does not show such a normal breast development, her voice may become lower in pitch, and she may fail to acquire an adult female body form.

In general, castrations performed on young females and young males of the infra-human species of mammals affect their physical development in ways which are comparable to those just noted for human females and males.

Some of the effects of castration may be partially or largely corrected by the administration of hormones from an outside source. This is true for both females and males, and for both the human and other mammalian species. But injections of hormones have their maximum effect if they are made at an early age. They cannot fully correct the damage done by a castration if they are not administered until some time after adolescence has begun, but they may still have some value even when the therapy is not started until the individual is essentially adult. But the corrective effects of hormonal administrations to young castrates can be maintained only if the treatment is continued throughout the growth period. Otherwise the individual may lapse into its castrated state. If hormonal treatments are continued until the individual has become completely adult—which in the human species means into the middle twenties or sometimes later—then the continued administration of hormones is not so necessary.[18]

Maintenance of Physical Characters in Adults. Damage to the gonads, or a complete castration of a human female or male after physical maturity has been acquired, prevents reproduction, but there are usually minimum effects on other physical characters. Some individuals (particularly some females) who have been castrated as adults may go for years and may even reach old age before they show any marked physical changes. Some females and males, on the contrary, may show more physical deterioration in the course of time. Clinicians commonly report characteristic aging effects on the genitalia of females who have had their ovaries removed.[19] It is generally believed that the deteriorations of old age come on sooner, although the specific data are inadequate on this point.

      congenitally absent or undeveloped ovaries are noted in Selye 1949:399–400.
[18] In the abundant literature on the correction of a gonadal insufficiency through the administration of hormones, see, for instance: Kenyon 1938:121–134. Vest and Howard 1938:177–182. McCullagh 1939. Hamilton and Hubert 1940:372. Escamilla and Lisser 1941. Biskind et al. 1941. Kearns 1941. Moore 1942:39. Heller, Nelson, and Roth 1943. Hurxthal 1943. Heller and Maddock 1947:414–418. Beach 1948:38–41, 45. Selye 1949:645–666. Howard and Scott in Williams 1950:328–333. H. H. Turner 1950:32–48.
[19] Menopausal-like involutional changes in the genitalia of castrated adult females, and their control by estrogen therapy, are noted, for example, in: Hoffman 1944:33–35. The lack of regressive changes in some hypogonadal males is discussed in: Heller and Maddock 1947:395.

Castrations may, however, have marked effects on the physiologic well-being of an adult.[20] Since gonadal secretions affect the levels of secretion of the pituitary, adrenal, and thyroid glands, all of which are important in the regulation of the general physiology of an animal, it is inevitable that castrations, even of adults, should have some effect; but this effect is usually minor among human females, and it is not clear that most human males have their physiologic well-being particularly modified by castration if the operation is performed after complete physical maturity has been acquired.

There seem to be more marked effects on the physical characters and on the physiologic well-being of males of lower mammalian species which are castrated as adults.[21]

The effects of castration on an adult animal, such as they are, may be more or less completely corrected by the administration of a sufficient supply of gonadal hormones. This has been demonstrated for laboratory animals, and hormones are often administered in clinical practice to middle-aged and older women who have had their ovaries removed. Usually testosterone is given to a male who is castrated as an adult, and estrogens to a female, but sometimes both hormones are given to individuals of both sexes. It is significant that the corrective administrations of hormones do not need to be kept up indefinitely in an adult, at least in an adult female. In some way the adult human body can adjust in a matter of months to a lack of gonadal hormones, and then it appears to be capable of more or less normal function, even though an important link in the endocrine chain has been eliminated. The capacity of the adult male body to adjust may not be as complete as that of the female. Some adult male castrates appear to adjust to a lack of male hormones for long periods of years; but others show some physical degeneration within a shorter period of years. Until there are further studies of long-time adult male castrates, we are uncertain how to interpret these contradictions in the reported data.[22]

LEVELS OF GONADAL HORMONES AND SEXUAL BEHAVIOR .

Much of the confusion concerning the function of the hormones which originate in the ovaries and the testes is a consequence of the

[20] The general physiologic effects of castration are discussed, for instance, in: Möbius 1903:28 ff. Laughlin 1922:435–436. Lipschütz 1924:12–14. Wolf 1934:257–268, 279. Lange 1934. Greenblatt 1947:254–257. Heller and Maddock 1947:393 ff. Ford and Beach 1951:221–225, 229–232.

[21] In animals, adult male castrates show a genital atrophy, an accumulation of fat, and a decrease in metabolism, sexual drive, and aggressiveness. See: Tandler and Grosz 1913:25–41. Rice and Andrews 1951:115–116.

[22] For a brief resumé of the studies on human male castrates, see pages 740–744 of this chapter.

unwarranted opinion that anything associated with reproduction must, *ipso facto*, be associated with an animal's sexual behavior and, contrariwise, that all sexual behavior is designed to serve a reproductive function. Since the glands which produce eggs and sperm also produce hormones, scientists and philosophers alike have considered it logical to believe that these must be the hormones which control sexual behavior. Reasoning thus, men throughout history have castrated criminals as punishment for sexual activity which their gonads were supposed to have inspired, and with the intention of controlling their further sexual activity.[23] In recent years, courts and state legislatures are again considering gonadal operations as a means of controlling sex offenders.[24] There has been some experimentation with hormone injections in an attempt to achieve that end. Castrations and the administration of sex hormones have been carried out under court order and under the direction of physicians and psychiatrists in various parts of the United States, in Denmark, in Holland, and still elsewhere in mental and penal institutions.[25] On the even more amazing assumption that anatomic defects in the genitalia may explain the social misuse of those organs, some of the medical and psychiatric officers in police courts and in penal and mental institutions routinely examine the genitalia of persons committed on sex charges.

But the fact that hormones are produced in the gonads is, without further evidence, no reason for believing that they are the primary

[23] For example, in the Middle Assyrian laws, which may date back to the 15th century B.C., tablet A, laws 15 and 20, provide castration as a punishment for certain sexual offenses. See: Pritchard 1950:181.

[24] There are statutes in some ten states providing for involuntary asexualization, or sterilization as a eugenic or therapeutic measure for criminal conduct. In several of these states (*e.g.*, Kansas and Oregon) the practice has been to allow castration as one of the permissible operations. Since the United States Supreme Court decision in Skinner v. Oklahoma 1941:316 U.S. 535, all operations have been "voluntary." In order to circumvent constitutional objections, the California Legislature at its Third Extraordinary Session in 1951 passed a bill, A.B.2367, providing for mandatory life imprisonment for persons convicted of certain types of criminal conduct who did not "consent" to being castrated. This bill was vetoed by Governor Warren on July 18, 1951. A milder version of the same bill, S.B.19, failed to pass in the 1952 regular session of the Legislature. We are informed that a similar proposal was introduced in the 1953 session of the Oregon Legislature. To our knowledge castration proposals were summarily rejected by the commissions on sex offenders in New Jersey and Illinois. See also, Michigan, Governor's Study Commission 1951:4, 6.

[25] For such a use of castration or the administration of hormones, see: Hirschfeld 1928:54. Lange 1934:44, 101. Wolf 1934:16–23. Böhme 1935:10–34. Sand and Okkels 1938:374. Kopp 1938:698–704. Hawke 1950. Tappan 1951:242–246 (Denmark). Bowman 1952:70 (cites Judge Turrentine of San Diego court as stating that "behavior disorders have been well controlled by castration in about 70 men," and that these men failed to become involved again with the law after castration. Since sex offenders are always among the lowest in their rate of recidivism, the Judge's criterion is inadequate.). Bowman 1952:79–80 gives a report from the Swedish authorities on 166 legal castrations between 1944 and 1950. Bowman and Engle 1953:10.

agents controlling those capacities of the nervous system on which sexual response depends. It is unfortunate, as we shall see, that these hormones were ever identified as sex hormones, and especially unfortunate that they were identified as male and female sex hormones, for the terminology inevitably prejudices any interpretation of the function of these hormones.

Estrogen Levels at Younger Ages. It should be borne in mind that estrogens (the female hormones), are to be found in the bodies of both females and males. The ovaries of the female are a chief source of her estrogens. The origins of the estrogens in the male are not so well established, but they seem to be produced, at least in part, by the testes.[26]

Estrogens are reported to occur in about equal amounts in the pre-adolescent human female and pre-adolescent human male until they reach the age of ten (Figure 152). But at about the time of adolescence, the estrogens increase abruptly in the female. There is only a slight increase in estrogens in the male at adolescence.[27] In the adult female there is, in consequence, a much higher estrogen level than in the adult male. There is, of course, wide individual variation in this matter.[28]

There is nothing, however, in the development of sexual responsiveness and activity, either in the female or in the male, which parallels these reported levels of estrogens in the human female or male. At the onset of adolescence there is no upsurge of sexual responsiveness and sexual activity in the female which parallels the dramatic rise in the levels of her estrogens (Figure 152). It is the male who suddenly becomes sexually active at adolescence, but his estrogens stay near their pre-adolescent levels.

Androgen Levels at Younger Ages. Androgens are also found in both females and males. The testes of the male are the chief source of his androgens; but the ovaries of the female apparently produce androgens as well as estrogens, and it is probable that the adrenal glands and still other structures in her body also produce androgens.[29]

---

[26] For data on the two estrogens, estradiol and estrone, found in the testes, see: Selye 1949:54, 84. For data on estrogenic substances from the adrenal cortex, see: Hartman and Brownell 1949:105. Selye 1949:80–81. Kepler and Locke in Williams 1950:203. Thorn and Forsham in Williams 1950:261.

[27] For levels of gonadal hormones in pre-adolescence and adolescence, see Nathanson et al. 1941.

[28] For androgen and estrogen levels in normal adult males and females, see: Gallagher et at. 1937:695–703. Koch 1938:228–230. Heller and Maddock 1947: 395–398. Dorfman in Pincus and Thimann 1948:496–508.

[29] The production of androgenic substances by the ovary is noted in: Hoffman 1944:47. C. D. Turner 1948:337. Burrows 1949:123–124. Selye 1949:627. Parkes 1950:108. The adrenal cortex as another source of androgens is noted

In the human species, from about age seven or eight until the middle teens, the androgen levels in the female and the male are about equal (Figure 153). Then the androgen levels begin to rise more markedly in the male, and less so in the female, and it is generally considered that older females have androgen levels that are about two-thirds as high as those of the males.

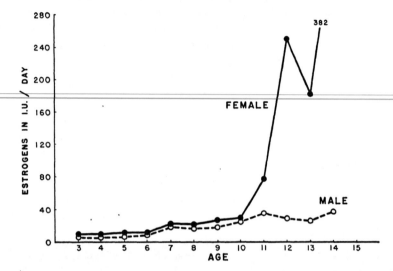

Figure 152. Estrogen levels in pre-adolescent and adolescent female and male

Averages from urinary assays reported by Nathanson et al. 1941.

Since we found a sudden upsurge of sexual responsiveness and overt sexual activity among human males at the beginning of adolescence, there may seem to be some correlation with the androgen picture; but the upsurge of sexual responsiveness in the male is much more abrupt than the steady rise in the levels of his androgens (compare Figures 143 and 153).

As for the female, there seems to be no correlation at all between the levels of her androgens and her slow and gradual development of sexual responsiveness and overt sexual activity (Figures 143 and 153). Although she has nearly as much androgenic hormone as the male in her pre-adolescent and early adolescent years, her levels of sexual response and overt sexual activity at that period are much lower than the levels in the average male. The near identity of the androgen levels in the female and male at the very age at which the two sexes develop

in: Koch 1938:218. Nathanson et al. 1941:862. C. D. Turner 1948:337. Burrows 1949:120. Selye 1949:80, 126–127. Kepler and Locke in Williams 1950:203. Parkes 1950:102. Perloff in Mazer and Israel 1951:124.

such strikingly different patterns of behavior, makes it very doubtful whether there is any simple and direct relationship between androgens and patterns of pre-adolescent and adolescent sexual behavior in either sex.

Levels of Gonadal Hormones in Older Adults. Unfortunately, levels of gonadal hormones seem not to have been established for any ade-

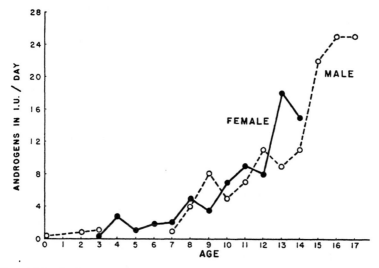

Figure 153. Androgen levels in pre-adolescent and adolescent female and male

Averages from urinary assays reported by Dorfman 1948.

quate series of older human adults, either female or male. There is some reason for assuming that the levels of male hormones drop in the male at advanced ages, and if this were proved to be so it would parallel the drop in the levels of sexual response and overt activity which we have found in the male. On the other hand, there is as much reason for assuming that the levels of male hormone similarly drop in the older female, but such a drop would not correlate with the fact that the frequencies of female response and sexual activity stay on a level from the teens into the fifties or sixties (Figure 143).[30]

GONADAL INSUFFICIENCIES AND SEXUAL BEHAVIOR

Although the importance of the gonadal hormones in respect to the physical growth and development of the young mammal is clearly established by castration experiments, it is more difficult to measure

[30] Dorfman in Pincus and Thimann 1948:502–504 differentiates the androgens from the 17-ketosteroid data and shows a decline in androgens with advancing age in both female and male. Ford and Beach 1951:227 state that it is generally agreed that testicular androgens decrease in later life.

the effects of castration on the capacity of an animal to respond sexually. In any case, it is difficult to know how many of the observed modifications of behavior represent the direct consequences of hormonal action, and how many are a product of the fact that the gonads influence other endocrine organs such as the pituitary and the thyroid which may affect the general metabolic level of all physiologic functions, including the functions of the nervous system. Finally, in the case of the human species, it should be noted that there may be pronounced psychologic effects from a castration. This is particularly true of the male because of the great importance which our culture attaches to his genital integrity and sexual potency. Many of the reported effects of castrations on sexual behavior are undoubtedly the product of the social maladjustments in which castrates often become involved. In those cultures where castration is observed as a religious duty, and in religious cults where the priests are regularly castrated, no social opprobrium is attached to such an operation, and the effects of castration do not seem as apparent as they are in our culture.[31]

Castration and Sexual Response in Young Females and Males. The behavioral effects of gonadal insufficiencies, whether they are the product of undeveloped or diseased ovaries or testes, or the product of complete castrations, are most evident in young animals. In many of the lower mammalian species, early castration more or less completely stops the development of all sexual responsiveness in both the female and male.

In the human male, pre-adolescent gonadal insufficiencies regularly delay the development of sexual responsiveness. The responses of a twenty-year-old male whose testes have degenerated because they have failed to descend into the scrotum, or of a male who has been castrated in pre-adolescence, may be on a level with those of the average pre-adolescent boy of eight or ten years of age. Erections and other signs of response occur less frequently in an adult who was an early castrate, arousal is not effected by as large a number of stimuli as in the normal male, and arousal by psychologic stimuli in particular may occur less frequently than in the normal male. Some degree of sexual responsiveness may develop in later years; but the levels of response in the few cases on which we have original data, and in the well known histories of eunuchs who were castrated at an early age, usually do not reach the levels which are typical of the average male.[32]

---

[31] For the social status of castrates in societies in which castration is socially approved, see: Möbius 1903:14, 16, 84.

[32] The effect of pre-adolescent castrations or gonadal under-development in lowering the sexual responsiveness of the human male is also recorded in: Lipschütz 1924:12–13. Commins and Stone 1932:497–499. Pratt in Allen et al. 1939:1268. Hoffman 1944:623, 625. Beach 1948:23–28. Selye 1949:646, 661. Ford and Beach 1951:231.

There is great need, however, for the accumulation of more data, for there appears to be considerable individual variation in such cases.

Early castrations of males of lower mammalian species similarly may prevent the development of any sexual responsiveness or reduce the levels of response.[33] But in some instances, castrations may have little effect on the development of sexual responsiveness. For instance, there are data on two male chimpanzees who were castrated at a very young age, and these castrations have not prevented the subsequent development of sexual responses comparable to those of normal pre-adolescent or adult chimpanzees.[34]

In lower mammalian females, early castrations have somewhat similar effects on the development of sexual responsiveness.[35] There are, however, practically no data on the effects of such early castrations on the sexual behavior of the human female.

The importance of the gonadal hormones in the development of sexual responses is further confirmed by the fact that when testosterone is administered to an early castrate, whether it is a human or a mammal of some lower species, and whether it is a female or male, the levels of sexual response may be raised to something approaching the normal. The administration of estrogens to a female who was castrated in pre-adolescence has a less marked effect on her behavior, and a still lesser effect (as far as the data are yet available) when administered to males who were castrated at an early age.[36]

Nevertheless, these demonstrations of the importance of the gonadal

---

[33] That an early castration depresses the sex drive of a male animal, although it does not abolish all indications of sexual responsiveness, is noted in: Stone in Allen et al. 1939:1219. Beach 1947a:34–35. Beach 1948:20–23. Rice and Andrews 1951:115. Ford and Beach 1951:229.

[34] At the Yerkes Laboratories at Orange Park, Florida, the male chimpanzee named Don, who is now almost 19 years old, was castrated at approximately 2 years of age (acc. Nissen, verbal communic.). Physically he lies within the norms for adult males. His frequency of erection and sexual interest in females are equal to those of an intact male, but he does not usually reach orgasm and ejaculate. Owing to this inability, he can copulate more frequently than an intact male. Under androgen therapy Don is capable of orgasm and ejaculation, and his behavior and coital frequency then are identical with those of an intact male. The case is also noted in: Clark 1945. Ford and Beach 1951: 231. Beach in Blake and Ramsey 1951:78.

Another male chimpanzee named Dag, now about 8 years old, was castrated at about 2 months of age. Physically Dag falls within the norms, and his frequency of erection and his incidental masturbation are the same as that of intact males of comparable age.

[35] When the females of mammalian species below the primates are castrated at an early age, they never develop sexual receptiveness, according to Beach 1947a: 35.

[36] According to Wilson and Young 1941:781–783, estrogens administered to pre-adolescent guinea pig castrates can produce estrus. Estrogens administered to castrated pre-adolescent male rats show less specific effects; see Ball 1939: 282.

hormones in the development of sexual responsiveness in the young
female and male, do not seem to warrant the conclusion that androgens
and estrogens have more specific effects upon the development and
functioning of the nervous system than they have upon the development and functioning of various other physical structures in the animal
body. Quite to the contrary, the retardation of sexual development in a
castrate is exactly what might be expected if the gonadal hormones
provide, as they appear to provide, simply one of the conditions necessary for the normal growth and development of the body as a whole.
Similar damage done to the pituitary, to the thyroid, or to some of the
other endocrine organs of a developing animal, may have similarly
disastrous effects on the normal course of its development and on the
development of its capacity to respond.[37] All of these endocrine organs,
as well as many of the other organs in the body, seem necessary for
the development of sexual responsiveness in the young animal.

Castration and Sexual Response in Adult Females. There are some
contradictions in the reported effects of the castration of a human adult.
There appear to be differences in the effects on different individuals.
Some of the recorded effects may depend upon the fact that the general level of all physiologic activities may be lowered by a gonadal insufficiency. Again, it is to be noted that the psychologic effects of a
castration on an adult, and especially on an adult male, may be more
severe than the psychologic effects on a pre-adolescent.

The effects of castration on the sexual behavior of a fully mature
female have generally been reported to be minor, or none at all. We
have the histories of 123 females who had had ovaries removed, and
our examination of these cases confirms the general opinion that there
is no modification of sexual responsiveness or capacity for orgasm, following an ovarian operation, which can be clearly identified as the
result of such an operation.[38] Some of the sexually most active females
in our sample were women in their fifties and sixties who were well

---

[37] Delayed sexual development consequent on insulin deficiencies, on thyroid over-
and under-development, on pituitary disturbances, and on adrenal malfunction,
are noted, for instance, in: Hoffman 1944:255, 268. Hartman and Brownell
1949:353. Selye 1949:124, 126–127, 524, 737. Williams 1950:42, 122, 173–
175. See also footnotes 12, 66.

[38] The lack of any consistent effect of castration of an adult female on her sex
drive is also noted in: Hegar 1878:71–72, 74. Canu 1897:ch. 1. Möbius 1903:
87 (sex drive remains after late castration). Glävecke acc. Kisch 1907:187–
188 (drive reported lessened in two-thirds of 27 castrated women). Commins
and Stone 1932:499–501. Havelock Ellis 1936(I,2):11–14. Hoskins 1941:234
(desire often not lessened). Filler and Drezner 1944:123–124 (41 female
castrates, 88 per cent reported no change in sex drive, majority under 35 years
of age). Huffman 1950:915–917 (68 adult female castrations, ages 26–43,
loss of responsiveness reported in 2 cases). Ford and Beach 1951:222–224.
Masters in Cowdry 1952:668 (atrophy of uterus, breasts, and vagina; more
rapid in younger females; usually decreased sex drive, although exceptions are
noted).

past the age of menopause. Some of them had had their ovaries removed ten to fifteen years before.

Of our total sample of 123 castrated (ovariectomized) females (Table 179), twenty-three had not experienced orgasm for a year or two before the operation, and did not experience orgasm after the operation. Of the remaining one hundred cases, forty-one appraised their own record as follows: 54 per cent had not recognized that their loss of ovaries had had any effect on their sexual responses or overt behavior. Some 19 per cent believed that their sexual responses had been increased by their operations, and 27 per cent believed that their sexual responses had been decreased. The record of specific activities on the full hundred cases showed that 42 per cent had not changed in their overt behavior, 15 per cent had increased their activity, and 43 per cent had decreased their activity. However, the median frequencies of orgasm calculated for the whole sample both before and after operation, indicate that over a ten-year period the drop had paralleled the drop in frequencies of total outlet in approximately that same age period, among the females in our total sample. It is to be recalled that the declining frequencies of socio-sexual activities among females are not primarily dependent on an aging process in the female, but upon an aging process in the male which reduces his interest in having frequent coitus (see pp. 353–354). Our cases, therefore, do not provide evidence that females deprived of their normal supplies of gonadal hormones have their levels of sexual responsiveness or their frequencies of overt activity lowered by ovarian operations.

The increases in sexual activity shown in some of our histories may have depended on the fact that some women who have gone through a natural or induced menopause feel more free to engage in sexual activity as soon as they are relieved of the possibility of becoming pregnant.

We have detailed data on 173 cases of females who had gone through natural menopause (Table 179). It is ordinarily considered that there is a considerable reduction in the amount of estrogen secreted by the ovaries after menopause. However, in our sample it would be difficult to identify any reduction of sexual response or activities which could be considered the consequence of any change at menopause. Out of the 173 cases, forty-six had not experienced orgasm for a year or two before menopause, and there was no change in their status following menopause. In the other 127 cases, thirty-one appraised their own record as follows: 39 per cent believed that their sexual responses and activities had not been affected by the menopause, 13 per cent believed that their responses had increased, and 48 per cent believed that their responses had decreased. The detailed record of the activities of the full 127 women, confirmed this distribution of cases (Table 179). Again, however, the decrease in median frequencies in this sample had merely paralleled the decrease in median frequencies in our total sample of females of approximately the same age. Note again that this decrease is primarily dependent upon the male's declining interest in socio-sexual activities.

Table 179. Effect of Castration and of Menopause on Sexual Response and Outlet

Effect	Castr.	Menop.	Castr.	Menop.
	Percent		*Cases*	
Subject's evaluation of effect				
No effect	*54*	*39*	22	12
Increase in response	*19*	*13*	8	4
Decrease in response	*27*	*48*	11	15
Number of cases	41	31	41	31
	Percent		*Cases*	
Effect on total outlet				
No effect	42	42	42	53
Increased frequency	15	5	15	7
Decreased frequency	43	53	43	67
Number of cases	100	127	100	127
	Freq. per week		*Cases*	
Median freq. of total outlet				
1–2 years before	1.0	1.5	100	127
During menopause		1.0		81
1–2 years after	0.7	0.7	100	127
3–4 years after	0.7	0.6	76	90
10 years after	*0.8*	*0.4*	28	38
20 years after		*0.4*		12

Table based on all females available in sample, whose marital status remained constant before and after castration or menopause and who had experienced orgasm within 1 to 2 years before and/or after castration or menopause.

The median age at castration was 38.6 years, the range was 17 to 53 years. The median age at the onset of menopause was 46.3 years, the range was 33 to 56 years.

Some of the decreased frequencies also depended upon the fact that some of these women had seized upon menopause or their ovarian operations as an excuse for discontinuing sexual relationships in which they were never particularly interested. Some of the cases of increased activity were, again, a product of the fact that some of these women had been relieved of their fear of pregnancy after going through menopause.

Castration of Adult Females of Lower Mammalian Species. This lack of effect of castration on an adult human female is not in accord with the reported effects of castrations on adult females of rats, guinea pigs, and some other species of mammals. It is reported that castrations in those species eliminate all evidence of sexual response.[39] Examination of the literature, however, indicates that the chief bases for these reports, persistent as they are, is the fact that the castration of a lower mammalian female puts an end to her periods of estrus—the period during which she will accept coitus from the male. It has always been assumed that she accepts coitus during estrus because she becomes sexually more responsive at that time. But we question whether the

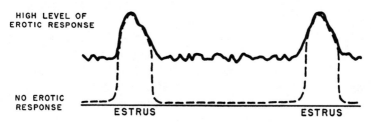

Figure 154. Relation of erotic response to estrus in infra-human females

Theoretic diagram. Lighter line shows previous interpretation. darker line shows present interpretation. In both instances, response is at a maximum during estrus; the present data indicate that there is also a considerable capacity for response in periods between estrus.

submission to a male is, in itself, sufficient evidence of erotic arousal. Moreover, we find considerable evidence that lower mammalian females who are not in estrus are frequently aroused erotically.

We have noted that a bull ordinarily gives evidence of his sexual arousal before he ever mounts a cow by showing a development of neuromuscular tensions throughout his body; the muscles on the sides of his abdomen become tensed in a corrugated design, his tail is arched as a result of tensions in that part of his body, he may show a partial erection, and his Cowper's secretions may start flowing before he has touched the cow. Usually a cow which is in estrus does not show similar evidences of sexual arousal until after she is mounted. Then the tensed muscles on the sides of her abdomen, her arched tail, her tumescent genital labia, and her vaginal mucous secretions provide

[39] In the earlier work, distinctions were made between the receptivity of estrus and sexual responsiveness, as in: Young and Rundlett 1939:449. But more recent scientific publications have been inclined to consider sexual responsiveness and estrus synonymous, just as most farmers and animal breeders do. See: Williams 1943:125. Ford and Beach 1951:221. Beach in Blake and Ramsey 1951: 75–76 ("Removal of the ovaries in lower mammals is followed by total and permanent loss of female sexual responses . . . the animal's tendency to become sexually aroused and to execute coital reactions is heavily dependent upon hormones from the ovaries." And again, "Removal of the ovaries promptly and permanently abolishes all sexual behavior"). See also Chapter 15, footnote 18.

evidence of her arousal and response. We have also noted (page 450) that cows quite regularly and frequently mount other cows, and that when this occurs, the cow that does the mounting is the one that first shows evidence of erotic arousal, although the cow which is mounted may not give such evidence until after she has been mounted. But the cows that do the mounting are usually not in estrus, while the cows that are mounted are almost always in estrus.

There is, of course, no question that an animal that is in estrus is capable of being aroused erotically, and it is common knowledge that some animals, like female dogs, may become more responsive and actively search for males when they are in estrus; but the data indicate that sexual arousal among infra-human females that are not in estrus may also occur with some frequency (Figure 154). There are also records of castrated adult female rats and dogs that will sometimes mount other females after castration, just as they did before castration; and this seems clear-cut evidence of sexual responsiveness after castration.[40]

Since reports on the effects of castrations in lower mammalian species usually do not describe those physiologic phenomena which are the best measures of sexual response (Chapter 15), and since there is so much evidence of sexual arousal outside of estrus, we doubt whether the presence or absence of estrus provides a sufficient measure of a lower mammalian female's capacity to respond sexually.

It has been said that the gonadal hormones are more important in controlling the sexual responses of the lower mammalian female, while cerebral controls are more important in the human female.[41] But this, again, seems to be based primarily on the fact that estrus stops after the castration of the lower mammalian female, while responsiveness after castration is retained in the adult human female. If reexamination of the experimental data or further experimental work shows that the lower mammalian female does not actually lose sexual responsiveness as a result of castration, it would mean that the role of the gonadal hormones in the lower mammalian female is, to this extent, about the same as in the human female.

Castration and Sexual Response in Adult Males. Reports on the sexual behavior of human males who have been castrated as fully grown adults have usually been very brief, and the various reports are contradictory in spite of the fact that such operations have been performed with some frequency throughout history. Castrations of adult

---

[40] That castrated female rats and dogs will sometimes mount other females is reported by: Beach and Rasquin 1942. Ford and Beach 1951:142.
[41] This "phylogenetic interpretation" of hormonal control is suggested in: Beach 1947b:293–294. Beach 1948:9–10. Beach 1950:261–269.

males were performed with religious objectives in various ancient groups, and have been performed for that purpose within the present century in certain groups.[31] In various cultures, castrations of adults were performed for the sake of obtaining eunuchs who could be used as household servants, or servants in harems, without the danger of their fathering offspring.[42] Castrations have been most frequently performed as indignities which were inflicted upon enemies captured or killed in battle, and as punishments for certain classes of criminals. Ancient Egyptian drawings, and the derived art of Northern Africa today, depict mounds of severed genitalia gathered from enemies destroyed in battle [43]; and during warfare in probably every part of the world, such mutilation has been considered the supreme subjugation which the conqueror could bestow upon the conquered. There is no doubt that the recurring interest in castration as a legal punishment today is, at least in part, a product of the same sadistic eroticism which has inspired genital mutilation throughout human history.

In more recent decades, both in Europe and in this country, castrations have been rationalized as attempts to modify some aspect of the individual's sexual behavior: to stop masturbation,[44] to transform homosexual into heterosexual patterns of behavior, to control exhibitionists and, in particular, to control adults who sexually "molest" children. Castrations have been used both in Europe and in this country to prevent feeble-minded, criminal, or irresponsible individuals from becoming parents; but simple sterilizations would satisfy that end if there were no other objective in a castration.[25]

Castrations have, of course, been necessarily performed when testes were diseased; and recently castrations of older males have become fairly frequent as a means of reducing their androgen levels, because these may influence cancerous growths or other hypertrophies of the prostate gland. In addition there are a fair number of males who have had to be castrated as a result of war injuries. There has, in consequence, been no shortage of cases for studying the effects of castration

---

[42] For historical accounts of castrations, see such references as: Möbius 1903:12–25. Tandler and Grosz 1913:45–46. Hirschfeld 1948:65–66.

[43] For a description of such mutilations in battle see: Möbius 1903:12–13. We have seen the ancient Egyptian and modern Ethiopian drawings.

[44] For an example of the use of castration to cure masturbation see: Flood [?1901] who reports the castration of 24 males, half of them under 14 years of age, for persistent masturbation and epilepsy, apparently in the Hospital for Epileptics at Palmer, Mass. (See Laughlin 1922:433). Flood concludes: "persistent masturbators . . . unpleasant for a refined woman to see . . . it seemed an absolute necessity to try something which we had not yet tried." See also the account of Dr. Pilcher's castrations of "confirmed masturbators" at the Kansas State Training School at Winfield, Kansas, in: Flood [?1901]:16. Cave 1911:123. Hawke 1950:1 ("Our castrations first started during the administration of Dr. Pilcher who conceived the idea that castration might help control excessive masturbation and pervert sexual acts"). Bowman 1952:69–70.

on adult males, but there have been very few detailed reports on the sexual behavior of such castrates.

A considerable proportion of the studies have reported, without specific data, that there was an improvement in health as a result of a castration, or an increase or decrease in sex drive, or a generally beneficial effect. Unfortunately some of the reports on which state legislators have recently relied have been in these same general terms, without specific data on the frequencies of response, the intensities of response, the number of items to which the castrated individual responded, the frequencies of erection, the frequencies of masturbatory and overt socio-sexual activities, or the frequencies of total activity to the point of orgasm.

Many of the reports have concerned castrations of older males past the age of fifty, and in many cases between sixty and eighty years of age. Males of such advanced ages normally have their rates so reduced that it would be difficult to determine how much of their inactivity should be credited to a castration. Even at fifty years of age there are 7 per cent of the males who are already impotent and unresponsive sexually, whether they are castrated or not.

The studies that do report frequencies of activity after castration, fail to allow for the fact that most males have their frequencies of sexual activity steadily reduced with advancing age. It means nothing to find that castrated males gradually, over some period of years, show diminished sexual interests and capacities, unless it is shown that the diminution of their activity occurs more rapidly than that which occurs in the population as a whole. It is to be recalled that in our total male sample we found (1948:226) that the average twenty-year-old, married male experienced orgasm with a median frequency of 3.2 per week, but that these frequencies dropped steadily through the years until they had reached 0.8 per week at sixty years of age among males who were not castrated.

Because of the general misunderstanding of the reliability of the evidence on the effects of adult castrations, and because some courts and state legislators have uncritically accepted the published records as justification for their consideration of castration as a means of controlling certain types of sex offenders, it seems appropriate to summarize briefly the data in the published studies. They are here arranged in chronologic order.

STUDIES ON CASTRATION OF ADULT HUMAN MALES

1. **Barr 1920.** A study of 6 male castrates, most of them with records of low intelligence. Ages from eleven to twenty. Results cannot be

evaluated because they are reported in nothing but general terms. They note "an improvement in general behavior" in most instances.

2. Commins and Stone 1932. An extensive review of the literature on the effects of castration on basal metabolism, on the nervous system, on reflex action, on voluntary activity, on sexual behavior in general, and on learning. Deals primarily with lower mammals.

3. Lange 1934. The most extensive and most specific study of the effects of male castration, based on 310 cases which included 242 complete castrates and 68 partial castrates. The data are drawn from a long-time study of 247 cases originating in war injuries, and 63 cases in which there had been a surgical removal of testes following tuberculosis. The following summary applies only to those cases in which there had been complete castrations.

At the time of castration, 10 per cent of the men were under twenty, 50 per cent under twenty-five, and 15 per cent over thirty-five years of age. All but 6 of the cases were observed for fifteen years or longer following castration. The physical changes, which were reported in detail, included regressive effects which appeared chiefly in the earliest and the latest years; but wide individual differences were noted, probably because of the wide age range of the subjects. Data on potency following castration are given on 99 complete castrates: 52 per cent lost potency immediately, 22 per cent lost potency gradually, and 26 per cent still retained their potency at the end of the period of observation. This loss was related to their ages at the time of castration (and hence to their ages at the final report). Some 73 per cent of those retaining potency were under the age of twenty-five at the time of castration. The author points out that the reports of defects due to injury or operation tended to be exaggerated in order to support claims for government compensation.

The sexual desire of many of the subjects had exceeded their potency. The author discusses the fallacy of castration as a cure for sex criminals, since such criminal violence is often the result of the conflict of weakened potency and strong sexual impulses (pp. 44, 101), and he questions the favorable results reported in Switzerland and Denmark on the basis of the selectivity of the groups and the short period of observation of those cases.

4. McCullagh and Renshaw 1934. A study of 12 subjects, 4 castrated between twenty-three and forty, 6 between forty and sixty, and 2 after sixty; observed from six months to twenty-seven years after castration. Partial responses remained in 3 cases, but were entirely lost in 9. There was a shrinking of the penis in 5 cases, a reduction of body and pubic hair in 11 cases, a decrease of energy and endurance in 10 cases, and changes in weight and general appearance. The specific data are, however, still insufficient for final analyses.

5. Wolf 1934. A German study summarizing 162 castrations of human males, many on the basis of fragmentary records from the older literature. Many of the subjects were feeble-minded or mentally deficient. There were 50 cases from the author's own data. The information on the sexual responses of the subjects after operation was quite incomplete. For 72 cases, the effects on responsiveness were minor in 35 per cent, the responses were much reduced in 28 per cent, the responses were completely gone in 37 per cent. However, the data are uninterpretable because there are no correlations with the ages at castration and no exact data on the subsequent frequencies of sexual activity.

6. Kopp 1938. A survey of the status of castration *and sterilization* of criminals in the United States and Europe; but the material is in very general terms, and chiefly historical in interest.

7. Feinier and Rothman 1939. A single male, castrated at twenty-three for tuberculosis of the testes, reported a normal married life and potency which had increased after castration and after recovery from the tuberculosis. At fifty-three (thirty years after the castration) he was having weekly coitus with his wife. The wife corroborated the story. The authors conclude that the only indispensable function of the testes in a fully grown adult is that of procreation, and that sexual responses and potency "are functions of and controlled by the pre-pituitary and psychic centers."

8. Tauber 1940. A good review of the literature, without original data. Reports studies on religious castrations; on castrations of criminals in Germany and Switzerland (where the author feels the data are not reliable); on castrations due to injuries; and on medical castrations. Feels the psychologic aspect of a castration is very important. Concludes that the sexual behavior of male castrates is highly variable, and that the range includes behavior which would be considered normal in non-castrates.

9. Huggins, Stevens, and Hodges 1941. A report on 21 patients castrated to control cancer of the prostate; 3 were between fifty-four and sixty years of age; the others ranged between sixty and eighty-four. Reports sexual drive and potency absent in all cases after operation; but all of the patients were so old that it is impossible to make any critical analysis of the reports on sexual behavior.

10. Engle in Cowdry 1942:489–491. Cites Rössle's study on 125 men castrated under German law for criminal sex offenses, in half of whom sexual responses were "weakened." Ages not given and no records of the specific frequencies of sexual activity. Cites a British report by Hammond of 7 males castrated at ages ranging from thirty to fifty-one, in whom coitus continued after castration, in one case for seventeen years. Concludes that "The evidence suggests then that in men in whom the psychic and neuromotor behavior patterns of sexual activity have been established, complete loss of the testes does not necessarily prevent participation in sexual activity."

11. **Hamilton 1943.** One male castrated because of cancer at twenty-five years demonstrated normal erections at age forty-three. One male castrated at twenty-six reported marital coitus two to three times a week, at age thirty-nine. Wife confirmed report. Androgen and estrogen levels low, so writer concludes that capacity for erection does not depend upon a supply of androgens from extra-gonadal sources.

12. **Stürup 1946a.** A psychiatrist's report on 123 males voluntarily castrated in the Danish asylum for psychopathic criminals at Herstedvester. Only general statements on sexual behavior after castration. "Some degree of sexuality is retained in certain cases, at any rate for a number of years, and some cases have been able to achieve a coitus satisfactory to their wives at intervals of about a month. To the majority of cases, however, this does not apply." Insufficient data on ages and on sexual responses and frequencies of activity before and after castration.

13. **Stürup 1946b.** A discussion of the psychiatric treatment of criminal psychopaths in Denmark. Only a general statement of "good results," without specific data which would allow critical analyses of the results. States that "The detainee must show hyper-sexuality beyond doubt or a stable sexually conditioned criminality, before we use this irreversible treatment."

14. **Beach 1948:23–27.** Surveys the studies on male castration. Concludes: "Despite the frequency and possibly the accuracy of generalizations regarding the depressing consequences of human castration, the literature is replete with references to complete retention of sexual function in individuals who have been castrated for many years." Then adds: "The frequency of accounts describing the survival of normal sexuality following castration need not obscure the fact that in many, if not the majority of cases, the human male exhibits a gradual loss of mating ability as a result of testicular removal." This last statement, however, ignores the fact that non-castrated males similarly show a gradual loss of mating ability with advancing age.

15. **Fuller 1950.** No original data. Points out that the medical profession is not agreed on the use of castration as a means of reducing

sexual responses. States that there is no assurance that "in man the sexual urge may not persist for years after the castration."

16. Hawke 1950. In a paper delivered before the Illinois Academy of Criminology, Hawke discusses the program of castration employed, under his direction, at the State Training School at Winfield, Kansas, where 330 male castrates furnished material for a nine-year research program. These cases were also the source of physiologic data on castrates in Hamilton 1948:257–322, who gives ages at castration in 57 of these cases. These ranged from eight to twenty-two years, and included 11 boys who were 12 years of age or younger. The cases were largely drawn from a defective delinquent group. The sweeping generalizations as to psychologic improvement, lack of inferiority feelings, increased stability, and lessening of the "social menace" in these cases are not substantiated in the paper by any sufficient data. Only three case histories of individuals, castrated at sixteen, eighteen, and twenty-four, are described in detail, and the only evidence given as to the satisfactory result of these castrations was the fact that they later adjusted to life outside the institution. Hamilton 1948:286–288 presents detailed data, however, showing that a group of these same subjects had difficulty in carrying out their motor activities. Hawke, nonetheless, states that the castrate is "physically a better organism." There do not seem to be any data on the sexual behavior of these subjects which allow reliable analyses of the sexual effects of the castrations.

On the basis of the more reliable of these published studies, and on the basis of the few cases we ourselves have seen, we may generalize as follows: Human males who are castrated as adults are, in many but not in all cases, still capable of being aroused by tactile or psychologic stimuli. They may still be capable of showing essentially all of the physiologic concomitants of sexual response, including the tumescence of all parts of the body and the specifically genital tumescence which may effect normal erection. They may still be capable of developing neuromuscular reactions which include rhythmic pelvic thrusts of the sort necessary for coitus, and they may still be capable of attaining orgasm. The frequency and intensity of response may or may not be reduced by the castration. The psychologic effects of such an operation may make it difficult for some of the males to make socio-sexual adjustments.

Ejaculation may or may not follow orgasm in a castrated human male. In the lower mammals, removal of the testes may cause degeneration of the prostate and seminal vesicles, which are the chief sources of the ejaculate, but there are some recorded instances of a reduced and modified ejaculate in some castrated human males. In most cases there is no ejaculation. The individual variation probably depends on the length of time which has elapsed since the castration and the stage of degeneration of the secreting glands. None of this, however, makes it impossible for a castrated male to have orgasm. We have the history of a male engaging in sexual activity with normal frequencies and with orgasm fifteen years after the castration, and it will be noted that in the literature cited above there are instances of the retention of sexual capacity for similarly long periods, including one case of a male who was normally active thirty years after castration.

Castrations of adult males of lower mammalian species produce, as we have already noted, a more general physical deterioration than is recorded for adult human male castrates. In general, male animals castrated as adults show diminished sexual responses, but the data are insufficient to allow critical analyses.[45] We have already noted that in chimpanzees there are records of adult male castrates who were as active sexually as males who had not been castrated.

In any event, the laboratory experiments on animals, and the data which are at present available on human male castrates, do not justify the opinion that the public may be protected from socially dangerous types of sex offenders by castration laws.

INCREASED SUPPLIES OF GONADAL HORMONES

The administration of an extra supply of male hormone to an animal, female or male, which has intact gonads, may increase its sexual responsiveness. This may appear to contradict the data on the effects of a castration, but the two bodies of data are not actually in conflict.

Excessive Gonadal Hormones in Young Animals. The administration of androgenic hormones to a young, non-castrated male, whether infrahuman or human, ordinarily speeds up its physical development and the development of its sexual responsiveness and overt sexual activity.[46] There are similar results when estrogens are given to young females.[47] There is considerable work on laboratory animals which establishes this fact. In the case of the human male, the clinical administration of testosterone to a pre-adolescent is ordinarily avoided because of the

---

[45] Data on the reduced sexual responses of castrated adult males of lower mammalian species may be found, for example, in: Stone 1927:369 (rats). Commins and Stone 1932:497 (rabbits and rats). Stone in Allen et al. 1939:1246 (rats). Beach 1944c:255 (rats). Beach 1948:21 (summarizes earlier studies). Rice and Andrews 1951:115–116 (farm animals).

[46] Experimental and clinical data on the administration of male hormones to young, non-castrated males are cited in: Beach 1942d. Hoffman 1944:157–159. Heller and Maddock 1947:417. Beach 1948:206–207. Burrows 1949:169. Howard and Scott in Williams 1950:339.

[47] Experimental data on the administration of female hormones to young, lower mammalian females, are cited in: Beach 1948:196, 203–204. Corner 1951: 57–58.

probability that it will start precocious development. But when adolescent development seems to be delayed, and particularly when there seems to be an under-development of the gonads, some physicians do administer testosterone. There may be complications, however, if more than the optimum dose is given, for an excessive supply of gonadal hormones may inhibit the secretory activity of the anterior lobe of the pituitary.

Precocious adolescent development sometimes, although rarely, may occur in children at five or six, or even at two or three years of age. In some of these cases there may be an endocrine imbalance which sometimes involves an androgen disturbance; but there are other cases in which clinical studies fail to show any sort of endocrine disturbance. Such children may show physical developments equal to those of a normal thirteen- or fourteen-year-old. However, the sexual responses and overt activities of such precocious children are ordinarily typical of those among normal pre-adolescent children of the same age. Investigators are inclined to emphasize that such a child masturbates, shows sexual curiosity, or engages in some form of socio-sexual play, and they are likely to conclude that these activities are a product of the precociousness. But it should not be forgotten that such activities are ordinarily found in the histories of normal pre-adolescent children (Chapter 4).

In the several cases which we have of precocious adolescent development, we have rarely found sexual activities which exceeded those ordinarily found among normal children of the same age.[48] We have the history of a five-year-old with the physical development of a fourteen-year-old, but with sexual responses which were normal or even lower than normal for a five-year-old. We have one group of four related cases in which hereditary factors seem to have been involved, for precociousness had appeared among the males of at least two generations in separated branches of the family. We have the history of another boy who turned adolescent at seven, and he was highly responsive and exceedingly active in socio-sexual contacts. Since cases of very early adolescent development are relatively rare, it is highly important that more extensive data be accumulated on the responses and overt sexual activity of such children.

The administration of androgens to a pre-adolescent human female may do considerable damage because of their over-stimulation of

---

[48] Summaries of published cases of precocious adolescent development in which there are data on sexual behavior, may be found in: Doe-Kulmann and Stone 1927:319 (adult sexuality not present in most precocious cases). Singer 1940: 19–22 (summary of published reports on 59 females, 48 males). Stotijn 1946:56 (a Dutch study). Dennis in Carmichael 1946:656–658. Sandblom 1948:110 (one case of a precocious three-year-old male, aggressive sexually).

physical development.[49] There seem to be no data on the effects of such an early administration of androgens on the sexual responsiveness and the overt sexual activity of the human female.

Excessive Androgens in Adults. When an extra supply of androgens is given an adult animal that has not been castrated, there may be an increase in the general level of its physical activity, its aggressiveness, and its frequency of sexual response and overt sexual performance.[50] This is true of laboratory and farm animals, and it is equally true of the human male. It is also true when androgens are given females, whether they are lower mammalian or human females.

When testosterone, for instance, is given the normal human male, there may be an increase in the frequency of his morning erections, the frequency of his erotic response to various stimuli, the frequency of his masturbation, and the frequency of his socio-sexual contacts.[51] This is ordinarily true of adult males of ages ranging at least from the twenties into the fifties or sixties. Testosterone has also been used clinically to increase the levels of sexual response in cases in which a failure to have offspring appears to depend on low rates of coitus. We have several histories of males who had had their coital frequencies increased by such clinical treatment. Sperm counts may also be increased by the administration of testosterone, and this may contribute to the relief of the sterility.[52] The indiscriminate use of testosterone, however, may involve some danger, for if the dose exceeds the amount necessary for optimum effectiveness, pituitary functions may be inhibited, and this may do damage to various structures, including the

---

[49] Discussion of dosages of androgens affecting precocious physical development in young females is found in: Selye 1949:614. Mazer and Israel 1951:126. Talbot et al. 1952:381, 440–441, 484.

[50] Effects of androgens on non-castrate adult male animals is described in: Beach 1942b:181–182, 193–194. Beach 1942c:227–247. Beach 1948:34, 36. Cheng et al. 1950:452 (increased sex activity in male rabbit).

[51] While there is a general use of androgen therapy on intact, normal males in clinical practice, there are few published accounts; but see: Miller, Hubert and Hamilton 1938:538–540. Kenyon et al. 1940:35. Barahal 1940:319 (increased drive in homosexual males). Heller and Maddock 1947:419 (says "normal" men fail to have an increase in sex drive, but on p. 422 agrees that such treatment seems to increase "the power of the sex drive in both normal and homosexual males . . ."). Burrows 1949:169 (excessive masculine urges in "normal" male adults). Perloff 1949:133 ff. H. H. Turner 1950: 38–43 (a survey; indicates contradictions in data).

[52] Surveys and studies on the effect of testosterone on sperm count include: Weisman 1941:240–242 (concludes small doses stimulate sperm production, large doses inhibit it). Heller and Maddock 1947:419 (temporary reduction of sperm). C. D. Turner 1948:344 (considers evidence for increase in sperm inconclusive; lack of sperm if dosage is excessive). H. H. Turner 1950:24–25, 53–54 (no change in sperm production, although recognizes that other reports are contradictory). Cheng et al. 1950:447–452 (increases volume of ejaculate in rabbit, but total number of sperm unchanged). Heckel and McDonald 1952:725–733 (reduces sperm production in 72 men; a return to a higher level after treatment stopped in twenty-three out of forty-five men).

gonads themselves, and there may be negative effects on sexual activities.[53] In laboratory animals, excessive doses of testosterone may reduce the gonads to more or less vestigial structures.

There is some clinical experience in administering testosterone to normal human females, and the results obtained are quite similar to those obtained in males.[54] Once again, the levels of physical activity may be increased, and the general level of aggressiveness may be increased. In the case of the lower mammals, this increased aggressiveness increases the frequency with which the female mounts other animals, either females or males. Because males are normally more aggressive than females, normal females usually find few opportunities to mount males; but females who have been given testosterone may become so aggressive that they succeed in mounting a larger number of males.[55]

The increased responsiveness of a normal female or male who has received an increased supply of testosterone has ordinarily been taken as evidence that the hormone plays a prime part in controlling sexual behavior. Such an interpretation, however, ignores the evidence of the castration experiments on adult females and males. It seems more correct to conclude that androgens, at every level which does not exceed the point of optimum effectiveness, are among the physiologic agents which step up the general level of metabolic activity in an animal's body, including the level of its nervous function and therefore of its sexual activity. For instance, as an example of the effect of testosterone on other physiologic activities, it may be noted that dairy breeders sometimes administer it to cattle in order to increase their food utilization. Thereby the breeder may increase the amount of meat which he secures when he gives the animal a given quantity of food.

There are so many other factors which affect the levels of metabolic activity in a fully mature animal that the loss of the usual supply of androgens, as in a castration, does not make it impossible for an animal to hold its metabolism at something approaching a normal level. This, however, does not preclude the possibility that supplies of male hormones in excess of those normally provided by the gonads, may raise

---

[53] The harmful effects of various doses of testosterone are pointed out in: Moore 1942:40–41. Heller and Maddock 1947:415–416. Ludwig 1950:453, 465 (small doses more injurious than large doses in male rats).

[54] For clinical use of androgens to increase sex drive in human females, see: Geist 1941. Salmon 1941 (survey of animal and clinical experiments). Salmon and Geist 1943. Greenblatt 1947:177–178. Hyman 1946(3):2406. Carter, Cohen and Shorr 1947:361–364 (a comprehensive survey). Beach 1948:62. Selye 1949:619.

[55] Increased aggressiveness among females of lower mammalian species after treatment with testosterone dosage are reported in: Stone 1939. Ball 1940:151–165. Young 1941:135. Birch and Clark 1946 (chimpanzee).

the metabolic levels and consequently the levels of sexual response and performance. But male hormones are not the only agents that step up the levels of activity, including the levels of sexual activity, for the administration of pituitary extracts, of thyroid, and of some other substances may have similar effects. In fact, good health, sufficient exercise, and plenty of sleep still remain the most effective of the aphrodisiacs known to man.

Excessive Estrogens in Adults. What effect the administration of an extra supply of estrogens may have on the sexual behavior of the human or lower mammalian female or male, is a matter which may not be asserted with assurance in our present state of knowledge. Some clinicians assert that they have raised the levels of sexual responsiveness in female patients by administering estrogens, while other clinicians make just as positive statements that they have never secured such a result.[56] Animal breeders and students experimenting with laboratory animals, female and male, give similarly contradictory reports.[57] The contradictions may mean that there is no simple and direct relationship between estrogens and sexual behavior, or they may mean that the effectiveness of an increased supply of estrogens depends upon the concomitance of a variety of factors, including such things as the general metabolic level, the general physical health, the levels of the other hormones in the body, the age, the point in the estrus cycle at which the estrogens are administered, and probably still other factors.

There is a theory that estrogens counteract the effectiveness of androgens. The theory is as yet unsubstantiated by adequate experiment,[58] and is confused by the fact that both androgens and estrogens occur simultaneously in both the female and male bodies. However, some clinicians in this country, in Denmark, and in Holland are using estrogens in an effort to reduce the levels of sexual responsiveness of males convicted as sex offenders.[59] It is possible that estrogens do re-

---

[56] Various effects of estrogen therapy are reported in: Frank 1940:1509. Soule and Bortnick 1941. Mazer and Israel 1946:34. Greenblatt 1947:177–178. Emmens and Parkes 1947:240–241. Kimbrough and Israel 1948:1217–1219. Perloff 1949:135. Ford and Beach 1951:225–226. Mazer and Israel 1951:33. Masters in Cowdry 1952:669.

[57] For estrogen effects on the behavior and physical structures of mammalian males, see: Clark and Birch 1945:328 (loss of dominance in chimpanzee). Emmens and Parkes 1947:236 ff. (atrophy of testis, prostate, Cowper's glands, spermatogenesis). Beach 1948:61–62. Selye 1949:360 (decreases sexual responsiveness). Goldzieher and Goldzieher 1949:1156 (male impotence). Howard and Scott in Williams 1950:342 (depresses spermatogenesis). Paschkis and Rakoff 1950:137–138 (testicular degeneration). Howard et al. 1950:134. Lynch 1952:734–741 (atrophy of rat testis, depresses spermatogenesis).

[58] The lack of any antagonistic effect between androgens and estrogens is noted in: Koch 1937:206–208. Hartman 1940:449–471 (female monkey). Hoffman 1944:42–43. Burrows 1949:122. Selye 1949:64–65.

[59] The use of estrogens on male sex offenders and others to reduce sex drive is discussed or reported in: Dunn 1940:2263–2264 (a single case, 96 day treat-

duce the amount of androgen which is secreted, but the attempt to control sexual behavior by lowering androgens depends, of course, on a misinterpretation of the role of the androgens in the sexual activities of the male. There are optimistic reports of "good results" from the estrogen injections, but, as usual, the reports are not supported by adequate details of what the "good results" are supposed to be. Since an excessive supply of estrogens may affect many body functions besides sexual behavior, and since an excessive supply may do irreparable damage to other glandular structures, several research endocrinologists assert that they consider the use of estrogens to lower the sexual responsiveness of a male nothing less than medical malpractice.

PITUITARY HORMONES IN SEXUAL BEHAVIOR

Because of the effects which secretions from the anterior lobe of the pituitary gland may have on all of the other endocrine organs, and particularly on the gonads and adrenal glands, the pituitary has often been described as a master gland dominating all of the other endocrine organs in the body. It seems, however, more correct to think of the gonads and the pituitary and adrenal glands as a chain of organs whose secretions have interlocking effects. The effect of the pituitary on the gonads, for instance, seems hardly more significant than the effect of the gonads on the pituitary.

Relation of Pituitary to Physical Characters. Damage to the anterior lobe of the pituitary in a young animal, or in the human female or male before the onset of adolescence, may affect physical development more seriously than a gonadal castration.[60] In the case of the human female or male, the individual remains immature or develops slowly and at a late age, and is abnormal in his or her physical proportions and functions. However, the effects of a pituitary insufficiency in a young animal may be relieved by the administration of pituitary hormones. Because of the general, regulatory function of the pituitary gland, some laboratory students and some clinicians are inclined to depend upon pituitary hormones to correct a gonadal insufficiency.

---

ment, resulting in loss of sex drive and degeneration of genitalia). Dunn 1941: 643 (later report on same case. Some effects disappeared after stopping therapy, but sexual levels still reduced). Rosenzweig and Hoskins 1941:87–88 (massive doses, no effect on one homosexual male). Greenblatt 1947:279–280. Golla and Hodge 1949 (13 males, uncontrollable sex urge, treated with estrone successfully). Perloff 1949:135. Brown 1950:52–53 (cites 300 male homosexuals, unsuccessful treatment with androgens). Tappan 1951:247–248 (in Holland).

[60] The physical effect of pituitary damage in young animals and human subjects, and subsequent hormonal therapy are discussed in: Pratt in Allen et al. 1939: 1294. Corner 1942:141 (females), 222 (males). Greenblatt 1947:183. Beach 1948:12–17, 22, 208. Selye 1949:262–317. Williams 1950:ch.2. Howard and Scott in Williams 1950:318–319. Ford and Beach 1951:167–169. Dempsey in Stevens 1951:221.

It is reported that the administration of pituitary hormones to a normal pre-adolescent in whom there is neither a pituitary nor a gonadal insufficiency, may induce a precocious adolescent development.[61]

Relation of Pituitary to Sexual Behavior. A pre-adolescent pituitary deficiency affects the sexual behavior of an animal in much the same way that an early castration affects its sexual behavior. Responsiveness develops slowly if at all, and it is probable that the levels of response which are ultimately reached by such an individual are below those which are normal.[62] Since the gonads are among the structures which do not develop normally when there is an early pituitary insufficiency, the effects of the pituitary hormones on sexual behavior may depend upon their regulation of the supply of gonadal hormones, and this interpretation is favored by the fact that the administration of androgens to an animal which has a pituitary deficiency may induce normal sexual behavior. On the other hand, it is also possible that the pituitary hormones directly affect the development and the physiologic function of the nervous system on which sexual behavior depends.[63]

Because the pituitary glands are not located near the organs of reproduction, there are only scattered references to the effects of pituitary deficiencies on sexual behavior. The opinion is generally held that such deficiencies in adults, or the administration of pituitary hormones in cases of deficiencies, or the administration of pituitary hormones to normal adults, does not have as pronounced an effect on sexual behavior as the administration of gonadal hormones.

Pituitary Secretions and Levels of Sexual Response. There are indications that the levels of pituitary secretion may correlate with the fundamental patterns of sexual behavior which we have found in the human female and male. Recently published studies on fowl show that the cells of the anterior lobe of the pituitary in a very young male contain a quite clear cytoplasm, which, however, begins to accumulate granular materials, the so-called mitochondria, soon after the animal

[61] Precocious adolescent development resulting from the pituitary treatment of young animals is described in: Smith and Engle 1927 (ablation of thyroid made no difference, showing it was not involved; mice, rats, cats, rabbits, guinea pigs). Young 1941:322. Beach 1948:12–16. Burrows 1949:22–23, 39. Ford and Beach 1951:168–170.

[62] That a lack of pituitary hormone depresses sex drive, is reported by: Burrows 1949:169. Selye 1949:271–282. Williams 1950:43, 46.

[63] Conversely, the effect of the nervous system on the secretory capacity of the pituitary is well-known. Examples of such neural control are in: Hoffman 1944:309–310 (suppression of menses). Hartman 1945:27 (rabbit ovulation). C. D. Turner 1948:364–365 (pseudopregnancy). Durand-Wever 1952:209–211 (gonads deteriorate as a result of fear). Talbot et al. 1952:302–303 (puberty depends on neural mechanisms operating via pituitary).

begins to grow.[64] These mitochondria are associated with the normal physiologic processes that go on in these cells, but as the male animal ages, the mitochondria steadily increase in quantity until the cells of the glands of the older male become more or less filled with the granular inclusions.

The secretory capacity of the cell is inversely proportional to the volume of the granular inclusions in the cell and therefore to the age of the animal, and in the older male bird the cells may, in consequence, lose most of their secretory capacity. The functional significance of the pituitary is therefore steadily lowered as the male becomes more advanced in age.

On the other hand, the cells of the anterior lobe of the pituitary of the female do not accumulate such a quantity of granular material, and cases are reported of ten- and twelve-year-old female birds in which the cells of the anterior lobe of the pituitary are practically as clear of mitochondria as they are in very young females. This means that the secretory capacities of the pituitaries of these females are maintained at a more or less constant level throughout the life of the bird.

This gradual loss of function in the cells of the pituitary of the males, among fowls, parallels the steady decline which we have found in the sexual activities of the human male. The maintenance of pituitary function on a more or less constant level in the females, among fowls, parallels the maintenance of sexual capacity which we have found in the human female. Among all of the biologic phenomena which we are aware of, these differences between female and male pituitary secretion, and the differences which we shall note below in the levels of the 17-ketosteroids among females and males, are the only phenomena that seem to parallel the differences which we find between the sexual activities of the human female and male.

Since cytologic and experimental studies of these differences in female and male pituitary functions are of very recent date, we have no idea whether such a differentiation applies to the human or to any other species of mammal. If the differentiation does hold in the human species, it does not necessarily imply that there is any simple, causal relationship between pituitary secretions and sexual behavior. The relationship may depend upon a whole chain of physiologic phenomena which are initiated by the pituitary hormones; or the pituitary picture may reflect some other physiologic condition of the female on which both pituitary and sexual function depend. In any event there is a parallel between these reported pituitary functions in the bird, and

---

[64] For the relation of age to the mitochondria in the cells of the anterior lobe of the pituitary, see: Payne 1949:197–198. Payne in Cowdry 1952:385.

the data which we have on the sexual behavior of the human female and male. We did not find such a correlation between the levels of gonadal hormones and the patterns of human sexual behavior.

THYROID HORMONES

Thyroid secretions have well known effects upon the general physiologic level of an animal's activities, including its nervous responses. Low levels of thyroid secretion lower the rates of most activities, and high levels raise the rates of activity; but there is no invariable relationship for, once again, the effects depend upon the concomitance of various other factors which may also influence metabolic activities. It is reported that male hormones tend to stimulate thyroid function, but female hormones are said to inhibit their function, thereby reducing metabolic levels.[65]

It is well known that early insufficiencies of thyroid secretions may delay or modify adolescent physical developments and lead to an infantilism of genital structures in both females and males.[66] Cases of such adolescent deficiencies which we have in our histories, show delayed or low levels of sexual response and activity.

There are few published data, but there is a rather widespread clinical understanding that thyroid deficiencies in an older individual may be associated with some lack of responsiveness, and an excess may be associated with rates of sexual activity which are above the average. Not a few physicians administer preparations of thyroid in order to increase the responsiveness of some of their patients. We have some histories of females and a fair number of histories of males to whom thyroid extracts were administered with that express purpose. In some cases, increased frequencies of coitus seem to have been the direct consequence of this therapy, in other cases the record did not seem to indicate that there was any modification of sexual activity, and in some cases a decrease in sexual activity was reported after the thyroid administration. It is, of course, difficult to know how much of the reported change in any particular case was the direct result of the administration of the hormone, how much was the consequence of a change in other physiologic functions, and how much depended on social situations.

Just as some physicians have used testosterone to increase the frequencies of coitus in cases of sterility in which the sterility appeared to be a consequence of low rates of coitus, so there are some who use

[65] Effects of male and female hormones on thyroid function are noted in: Hoffman 1944:240–241 (reciprocal relationship). Williams 1950:115 (estrogen antagonizes thyroid function; testosterone stimulates thyroid and increases metabolic rate).

[66] Delayed adolescence resulting from lack of thyroid is described in: Lisser 1942. Hoffman 1944:239. Williams 1950:173–175. Talbot et al 1952:21, 31.

thyroid to accomplish the same end. Just as with the administration of gonadal hormones, there are reports that such thyroid therapy increases the sperm count and the motility and viability of the sperm, thereby increasing the chances for conception to occur.[67] There is, however, a considerable need for the accumulation of more specific data on these matters, and more carefully controlled experimentation. Since the administration of thyroid extracts may affect a number of other body functions, there is need for caution in the use of these hormones in any attempt to control sexual responses.

ADRENAL HORMONES

The adrenal glands are paired organs lying above the kidneys, which places them in the small of the back of the human animal. They are known to produce a considerable number—perhaps twenty or thirty or more—chemical compounds (steroids) which function as hormones. The possibility that some of these are related to sexual behavior has been considered for some time, but there are still few specific data to establish such a relationship.

Adrenaline. The best known of the secretions of the adrenal glands is adrenaline. This is a product of the central core, the medulla, of the adrenal organ. The role of adrenaline in various types of emotional response, and particularly in the case of fear and fright, was one of the first things known about the glands of internal secretion (p. 692).

Because many of the gross physiologic changes which occur during sexual response are similar to those which may be induced in an animal by injecting it with adrenaline, or by stimulating it so its adrenal glands secrete an extra supply of adrenaline, it has frequently been suggested that adrenaline is responsible for the initiation of sexual response or for some portion of it.

On the other hand, there is a longstanding theory that adrenaline interferes with sexual responses.[68] The theory may have originated in the observation that persons who are frightened during sexual activity may have their activity interrupted. There are, however, no experimental data to establish this inhibitory effect of adrenaline on sexual response.

We have already emphasized (Chapter 17) that sexual reactions may so quickly involve the whole animal body that it is difficult to believe that a blood-circulated agent such as adrenaline could be re-

---

[67] The administration of thyroid in sterility cases is noted in: Siegler 1944:311–313. Smith in Williams 1950:430. Mazer and Israel 1951:461, 488 (to increase sperm count).

[68] The opinion that adrenaline interferes with sexual response, or inhibits erection, is implied or specifically stated in: Cannon 1920:270–271. Kuntz 1945:311.

sponsible for the major body of physiologic changes which constitute sexual response. Nevertheless, many of the physiologic phenomena involved in sexual response, and particularly the activity of the autonomic nervous system during sexual response, may lead to the secretion of adrenaline, and it is quite possible that some of the phenomena seen in the later stages of sexual response may be products of adrenaline secretion (p. 693).

Cortical Hormones. The outer portion of the adrenal organ, its cortex, is the source of a variety of hormones, including androgens, estrogens, and the various compounds known as the 17-ketosteroids. It is understandable that the adrenal cortex should have some of the same functions as the gonads, for the testes, the ovaries, and the cortex of the adrenals originate from a common embryonic cell mass, even though the organs into which they develop may move apart in the adult anatomy. It has generally been considered that the effects of a gonadal castration are not more severe than they are because the adrenal cortex shares some of the responsibility for the production of the same hormones.[69] The importance of the adrenals in this regard, however, may have been overemphasized in the earlier literature.

The 17-ketosteroids are so called because they have a ketone group at the 17 position on the molecule. Some, but not all, of the 17-ketosteroids are androgens. The 17-ketosteroids are supposed to originate in various organs, including the gonads and the adrenal cortex. It has been estimated that the adrenal cortex may produce as much as 60 or 70 per cent of the 17-ketosteroids found in the mammalian body; but the removal of both the testes and the adrenal cortex in a laboratory animal does not wholly deprive it of its 17-ketosteroids, and this indicates that other organs also produce these compunds.[70]

There are suggestions in the literature that the 17-ketosteroids are in some fashion related to levels of sexual responsiveness, and to frequencies of overt sexual activity in a mammal. The suggestions, however, have been theoretic, and there seems to have been no demonstration of such a relationship in either experimental or clinical data. We do find, however, that the reported levels of the 17-ketosteroids in the

---

[69] The compensatory function of the adrenal cortex in castrates is described in: Burrows 1949:189. Dempsey in Stevens 1951:216 (potency of male depressed more severely by destroying adrenals than by castration).

[70] For the relative importance of the cortex in supplying 17-ketosteroids in the body see: Fraser et al. 1941:255 (two-thirds adrenal in male, all from adrenal in female). Dorfman et al. 1947:487 (17-ketosteroids still significant in male and female monkeys 40 days after removal of both gonads and adrenals). Hamburger 1948:31–32. Dempsey in Stevens 1951:215 (one-third gonadal, remainder from adrenal cortex). Engle in Cowdry 1952:712 (two-thirds from adrenal). Maddock et al. 1952:668 (both testes and cortex are important).

human male differ from the reported levels of the 17-ketosteroids in the human female in a manner which more or less parallels the differences which we have found between the median frequencies of orgasm at various ages in the two sexes (compare Figures 143 and 155).[71]

It is reported that the 17-ketosteroids increase sharply in the urine of the developing human male at the approach of adolescence and during early adolescence (Figure 155). They reach a peak in the late teens

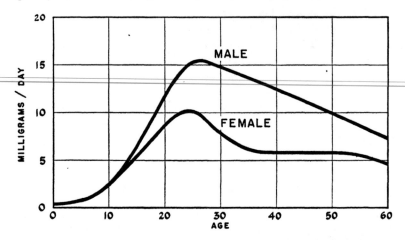

Figure 155. Levels of 17-Ketosteroids in Human Female and Male

Showing aging effect in male, and a prolonged plateau in the female, somewhat similar to average median frequencies of total outlet among females and males shown in Figure 143. Data from Hamburger 1948.

or early twenties in the male, but from that point they drop steadily into old age. Consequently, the levels of sexual activity in the human male more closely match the levels of the 17-ketosteroids than the levels of gonadal hormones. We have already noted that there is also a parallel to the human male behavior curves, in the secretory capacity of the anterior lobe of the pituitary in one species of laboratory animal.

The levels of the 17-ketosteroids which are reported for the human female are quite different from those reported for the male (Figure 155). During adolescence and the later teens, there is a sharp rise in the ketosteroid levels of the female. The curve reaches its peak somewhere in the middle twenties, shows a rather sharp drop in the next

[71] Data on levels of 17-ketosteroids are found in the following sources: Fraser et al. 1941:237–255 (shows decrease in older ages, both male and female). Talbot et al. 1943:364. Hamburger 1948:34. Hamilton and Hamilton 1948: 438. Howard et al. 1950:127 ff. Salter, Humm, and Oesterling 1948:302–303, fig. 3 (decrease with age in male after twenties). Kenigsberg, Pearson, and McGavack 1949:426–427 (in males increase up to thirties, decrease thereafter. In contrast to other studies, notes no changes in twenty females aged 17 to 64).

ten years, and then lies on a plateau for some years. It continues on this level through menopause and into the middle fifties, and does not show a further drop until the late fifties or sixties. Another study shows a more or less continuous plateau from ages seventeen to sixty-four, without the teen-age rise which the first-mentioned study showed. This plateau parallels the plateau which we have found in the sexual activities of the female (Figure 143). The chief difference lies in the rise in the 17-ketosteroids reported in one study during the late teens and early twenties, and this rise is not found in the female's sexual performance. This may raise a question as to the significance of the correlation; but it is also clear that a part of the lack of sexual response in the younger American female is a product of cultural restraints on her sexual performance. We have presented evidence for this throughout the present volume, particularly in connection with the data showing the effect of the religious tradition on her behavior.

These correlations between ketosteroid levels and the sexual activities of the human female and male may represent a simple causal relationship, or each of these phenomena may be a product of some other basic physiologic phenomenon, or of a whole complex of physiologic phenomena which are responsible for adrenal function, for pituitary function, and for the capacities of the nervous system to be affected by what we call sexual stimuli.

HORMONES AND PATTERNS OF SEXUAL BEHAVIOR

While the data which we have presented show that hormones may affect the capacity of the central nervous system to be stimulated sexually, and therefore may affect the levels of sexual response, there are no data which warrant the conclusion that an individual's choice of a sexual partner or of a particular type of sexual activity may be directly influenced by any of the hormones. As far as sexual performance is concerned, it is incorrect to think of endocrine organs as the glands of personality.

We have already pointed out that the sexual capacity of an animal depends upon its inherited anatomy and particularly upon its nervous structures and the physiologic capacities of those structures. The capacity of an animal to be aroused by tactile stimuli or psychologic stimuli, and to respond with a development of neuromuscular tensions which produce, among other things, the rhythmic pelvic thrusts which are typical of mammalian coitus, is a capacity which is inherent at birth (Chapter 15).

We have also emphasized (Chapter 16) that all organisms, and particularly such highly evolved organisms as the human animal, are conditioned by their experience, and consequently in the course of time come to respond more readily to particular types of stimuli. Thus, they develop preferences for the repetition of those types of sexual experience which have been most satisfactory, and become conditioned against the repetition of those types of sexual activity which have been unsatisfactory. In addition to being affected by firsthand experience, the individual may be conditioned by the reported experience of his or her friends, and by the social tradition (Chapter 16). Such conditioning seems a sufficient explanation of a preference for a sexual partner who is a blonde or a brunette, or a preference for a sexual partner of the same or of the opposite sex, or for the utilization of one type of technique rather than another in the course of a sexual relationship. It is curious that psychologists and psychiatrists, who are the ones who most often emphasize the importance of psychologic conditioning, so often look for hormonal explanations of any type of sexual behavior which departs from the Hittite and Talmudic codes.

On the unwarranted assumption that homosexuality represents some sort of femininity in an anatomic male, and masculinity in an anatomic female, there have been clinical attempts to redirect the sexual behavior of homosexual males, and occasionally of homosexual females, by giving them gonadal hormones. Recent suggestions concerning the importance of the 17-ketosteroids have led to investigations of the possibilities of using those substances to control behavior. Certain clinicians have developed a lucrative business supplying hormones to patients with homosexual histories, and androgens have been given males who have been discovered in homosexual activity in penal institutions, in the Armed Forces, and elsewhere. Certain of the drug companies have encouraged the use of male hormones for this purpose.

The possibility that hormones might modify homosexual patterns of behavior found some encouragement in early reports on the hormonal injections of laboratory animals. These reports, unfortunately, involved an erroneous use of the term *homosexual*.[72] Among most of the mammals a female in heterosexual coitus typically assumes a crouched position, often with her head drawn down and her haunches raised in a manner which allows the male to mount in coitus. The female rat arches her back convexly (in a *lordosis*) and her ears go into a tremor. The male, on the other hand, typically sits or rises on his hind legs while he effects coitus. The assumption of the crouched position by a male during a sexual relationship, or the assumption of an upright posi-

---

[72] The concept of a "reversal" of "masculinity" or "femininity" in sexual behavior among lower mammals is found, for instance, in: Ball 1937, 1939, 1940. Beach 1938, 1941, 1942b, 1945, 1947b. Young and Rundlett 1939:449 (labelling such a reversal homosexual). Beach corrects this interpretation in: Beach 1948:68. Ford and Beach 1951:41–42.

tion by a female, was recorded in the earlier literature on animal behavior as a *sex reversal* or *homosexual* activity. But this use of the term had no relation to its use in human psychology, where *homosexual* refers to the choice of a partner of the same sex.

We have already noted (p. 747) that rats which are injected with testosterone become more aggressive in all of their behavior, including their sexual behavior, and that females who are given testosterone mount other females and males more frequently than they did before they were given the hormones. But they do this because their aggressiveness has been increased; and when they mount males their behavior is still heterosexual and not homosexual, even though the positions assumed in the coitus have been reversed.

In connection with the clinical treatment of homosexuality there are, of course, reports of "good results," but no adequate data to show that the behavioral patterns of such subjects have ever been modified by hormonal treatments.[73] We have the histories of an appreciable number of males who had received hormonal treatments, but we have never seen an instance in which a homosexual pattern had been eliminated by such therapy. Not infrequently, however, the levels of sexual responsiveness had been raised by the treatment with male hormones, and this, in many instances, had complicated the situation by increasing the capacity of the homosexual individual to be aroused by *homosexual* stimuli. Such failures to attain the desired results in therapy had led to social tragedies in a number of the cases.

SUMMARY AND CONCLUSIONS

Our present understanding of the relations between hormones and the sexual behavior of the human female and male may, then, be summarized as follows:

1. The early development of sexual responsiveness in the human male and its later development in the female, the location of the period of maximum responsiveness for the male in the late teens and early twenties and for the female in the late twenties, the subsequent decline of the male's sexual capacities from that peak into old age, and the maintenance of female responsiveness on something of a level throughout most of her life, are patterns which are not explained by the known

---

[73] For the hormonal treatment of homosexual cases, see: Barahal 1940:329 (7 cases, increased the sex drive of homosexuals, failed to change its direction). Glass, Deuel and Wright 1940. Kinsey 1941 (a critical appraisal of earlier data). Glass and Johnson 1944 (some benefit in 3 cases; no benefit in 8). T. V. Moore 1945:80. Heller and Maddock 1947:419–422 (a well balanced discussion). Perloff 1949:136–139. Howard and Scott in Williams 1950:333. Brown 1950: 52–53 (300 male homosexuals, unsuccessful treatment with androgens). See also footnote 59.

anatomy or physiology of sexual response. Neither are they explained by the known differences in the capacities of females and males to be aroused by psychosexual stimuli.

2. Various hormones may affect the levels of sexual responsiveness in the human female and male because of their effects on the levels of all physiologic activities, including those capacities of the nervous system on which sexual behavior depends.

3. Estrogens, the so-called female hormones, originate primarily in the ovaries and the adrenal cortex. Androgens, the so-called male hormones, originate in the testes, the adrenal cortex, and still other organs. Both estrogens and androgens are present in both females and males. They occur in about equal abundance in human females and males in the pre-adolescent years, but during adolescence the androgens rise to a somewhat higher level in the male, and the estrogens rise to a much higher level in the female. In neither sex, however, does this rise correlate with the levels of sexual responsiveness in the adolescent or older female or male.

4. Gonadal hormones are among the elements necessary for normal physical development in young animals, both female and male. Consequently they are necessary for the development of the capacities of the nervous system and therefore may, indirectly, affect the development of sexual responsiveness in the female and male.

5. In human adults, gonadal hormones may be withdrawn (as in a castration) with minimum effects on physical characters, with a minimum or no effect on the sexual responsiveness of the female, and with little or no effect on the sexual responsiveness of half or more of the males. In a smaller number of cases there may be a slow physical degeneration in the male which may be accompanied by a reduction in his sexual responsiveness.

6. Estrogen and androgen levels, as far as they are known in the normal human female and male, do not correlate with the frequencies of sexual activity in the two sexes. There seem to be no correlations with the aging patterns of the male in his sexual performance, and with the female's maintenance of a level of response throughout most of her life.

7. Increased supplies of androgens given to adult human females or males do raise their levels of sexual response. There is serious danger, however, in the administration of such hormones except under the direction of an experienced clinician.

8. Pituitary hormones are also among the elements necessary for the normal physical development of the female and male, and for the

development of many physiologic capacities, including those capacities of the nervous system upon which sexual behavior depends.

9. The levels of pituitary secretion are known to drop steadily among the males of one species of laboratory animal, while the levels of pituitary secretion in the female of the same species may remain constant for some years. If this is also true of the human species, this would parallel the differences which we find in levels of sexual responsiveness among human females and males. The correlations might represent a causal relationship, or both phenomena might be independent products of more basic physiologic factors.

10. The levels of thyroid secretion sometimes show a direct correlation with the levels of sexual responsiveness. Higher thyroid levels sometimes correlate with higher levels of response, and lower thyroid levels sometimes correlate with lower levels of response; but the correlations are not invariable, for there appear to be other factors which determine the effectiveness of the hormone from the thyroid.

11. Adrenaline, from the medulla of the adrenal glands, does not seem to be involved in the initiation of sexual response. However, as sexual activities progress, the adrenal glands may be stimulated into secretion, and this secretion may be responsible for certain of the phenomena which appear in the later stages of sexual activity. There is no evidence that adrenaline actually inhibits sexual response.

12. The levels of the 17-ketosteroids, which originate largely in the adrenal cortex but also develop in the testes and ovaries and apparently in other organs of the body, show rather striking correlations with the levels of sexual response in the human female and male. In the human male, the 17-ketosteroids drop steadily from a peak in the late teens or twenties into old age, while they remain on a plateau for some period of years in the female. Again, this may represent a direct causal relationship, or (what is more likely) both the 17-ketosteroids and the levels of sexual responsiveness may reflect a more basic and probably more complex physiologic situation.

13. While hormonal levels may affect the levels of sexual response—the intensity of response, the frequency of response, the frequency of overt sexual activity—there is no demonstrated relationship between any of the hormones and an individual's response to particular sorts of psychologic stimuli, an individual's interest in partners of a particular sex, or an individual's utilization of particular techniques in his or her sexual activity. Within limits, the levels of sexual response may be modified by reducing or increasing the amount of available hormone, but there seems to be no reason for believing that the patterns of sexual behavior may be modified by hormonal therapy.

BIBLIOGRAPHY

The following bibliography includes the items which are cited in the footnotes in the present volume, and a limited number of additional references to important general studies on sex. It does not purport to be a general bibliography on human sexual behavior.

Anon. (Madame B——le.). **1707.** The fifteen plagues of a maiden-head. London, F. P., 8p. [reprint].

Anon. 1772. The fifteen plagues of a maidenhead. *In:* Anon., comp. The merry Andrew, pp. 136–142.

Anon. (Buck, M. S., ed.). **1916** (orig. 6th cent. A.D.). The Greek anthology (Palatine ms). The amatory epigrams. Priv. print., 142p.

Anon., comp. (Paton, W. R., trans.). **1917, 1918** (orig. 1st cent. A.D.). The Greek anthology. In five volumes. Cambridge, Mass., Harvard University Press, v. 3, 456p. v. 4, 422p.

Anon. (Tomita, K., trans.). ca. **1845.** Hikatsu-sho. Ms, typed, 1 + 83p.

Anon. (Alix (Petrus)). **n.d.** Histoire et psycho-physiologie du vice à travers les siècles et les êtres. Paris, Librairie Mondaine, 160p.

Anon. (Azama Otoko, pseud.) (Tomita, K., trans.). ca. **1845.** Ikkyū zenshi shoshoku monogatari. Ms, typed, 1 + 24p.

Anon. (Tomita, K., trans.). ca. **1850.** Jiiro haya-shinan. Ms, typed, 10p.

Anon. (Tomita, K., trans.). ca. **1840.** Makurabunko. Ms, typed, 11p.

Anon. (?Carrington, Charles). **1896.** Marriage, love and woman amongst the Arabs. Otherwise entitled The book of exposition . . . Translated . . . by an English Bohemian. Paris, Charles Carrington, 1 pl. + xlviii + 285p.

Anon., comp. (Ferdinando Funny, pseud.). **1772.** The merry Andrew: being the smartest collection ever yet published . . . to which is added . . . The fifteen plagues of a maidenhead. London, P. Wickes, 1 pl. + 144p.

Anon. [Burton, R. F., and Smithers, L. C. trans.]. **1888.** Priapeia, or the sportive epigrams of divers poets on Priapus now first completely done into English prose from the original Latin, with notes . . . to which is appended the Latin text. Athens, Erotika Biblion Society, priv. print., xxxiv + 238p.

Anon. (Tomita, K., trans.). ca. **1885.** Takarabunko. Ms, typed, 20p.

Anon. 1829. Über die Behandlung der Unarten, Fehler und Vergehungen der Jugend. Graudenz, C. O. Röthe; Berlin, Enslin, [6] + 116p.

Anon. 1772. The virgin's dream. *In:* Anon., comp. The merry Andrew, pp. 143–144.

Abernethy, E. M. 1925. Correlations in physical and mental growth. J. Educ. Psychol. 16:458–466, 539–546.

Abraham, K. 1948 (orig. **1927**). (Bryan, D., and Strachey, A., trans.). Selected papers of Karl Abraham, M.D. London, Hogarth Press and Institute of Psychoanalysis, 527p.

Achilles, P. S. 1923. The effectiveness of certain social hygiene literature. New York, American Social Hygiene Association, 116p.

Ackerson, L. 1931. Children's behavior problems. A statistical study based upon 5000 children examined consecutively at the Illinois Institute for Juvenile Research. I. Incidence, genetic and intellectual factors. Chicago, University of Chicago Press, xxi + 268p.

Adams, C. R. 1946. Building for successful marriage and family living. *In:* Chivers, W. R., ed. Successful marriage and family living, pp. 30–35.

Adler, O. 1911. Die mangelhafte Geschlechtsempfindung des Weibes. Anaesthesia sexualis feminarum. Anaphrodisia. Dyspareunia. Berlin, H. Kornfeld, xiv + 231p.

    1912. Die frigide Frau. Sexual-Probleme 8:5–17.

Aeschines. 1919 (orig. 4th cent. B.C.). (Adams, C. D., trans.). The speeches: against Timarchus, on the embassy, against Ctesiphon. Cambridge, Mass., Harvard University Press, xxiii + 528p.

Alibert. n.d. [ca. 1900?]. Onanism. Onanism amongst men. Its causes, 'methods, and disorders. Masturbation amongst women. Its causes. Different methods of masturbation. Symptoms. Consequences. Treatment. Paris, Medical Library, 95p.

    1926. Tribadism and saphism. Paris, priv. print., 86p.

Allen, C. 1949. The sexual perversions and abnormalities. A study in the psychology of paraphilia. London, Oxford University Press, x + 346p.

Allen, E., et al., ed. 1939. Sex and internal secretions. A survey of recent research. Baltimore, Williams & Wilkins Co., xxxvi + 1346p.

Allen, F. L. 1931. Only yesterday. An informal history of the nineteen-twenties. New York, Bantam Books, 413p.

Allendy, R., and **Lobstein, H.** 1948. (Larsen, E., trans.). Sex problems in school. London, New York, Staples Press, 182p.

Alpert, L. M. 1938. Judicial censorship of obscene literature. Harvard Law Rev. 52:40–76.

American Association of Marriage Counselors. 1952. Premarital sex relations: The facts and the counselor's role in relation to the facts. (Report of a round table meeting held by the Section on Marriage and Family Counseling of the National Council on Family Relations at Lake Geneva, Wisc., Aug. 30, 1951.) Marr. & Fam. Liv. 14:229–238.

Ananga-Ranga. 1935 (orig. 12th cent. A.D.?). (Burton, R. F., trans.). The secret places of the human body. Known as the Ananga-Ranga, or the Hindu art of love. Priv. print., 218p., 10 pl.

Andrews, F. N., and **McKenzie, F. F.** 1941. Estrus, ovulation, and related phenomena in the mare. Columbia, Mo., University of Missouri (Agricultural Experiment Research Bull. 329), 117p.

Andrews, T. G., ed. 1948. Methods of psychology. New York, John Wiley & Sons; London, Chapman & Hall, xiv + 716p.

Anthropophyteia (Krauss, F. S., ed.). 1904–1913. Jahrbücher für folkloristische Erhebungen und Forschungen zur Entwicklungsgeschichte der geschlechtlichen Moral. Leipzig, Deutsche Verlagsaktiengesellschaft & Ethnologischer Verlag, v. 1–10.

Apfelberg, B., Sugar, C., and **Pfeffer, A. Z.** 1944. A psychiatric study of 250 sex offenders. Amer. J. Psychiat. 100:762–770.

Apuleius. 1822 (orig. 2nd cent. A.D.). (Taylor, T., trans.). The metamorphosis, or golden ass, and philosophical works. London, Robert Triphook etc., xxiv + 400 + 10p.

    1915. (Adlington, W., trans.). The golden ass. Being the Metamorphoses of Lucius Apuleius. London, William Heinemann; New York, G. P. Putnam's Sons, xxiv + 608p.

    1923. (Adlington, W., trans.). The golden ass. London, John Lane, xxxv + 282p., 8 pl., 8 col. pl.

Aretaeus. 1856 (orig. 2nd and 3rd cent. A.D.). (Adams, F., trans.). The extant works of Aretaeus, the Cappadocian. London, priv. print., Sydenham Society, xx + 510p.

Arey, L. B. 1946. Ed. 5. Developmental anatomy. A textbook and laboratory manual of embryology. Philadelphia and London, W. B. Saunders Co., ix + 616p.

Arieff, A. J., and **Rotman, D. B.** 1942. One hundred cases of indecent exposure. J. Nerv. & Ment. Dis. 96:523–528.

Aristophanes. 1912 (orig. 5th–4th cent. B.C.). Aristophanes. The eleven comedies. London, Athenian Society, priv. print., 2v. v. 1, 392p. v. 2, 476p.

1924. (Rogers, B. B., trans.). Aristophanes. In three volumes. I. The Acharnians. The clouds. The knights. The wasps. II. The peace. The birds. The frogs. Cambridge, Mass., Harvard University Press, 2v. v. 1, xvi + 555p. v. 2, 1 pl. + 443p.

Aristotle. 1910 (orig. 4th cent. B.C.). (Thompson, D. W., trans.). The works of Aristotle. Volume IV. Historia animalium. Oxford, Clarendon Press, xv + n.pag.

1936. (Hett, W. S., trans.). Problems. London, William Heinemann, 2v. v. 1, x + 461p. v. 2, vi + 456p.

Arregui, A. M. 1927. Summarium theologiae moralis, ad recentem codicem iuris canonici accommodatum. Bilbao, El Mensajero Del Corazón de Jesús, xix + 665p.

Asayama, Sin-iti. 1949. Gendai gakusei no seikōdō. (Sex behavior of present-day Japanese students.) Kyoto, Usui Shobo, 346p.

Asdell, S. A. 1946. Patterns of mammalian reproduction. Ithaca, N. Y., Comstock Publishing Co., xi + 437p.

Aurand, A. M., Jr. 1938a. Little known facts about bundling in the New World. Harrisburg, Penna., priv. print., 32p.

1938b. Slants on the origin of bundling in the Old World. Harrisburg, Penna., priv. print., 32p.

Azama Otoko. See Anon. Ikkyū zenshi shoskoku monogatari.

Baber, R. E. 1936. Some mate selection standards of college students and their parents. J. Soc. Hyg. 22:115–125.

1939. Marriage and the family. New York and London, McGraw-Hill Book Co., xii + 656p.

Back, G. 1910. Sexuelle Verirrungen des Menschen und der Natur. Berlin, John Pohl, 2v. v. 1, 7 + x + 544p. v. 2, 545–973p.

Backhouse, E., and Scott, W. B. 1888. Martyr scenes of the sixteenth and seventeenth centuries. London, Hamilton, Adams & Co., 86p., 12 pl.

Bailey, F. R. 1948. Ed. 12. (Smith, P. E., and Copenhaver, W. M., ed.). Textbook of histology. Baltimore, Williams & Wilkins Co., xix + 781p.

Baker, J. R. 1926. Sex in man and animals. London, George Routledge & Sons, xvi + 175p.

Baldwin, B. T. 1921. The physical growth of children from birth to maturity. Iowa City, Ia., The University of Iowa Studies in Child Welfare, Monogr. v. 1, no. 1, 411p.

Ball, Josephine. 1937. Sex activity of castrated male rats increased by estrin administration. J. Comp. Psychol. 24:135–144.

1938. Partial inhibition of sex activity in the intact female rat by injected estrin. Endocrinol. 23:197–199.

1939. Male and female mating behavior in pre-pubertally castrated male rats receiving estrogens. J. Comp. Psychol. 28:273–283.

1940. The effect of testosterone on the sex behavior of female rats. J. Comp. Psychol. 29:151–165.

Ball, J., and Hartman, C. G. 1935. Sexual excitability as related to the menstrual cycle in the monkey. Amer. J. Obstet. & Gynec. 29:117–119.

Ballerini, A. 1890. Opus theologicum morale. . . . Prati, Giachetti, Filii et Soc., v. 2, pp. 678–739, v. 6, var.pag., v. 7, 421–442p.

Banay, R. S., and Davidoff, L. 1942. Apparent recovery of a sex psychopath after lobotomy. J. Crim. Psychopath. 4:59–66.

Banning, Margaret C. 1937. The case for chastity. Readers Digest 31:(Aug.), 1–10.

Barahal, H. S. 1940. Testosterone in psychotic male homosexuals. Psychiat. Quart. 14:319–330.

Barash, M. 1926. Sex life of the workers of Moscow. J. Soc. Hyg. 12:274–288.

Bard, P. 1934. On emotional expression after decortication with some remarks on certain theoretical views. Parts I and II. Psychol. Rev. 41:309–329, 424–449.

1936. Oestrual behavior in surviving decorticate cats. Amer. J. Physiol. 116:4–5.

1939. Central nervous mechanisms for emotional behavior patterns in animals. Res. Publ. Assoc. Nerv. & Ment. Dis. 19:190–218.

1940. The hypothalamus and sexual behavior. In: Fulton, J. F., et al., ed. The hypothalamus and central levels of autonomic function, pp. 551–579.

1942. Neural mechanisms in emotional and sexual behavior. Psychosom. Med. 4:171–172.

Barker, R. G., and Stone, C. P. 1936. Physical development in relation to menarcheal age in university women. Human Biol. 8:198–222.

Barr, M. W. 1920. Some notes on asexualization; with a report of eighteen cases. J. Nerv. & Ment. Dis. 51:231–241.

Barré, M. L., ed. 1839, 1840, 1861. Herculanum et Pompéi. Recueil général des peintures, bronzes, mosaïques, etc. . . . Paris, Firmin Didot Frères, 8v., var.pag.

Barton, G. A. 1925. Ed. 4. Archaeology and the Bible. Philadelphia, American Sunday-School Union, xv + 561p., 122 pl.

Bauer, B. A. 1927. (Haire, N., Jerdan, E. S., and Paul, E. and C., trans.). Woman and love. New York, Liveright Publishing Corp., 2v. v. 1, 7 + 360p. v. 2, xxvii + 396p.

1929. Wie bist Du, Weib? Mit einem besonderen Anhang: Weib, bleibe jung und schön! Hygiene der modernen Frau. Zürich, Viktoria-Verlag, xxiii + 631p.

Bauer, M. 1924. Liebesleben in deutscher Vergangenheit. Berlin, P. Langenscheidt, 390p.

Beach, F. A. 1938. Sex reversals in the mating pattern of the rat. J. Genet. Psychol. 53:329–334.

1940. Effects of cortical lesions upon the copulatory behavior of male rats. J. Comp. Psychol. 29:193–245.

1941. Female mating behavior shown by male rats after administration of testosterone propionate. Endocrinol. 29:409–412.

1942a. Central nervous mechanisms involved in the reproductive behavior of vertebrates. Psychol. Bull. 39:200–226.

1942b. Analysis of factors involved in the arousal, maintenance and manifestation of sexual excitement in male animals. Psychosom. Med. 4:173–198.

1942c. Effects of testosterone propionate upon the copulatory behavior of sexually inexperienced male rats. J. Comp. Psychol. 33:227–247.

1942d. Sexual behavior of prepuberal male and female rats treated with gonadal hormones. J. Comp. Psychol. 34:285–292.

1943. Effects of injury to the cerebral cortex upon the display of masculine and feminine mating behavior by female rats. J. Comp. Psychol. 36:169–199.

1944a. Effects of injury to the cerebral cortex upon sexually-receptive behavior in the female rat. Psychosom. Med. 6:40–55.

1944b. Experimental studies of sexual behavior in male mammals. J. Clin. Endocrinol. 4:126–134.

1944c. Relative effects of androgen upon the mating behavior of male rats subjected to forebrain injury or castration. J. Exper. Zool. 97:249–295.

1945. Bisexual mating behavior in the male rat: Effects of castration and hormone administration. Physiol. Zool. 18:390–402.

1947a. Hormones and mating behavior in vertebrates. Rec. Prog. in Hormone Res. 1:27–63.

1947b. A review of physiological and psychological studies of sexual behavior in mammals. Physiol. Rev. 27:240–307.

1948. Hormones and behavior. A survey of interrelationships between endocrine secretions and patterns of overt response. New York and London, Paul B. Hoeber, xiv + 368p.

1949. A cross-species survey of mammalian sexual behavior. *In:* Hoch, P. H., and Zubin, J., ed. Psychosexual development in health and disease, pp. 52–78.

1950. Sexual behavior in animals and men. Springfield, Ill., Charles C Thomas, from the Harvey Society Lectures (1947–1948), Series 43: 254–280.

1951. Body chemistry and perception. *In:* Blake, R. R., and Ramsey, G. V. Perception, pp. 56–94.

Beach, F. A., and Gilmore, R. W. 1949. Response of male dogs to urine from females in heat. J. Mammalogy 30:391–392.

Beach, F. A., and Rasquin, P. 1942. Masculine copulatory behavior in intact and castrated female rats. Endocrinol. 31:393–409.

Beauvoir, Simone de. 1949. Le deuxième sexe. I. Les faits et les mythes. II. L'expérience vécue. Paris, Gallimard, v. 1, 399p. v. 2, 581p.

1952. (Parshley, H. M., trans.). The second sex. New York, Alfred A. Knopf, xxx + 746p.

Becker, H., and Hill, R., ed. 1942. Marriage and the family. Boston, D. C. Heath & Co., xxxii + 663p.

Becker, J. E. de. 1899. The nightless city, or the history of the Yoshiwara yukwaku. Yokohama, Z. P. Maruya & Co., 441 + xxv p., 10 pl., 6 col. pl.

Bekker, ?B. 1741 (ex Eng. ed. 15, orig. 1722?). (Crouch, N., ed.). Onania, oder die erschreckliche Sünde der Selbst-Befleckung . . . aus dem Englischen ins Deutsche übersetzet. Leipzig, Johann George Löwe, 16 + 440p.

Belknap, G., and Campbell, A. 1951. Political party identification and attitudes toward foreign policy. Pub. Opin. Quart. 15:601–623.

Bell, R. H. 1921. Some aspects of adultery. A study. New York, Critic and Guide Co., 54p.

Bell, S. 1902. A preliminary study of the emotion of love between the sexes. Amer. J. Psychol. 13:325–354.

Bender, Lauretta. 1939. Mental hygiene and the child. Amer. J. Orthopsychiat. 9:574–582.

1952. Child psychiatric technics. Diagnostic and therapeutic approach to normal and abnormal development through patterned, expressive, and group behavior. Springfield, Ill., Charles C Thomas, xi + 335p., 16 pl., 4 col. pl.

Bender, Lauretta, and Blau, A. 1937. The reaction of children to sexual relations with adults. Amer. J. Orthopsychiat. 7:500–518.

Bender, Lauretta, and Cramer, J. B. 1949. Sublimation and sexual gratification in the latency period of girls. *In:* Eissler, K. R., et al., ed. Searchlights in delinquency, pp. 53–64.

Benedek, Therese. 1952. Psychosexual functions in women. New York, Ronald Press, x + 435p.

Benedek, T., and Rubenstein, B. B. 1942. The sexual cycle in women. Washington, D. C., Nat. Res. Council (Psychosom. Med. Mono. v. 3, no. 1 and 2), viii + 307p.

Bénézit, E. 1948, 1949, 1950, 1951, 1952. Dictionnaire critique et documentaire des peintres, sculpteurs, dessinateurs et graveurs. . . . Librairie Gründ, v. 1, xvi + 770p., 32 pl. v. 2, 776p., 32 pl. v. 3, 815p., 32 pl. v. 4, 773p., 32 pl. v. 5, 843p., 32 pl.

Bensing, R. G. 1951. A comparative study of American sex statutes. J. Crim. Law & Criminol. 42:57–72.

Benvenuti, M. 1950. L'ipersessualità come fattore degenerogeno. Pisa, "Omnia Medica," 209p.

Bergler, E. 1943. The respective importance of reality and phantasy in the genesis of female homosexuality. J. Crim. Psychopath. 5:27–48.

1944. The problem of frigidity. Psychiat. Quart. 18:374–390.

1948. Lesbianism, facts and fiction. Marr. Hyg. 1:197–202.

1951. Neurotic counterfeit-sex. Impotence, frigidity, "mechanical" and pseudo-sexuality, homosexuality. New York, Grune & Stratton, xii + 360p.

Bernard, J. 1935. Some biological factors in personality and marriage. Human Biol. 7:430–436.

Bernard, W. S. 1938. Student attitudes on marriage and the family. Amer. Sociol. Rev. 3:354–361.

Berrien, F. K. 1935. A study of the drawings of abnormal children. J. Educ. Psychol. 26:143–150.

Bertram, J. G. (Cooper, W. M., pseud.). 1869. A history of the rod, in all countries, from the earliest period to the present time. (Flagellation & the flagellants). London, John Camden Hotten, x + 544p., xx pl.

Biederich, P. H., and Dembicki, L. 1951. Die Sexualität des Mannes. Darstellung und Kritik des "Kinsey-Report." Regensburg, Wien, Franz Decker, 232p.

Bienville, D. T. de. 1771. La nymphomanie ou traité de la fureur utérine . . . Amsterdam, Marc-Michel Rey, xx + 168p.

Bilder-Lexikon (Schidrowitz, L., ed.). **1928, 1929, 1930, 1931.** 1. Kulturgeschichte, 943p. 2. Literatur und Kunst, 944p. 3. Sexualwissenschaft, 917p. 4. Ergänzungsband, 877p. Wien und Leipzig, Verlag für Sexualforschung.

Bingham, A. T. 1922. Determinants of sex delinquency in adolescent girls based on intensive studies of 500 cases. J. Crim. Law & Criminol. 13:494–586.

Bingham, H. C. 1928. Sex development in apes. Baltimore, The Johns Hopkins Press (Comp. Psychol. Monogr. v. 5, Ser. No. 23), 165p.

Birch, H. G., and Clark, G. 1946. Hormonal modification of social behavior. II. The effects of sex-hormone administration on the social dominance status of the female-castrate chimpanzee. Psychosomat. Med. 8:320–331.

Biskind, G. R., Escamilla, R. F., and Lisser, H. 1941. Treatment of eunuchoidism. Implantation of testosterone compounds in cases of male eunuchoidism. J. Clin. Endocrinol. 1:38–49.

Bissonnette, T. H. 1950. Ferrets. In: Farris, E. J., ed. The care and breeding of laboratory animals, pp. 234–255.

Blackwood, B. 1935. Both sides of Buka passage. An ethnographic study of social, sexual, and economic questions in the north-western Solomon Islands. Oxford, Clarendon Press, xxiii + 624p., 80 pl.

Blake, R. R., and Ramsey, G. V. 1951. Perception. An approach to personality. New York, Ronald Press Co., viii + 442p.

Blanchard, Phyllis. 1929. Sex in the adolescent girl. In: Calverton, V. F., and Schmalhausen, S. D., ed. Sex in civilization, pp. 538–561.

Blanchard, P., and Manasses, C. 1930. New girls for old. New York, Macaulay Co., xii + 281p.

Bleuler, E. 1949. Ed. 8 (orig. 1916). Lehrbuch der Psychiatrie. Berlin, Springer, x + 504p.

Bloch, I. (Dühren, E., pseud.). **1900.** Der Marquis de Sade und seine Zeit. Ein Beitrag zur Cultur- u. Sittengeschichte des 19. Jahrhunderts. Berlin und Leipzig, H. Barsdorf, vi + 502p.

1902–1903. Beiträge zur Aetiologie der Psychopathia sexualis. Pt. I and II. Dresden, H. R. Dohrn, 2v. in 1, xvi + 272 + xviii + 400p.

1908. (Paul, M. E., trans.). The sexual life of our time in its relations to modern civilization. London, Rebman, xviii + 790p.

1909 (orig. 1906). Das Sexualleben unserer Zeit in seinen Beziehungen zur modernen Kultur. Berlin, Louis Marcus, xii + 850 + xix p.

1933. (Wallis, Keene, trans.). Anthropological studies in the strange sexual practices of all races in all ages . . . New York, Anthropological Press, priv. print., ix + 246 + xxiii p.

Blumenthal, A. 1932. Small-town stuff. Chicago, University of Chicago Press, xvii + 416p.

Boas, E. P., and Goldschmidt, E. F. 1932. The heart rate. Springfield, Ill., Charles C Thomas, xi + 166p.

Boas, F. 1932. Studies in growth. Human Biol. 4:307–350.

Boas, F., ed. 1938. General anthropology. Boston, D. C. Heath and Co., xi + 718p.

Böhme, A. 1935. Psychotherapie und Kastration. München, J. F. Lehmann, 183p.

Boileau, Jacques. 1700. Historia flagellantium de recto et perverso flagrorum usu apud Christianos, ex antiquis scripturae, patrum, pontificum, conciliorum, & scriptorum profanorum monumentis cum cura & fide expressa. Paris, Joannes Anisson, 18 + 341 + 19p.
  1732 (ed. 1, 1701). Histoire des flagellans, ou l'on fait voir le bon & le mauvais usage des flagellations parmi les chrétiens. Amsterdam, Henry du Sauzet, xxxii + 316p.

Boling, J. L., et al. 1939. Post-parturitional heat responses of new born and adult guinea pigs. Data on parturition. Proc. Soc. Exper. Biol. & Med. 42:128–132.

Bölsche, W. 1926. (Brown, C., trans.). Love-life in nature. The story of the evolution of love. New York, Albert & Charles Boni, 2v. v. 1, 304p. v. 2, 726p.

Bonaparte, M. 1935. Passivity, masochism and femininity. Int. J. Psychoanal. 16:325–333.

Bonnar, A. 1941. c. 1939. The Catholic doctor. New York, P. J. Kenedy & Sons, xvii + 184p.

Bors, E., et al. 1950. Fertility in paraplegic males. A preliminary report of endocrine studies. J. Clin. Endocrinol. 10:381–398.

Boswell, J. 1950. (Pottle, F. A., ed.). London journal, 1762–1763. New York, McGraw-Hill Book Co., xxix + 370p.

Bowman, H. A. 1942. Marriage for moderns. New York and London, McGraw-Hill Book Co., ix + 493p.

Bowman, K. M. 1951. Report of Karl M. Bowman, medical superintendent of the Langley Porter Clinic. In: Brown, R. M. Progress report to the Calif. legislature . . . , pp. 148–160.
  1952. Sexual deviation research. [Report to Assembly Judiciary Committee Subcommittee of Sex Research.] Sacramento, Calif., Assembly of the State of California, 80p.

Bowman, K. M., and Engle, B. 1953. The problem of homosexuality. J. Soc. Hyg. 39:2–16.

Boys' Clubs of America. 1946. Social hygiene in a boys' club. New York, priv. print., 19p.

Brantôme, P. de B. de. 1901, 1902 (orig. 16th cent., ed. 1, 1666). (Allinson, A. R., trans.). Lives of fair and gallant ladies. Paris, Charles Carrington, 2v. in 1. v. 1, xliv + 379p. v. 2, xxiii + 464p.

Brattgård, Sven-Olof. 1950. Personlighetsattityd och sexuellt beteende. En studie på sjukhusmaterial. Svenska Läkartidningen 47:1677–1682.

Braude, J. M. 1950. The sex offender and the court. Fed. Probat. 14(3):17–22.

Brettschneider, R. 1931. Die Rolle der erotischen Photographie in der Psychopathia Sexualis. In: Wulffen, E., et al. Die Erotik in der Photographie, pp. 89–138.

Briffault, R. 1927. The mothers. A study of the origins of sentiments and institutions. New York, The Macmillan Co., 3v. v. 1, xix + 781p. v. 2, xx + 789p. v. 3, xv + 841p.

Brill, A. A. 1946. Lectures on psychoanalytic theory. New York, Alfred A. Knopf, viii + 292 + xiii p.

Brody, M. W. 1943. An analysis of the psychosexual development of a female—with special reference to homosexuality. Psychoanal. Rev. 30:47–58.

Bromley, D. D., and Britten, F. H. 1938. Youth and sex. A study of 1300 college students. New York and London, Harper & Brothers, xiii + 303p.

Brookhart, J. M., and Dey, F. L. 1941. Reduction of sexual behavior in male guinea pigs by hypothalamic lesions. Amer. J. Physiol. 133:551–554.

Brookhart, J. M., Dey, F. L., and Ranson, S. W. 1941. The abolition of mating behavior by hypothalamic lesions in guinea pigs. Endocrinol. 28:561–565.

Brooks, C. McC. 1937. The role of the cerebral cortex and of various sense organs in the excitation and execution of mating activity in the rabbit. Amer. J. Physiol. 120:544–553.

Brower, D., and Abt, L. E., ed. 1952. Progress in clinical psychology. Volume I (Section 1), xi + 328p. (Section 2), xii–xxiii + 329–564p. New York, Grune & Stratton.

Brown, F., and Kempton, R. T. 1950. Sex questions and answers. A guide to happy marriage. New York, McGraw-Hill Book Co., xiv + 264p.

Brown, J. F. 1940. The psychodynamics of abnormal behavior. New York and London, McGraw-Hill Book Co., xvi + 484p.

Brown, R. M. 1950. Preliminary report of the Subcommittee on Sex Crimes of the Assembly Interim Committee on Judicial System . . . Sacramento, California State Legislature, 269p., 1 fold. pl.

 1951. Progress report to the legislature . . . by the Assembly Interim Committee on Judicial System and Judicial Process . . . Sacramento, California, Assembly, 170p.

Bruckner, P. J. 1937. How to give sex instructions. A guide for parents, teachers and others responsible for the training of young people. St. Louis, Mo., The Queen's Work, 64p.

Brusendorff, O. 1938. (Rimestad, C., ed.). Erotikens historie, fra Graekenlands oldtil til vore dage. København, Universal-Forlaget, 3v. v. 1, 423p. v. 2, 598p. v. 3, 475p.

Bryk, F. 1933 (orig. 1928). (Sexton, M. F., trans.). Voodoo-Eros. Ethnological studies in the sex of the African aborigines. New York, priv. print., 351p.

 1934. (Berger, D., trans.). Circumcision in man and woman. Its history, psychology, and ethnology. New York, American Ethnological Press, 342p.

Buck, W. 1936. Measurements of changes in attitudes and interests of university students over a ten-year period. J. Abnorm. & Soc. Psychol. 31:12–19.

Budge, E. A. W. 1895. The book of the dead. The papyrus of Ani in the British Museum. The Egyptian text with interlinear transliteration and translation, a running translation, introduction, etc. London, British Museum, clv + 377p.

Bühler, C. 1927. Vergleich der Pubertätsentwicklung bei Knaben und Mädchen. In: Stern, E., ed. Die Erziehung und die sexuelle Frage, pp. 155–169.

 1928. Männliche und weibliche Pubertätsentwicklunk. In: Marcuse, M., ed. Verhandlungen des I. International Kongresses für Sexualforschung, v. 3, pp. 35–41.

 1931. Zum Problem der sexuellen Entwicklung. Ztschr. f. Kinderheilkunde 51:612–641.

Bulliet, C. J. 1932. Modern painting. In: McDermott, J. F., and Taft, K. B., ed. Sex in the arts, pp. 233–252.

Bundesen, H. N. 1951. Toward manhood. Philadelphia and New York, J. B. Lippincott Co., 175p.

Burgess, E. W., and Cottrell, L. S., Jr. 1939. Predicting success or failure in marriage. New York, Prentice-Hall, xxiii + 472p.

Burrows, H. 1949. Ed. 2. Biological actions of sex hormones. Cambridge University Press, xiii + 615p.

Butterfield, O. M. 1939. Love problems of adolescence. New York, Emerson Books, viii + 212p.

 1940. Sex life in marriage. New York, Emerson Books, 192p.

Calverton, V. F., and Schmalhausen, S. D., ed. 1929. Sex in civilization. New York, Macaulay, 719p., 1 pl.

Campbell, E. H. 1939. The social-sex development of children. Genetic Psychol. Monogr. 21:461–552.

Canby, H. S. 1934. Sex and marriage in the nineties. Harper's Magazine, Sept., pp. 427–436.

Cannon, W. B. 1920. Bodily changes in pain, hunger, fear and rage. New York & London, D. Appleton and Co., xiii + 311p.

Canu, É. 1897. Résultats thérapeutiques de la castration chez la femme . . . Paris, Ollier-Henry, 188p.

Carmichael, L., ed. 1946. Manual of child psychology. New York, John Wiley & Sons; London, Chapman & Hall, viii + 1068p.

Carpenter, C. R. 1942. Sexual behavior of free ranging rhesus monkeys (Macaca mulatta). J. Comp. Psychol. 33:113–162.

Carpenter, E. 1930 (orig. 1908). The intermediate sex. A study of some transitional types of men and women. London, George Allen & Unwin, 176p.

Carpenter, E., ed. 1902. Ioläus. An anthology of friendship. London, Swan Sonnenschein & Co.; Manchester, priv. print.; Boston, Charles E. Goodspeed, vi + 191p.

Carpenter, N. 1932. Courtship practices and contemporary social change in America. Ann. Amer. Acad. Polit. & Soc. Sci. 160:38–44.

Carter, A. C., Cohen, E. J., and Shorr, E. 1947. The use of androgens in women. Vitamins and Hormones 5:317–391.

Cary, H. N. 1948, ex ?1916. Sexual vocabulary . . . Compiled from manuscript, carbon, and printed materials . . . [from] the files of H. N. Cary . . . Bloomington, Ind., [Institute for Sex Research], Ms, typed, 5v. v. 1–3, n.pag., ca. 262p. ea. v. v. 4, 788–1049p. v. 5, 1050–1396p.

Casañ, V. S. n.d. (ca. ?1900). Conocimientos para la vida privada. I. La prostitución. II. Secretos del lecho conyugal. III. La virginidad. IV. Onanismo conyugal. V. Los vicios solitarios. Barcelona, Maucci, 5v. in 1, 126 + 111 + 95 + 92 + 95p.

Catullus. 1894 (orig. 1st cent. B.C.). (Burton, R. F., and Smithers, L. C., trans.). The Carmina of Caius Valerius Catullus. London, priv. print., xxiii + 313p.

Catullus, et al. 1913 (orig. 1st cent. B.C.). (Cornish, F. W., et al., trans.). Catullus, Tibullus and Pervigilium Veneris. Cambridge, Mass., Harvard University Press, xi + 376p.

Caufeynon, Dr., pseud. of Fauconney, J. [1902.] La masturbation et la sodomie feminine. Paris, Administration de la Vie en Culotte Rouge, 112p.

1903. Orgasme. Sens genital jadis et aujourd'hui . . . Paris, Charles Offenstadt, 237p.

1934. Unisexual love. A documentary study of the sources, manifestations, the physiology and psychology of sexual perversion in the two sexes. New York, New Era Press, 164p.

Cave, F. C. 1911. Report of sterilization in the Kansas State Home for Feebleminded. J. Psycho-Asthenics 15:123–125.

Cheng, Peilieu, et al. 1950. Different intensities of sexual activity in relation to the effect of testosterone propionate in the male rabbit. Endocrinol. 46:447–452.

Chesser, E. 1947. Love without fear. How to achieve sex happiness in marriage. New York, Roy, 307p.

Chideckel, M. 1935. Female sex perversion; the sexually aberrated woman as she is. New York, Eugenics Publishing Co., xviii + 331p.

Childers, A. T. 1936. Some notes on sex mores among Negro children. Amer. J. Orthopsychiat. 6:442–448.

Chivers, W. R., ed. 1946. Successful marriage and family living. Atlanta, Ga., Morehouse College, Department of Sociology Publication Number 1, 8 + 47p.

Chlenov. See Feldhusen, F.

Christensen, H. T. 1950. Marriage analysis. Foundations for successful family life. New York, Ronald Press Co., viii + 510p.

1952. Dating behavior as evaluated by high-school students. Amer. J. Sociol. 57:580–586.

Clark, G. 1942. Sexual behavior in rats with lesions in the anterior hypothalamus. Amer. J. Physiol. 137:746–749.

1945. Prepubertal castration in the male chimpanzee, with some effects of replacement therapy. Growth 9:327–339.

Clark, G., and Birch, H. G. 1945. Hormonal modifications of social behavior. I. The effect of sex-hormone administration on the social status of a male-castrate chimpanzee. Psychosom. Med. 7:321–329.

Clark, L. C., and Treichler, P. 1950. Psychic stimulation of prostatic secretion. Psychosom. Med. 12:261–263.

Clark, LeMon. 1937. Emotional adjustment in marriage. St. Louis, C. V. Mosby Co., 261p.

1949. Sex and you. Indianapolis and New York, The Bobbs-Merrill Co., 203p.

1952. A further report on the virginity of unmarried American women. Int. J. Sexol. 6:27–32.

Clark, W. E., ed. 1937. Two lamaistic pantheons. Cambridge, Mass., Harvard University Press, 2v. v. 1, [Indexes] xxiv + 169p. v. 2, [Plates] 314p.

Clark, W. L., and Marshall, W. L. (Kearney, J. J., ed.). 1940 (ed. 1, 1900). A treatise on the law of crimes. Chicago, Callaghan and Co., lxiii + 724p.

Clement of Alexandria. 1885 (orig. 2nd cent. A.D.). The instructor (Paedagogus). In: Roberts, A., Donaldson, J., and Coxe, A. C., ed. The Ante-Nicene fathers, pp. 209–296.

Cleugh, J. 1951. The Marquis and the chevalier. A study of sex as illustrated by the lives . . . of the Marquis de Sade . . . and the Chevalier von Sacher-Masoch . . . London, Andrew Melrose, 255p., 2 pl.

Clinton, C. A. 1935. Sex behavior in marriage. New York, Pioneer Publications, 159p.

Cochran, W. G. 1953. Sampling techniques. New York, John Wiley & Sons; London, Chapman & Hall, xiv + 330p.

Cohen, A. 1949. Everyman's Talmud. New York, E. P. Dutton & Co., xli + 403p.

Colmeiro-Laforet, C. 1952. Lack of orgasm as a cause of sterility. Int. J. Sexol. 5:123–127.

Comfort, A. 1950. Sexual behavior in society. London, Gerald Duckworth & Co., 158p.

Commins, W. D., and Stone, C. P. 1932. Effects of castration on the behavior of mammals. Psychol. Bull. 29:493–508.

Conn, J. H. 1939. Factors influencing development of sexual attitudes and sexual awareness in children. Amer. J. Dis. Child. 58:738–745.

1940. Sexual curiosity of children. Amer. J. Dis. Child. 60:1110–1119.

Cooper, W. M. See Bertram, J. G.

Coppens, C. (Spalding, H. S., ed.). 1921. Moral principles and medical practice. The basis of medical jurisprudence. New York, Benziger Brothers, 320p.

Corner, G. W. 1942. The hormones in human reproduction. Princeton, N. J., Princeton University Press, xix + 265p.

1951. The reproductive cycle of the rhesus monkey. Amer. Scientist 39: 50–74, 109.

1952. Attaining womanhood. A doctor talks to girls about sex. New York, Harper & Brothers, xi + 112p.

Corner, G. W., and Landis, C. 1941. Sex education for the adolescent. Hygeia (July), 18p.

Corpus inscriptionum latinarum, consilio et auctoritate Academiae litterarum regiae borussicae editum . . . 1862–1952. 16v. Berlin, G. Reimer.

Cory, D. W., pseud. 1951. The homosexual in America. A subjective approach. New York, Greenberg, xvii + 326p.

Cowdry, E. V. (Lansing, A. I., ed.). 1952. Ed. 3 (ed. 1, 1939). Problems of ageing. Biological and medical aspects. Baltimore, Williams & Wilkins Co., xxiii + 1061p.

Crawley, E. 1927. Ed. 2. (Besterman, T., ed.) The mystic rose. A study of primitive marriage and of primitive thought in its bearing on marriage. New York, Boni and Liveright, 2v. v. 1, xx + 375p. v. 2, vii + 340p.

Crespi, L. P. 1951. Germans view the U. S. reorientation program. I. Extent of receptivity to American ideas. II. Reactions to American democratization efforts. Int. J. Opin. & Attit. Res. 5:179–190, 335–346.

Crisp, K. B. 1939. Growing into maturity. Chicago, J. B. Lippincott Co., 38p.

Crossley, H. M., and Fink, R. 1951. Response and non-response in a probability sample. Int. J. Opin. & Attit. Res. 5:1–19.

Crouch, N. See Bekker, ?B.

Cuber, J. F., and Pell, B. 1941. A method for studying moral judgments relating to the family. Amer. J. Sociol. 47:12–23.

Curtis, A. H. 1946. Ed. 5. A textbook of gynecology. Philadelphia and London, W. B. Saunders Co., 755p.

Daly, M., ed. 1951. Profile of youth. By members of the staff of the Ladies Home Journal. Philadelphia and New York, J. B. Lippincott Co., 256p.

Daniels, G. E., and Tauber, E. S. 1941. A dynamic approach to the study of replacement therapy in cases of castration. Amer. J. Psychiat. 97:905–918.

Davis, C. D. 1939. The effect of ablations of the neocortex on mating, maternal behavior and the production of pseudopregnancy in the female rat and on copulatory activity in the male. Amer. J. Physiol. 127:374–380.

Davis, H. 1946. Ed. 5 (ed. 1, 1935). Moral and pastoral theology, in four volumes. Volume one. Human acts, law, sin, virtue, xix + 361p. Volume two. Commandments of God. Precepts of the Church, x + 463p. Volume three. The sacraments in general, baptism, confirmation, holy eucharist, penance, indulgences, censures, xviii + 504p. Volume four. Extreme unction, holy orders, marriage, the clerical state, the religious state, duties of laypeople, xiii + 432p. New York, Sheed and Ward.

Davis, Katharine B. 1929. Factors in the sex life of twenty-two hundred women. New York and London, Harper & Brothers, xx + 430p.

Dearborn, L. W. 1947. Extramarital relations. In: Fishbein, M., and Burgess, E. W., ed. Successful marriage, pp. 163–173.
    1947. Masturbation. In: Fishbein, M., and Burgess, E. W., ed. Successful marriage, pp. 356–367.
    1952. The problem of masturbation. Marr. & Family Liv. 14:46–55.

Decurtins, F. 1950. Ärztliches über die Enthaltsamkeit. In: Hornstein, X. von, and Faller, A., ed. Gesundes Geschlechts Leben, pp. 132–142.

Dell, F. 1930. Love in the machine age. A psychological study of the transition from patriarchal society. New York, Farrar & Rinehart, 428p.

Deming, W. E. 1950. Some theory of sampling. New York, John Wiley & Sons, xvii + 602p.

Dempsey, E. W. 1951. Homeostasis. In: Stevens, S. S., ed. Handbook of experimental psychology, pp. 209–235.

Dempsey, E. W., and Rioch, D. McK. 1939. The localization in the brain stem of the oestrus responses of the female guinea pig. J. Neurophysiol. 2:9–18.

Dennis, W. 1946. The adolescent. In: Carmichael, L., ed. Manual of child psychology, pp. 633–666.

DeRiver, J. P. 1949. The sexual criminal. A psychoanalytical study. Springfield, Ill., Charles C Thomas, xvii + 281p.

Deutsch, Helene. 1933. Homosexuality in women. Int. J. Psychoanal. 14:34–56.
    1944, 1945. The psychology of women. A psychoanalytic interpretation. Volume one, vii + 399p. Volume two, Motherhood, vi + 498p. New York, Grune & Stratton.

Devereux, G. 1936. Sexual life of the Mohave Indians. An interpretation in terms of social psychology. (Ph.D. dissertation.) Berkeley, Calif., University of California, Ms, ii + 116 + vi p.

1937. Institutionalized homosexuality of the Mohave Indians. Human Biol. 9:498–527.

Dey, F. L., Leininger, C. R., and Ranson, S. W. 1942. The effect of hypophysical lesions on mating behavior in female guinea pigs. Endocrinol. 30:323–326.

Dickerson, R. E. 1931 (c. 1930). So youth may know. New viewpoints on sex and love. New York, Association Press, x + 255p.

1947. Home study course [in] social hygiene guidance. Lessons I–VI. Portland, Ore., E. C. Brown Trust for Social Hygiene Education, 72p.

Dickinson, R. L. 1933. Human sex anatomy. Baltimore, Williams and Wilkins Co., vii + 145p. + 175 pl.

1940. Chapter 16. Autoerotism. [From unpublished manuscript entitled The doctor as marriage counselor.] Ms, 27p.

1949. Ed. 2. Human sex anatomy. A topographical hand atlas. Baltimore, Williams & Wilkins Co., 21 + 145p., 204 pl.

Dickinson, R. L., and Beam, L. 1931. A thousand marriages. A medical study of sex adjustment. Baltimore, Williams & Wilkins Co., xxv + 482p.

1934. The single woman. A medical study in sex education. Baltimore, Williams & Wilkins Co., xix + 469p.

Dickson, H. R. P. 1949. The Arab of the desert. A glimpse into Badawin life in Kuwait and Sau'di Arabia. London, George Allen & Unwin, 664p., 49 pl., 8 col. pl., + 3 maps + 6 geneal. tab.

Diehl, E. L. T., ed. 1910. Pompeianische Wandinschriften und Verwandtes. Bonn, A. Marcus und E. Weber, 60p.

Dillon, M. S. 1934. Attitudes of children toward their own bodies and those of other children. Child Develop. 5:165–176.

Dinerman, H. 1949. 1948 votes in the making—a preview. Pub. Opin. Quart. 12:585–598.

Döderlein, A., and Krönig, B. 1907. Operative Gynäkologie. Leipzig, Georg Thieme, xvi + 721p.

Doe-Kulmann, L., and Stone, C. P. 1927. Notes on the mental development of children exhibiting the somatic signs of puberty praecox. J. Abnorm. & Soc. Psychol. 22:291–324.

Dollard, J. 1949. (Ed. 1, 1937.) Caste and class in a southern town. New York, Harper & Brothers, xvi + 502p.

Donohue, J. F. 1931. The impediment of crime; an historical synopsis and commentary. Washington, D. C., The Catholic University of America, viii + 110p.

Dorfman, R. I. 1948. Biochemistry of androgens. In: Pincus, G., and Thimann, K. V., ed. The hormones, pp. 467–548.

Dorfman, R. I., et al. 1947. Metabolism of the steroid hormones: Studies on 17-ketosteroids and androgens. Endocrinol. 41:470–488.

Dragoo, D. W. 1950. Transvestites in North American tribes. Ms, 12p.

Dublin, L. I., and Spiegelman, M. 1951. The facts of life, from birth to death. New York, The Macmillan Co., 461p.

Du Bois, C. 1944. The people of Alor. A social-psychological study of an East Indian Island. Minneapolis, University of Minnesota Press, xvi + 654p., 32 pl.

Dubois, J. 1848. The secret habits of the female sex . . . New York, priv. print., 185p., 4 pl.

Dubois-Desaulle, G. 1933. (A. F. N., trans. and ed.) Bestiality. An historical, medical, legal and literary study. New York, Panurge Press, priv. print., 300p.

Dück, J. 1914. Aus dem Geschlechtsleben unserer Zeit. Eine kritische Tatsachenschilderung. II. Der erste Geschlechtsverkehr. III. Die Masturbation. Sexual-Probleme 10:545–556, 713–766.

1941. Virginität und Ehe. Reihenuntersuchung aus der allgemeinen ärztlichen Praxis. München, priv. print., 11p.

1949. [Two charts received through correspondence:] Sources of first outlet of sexuality. First outlet of sexuality [by age].

Dühren, E. (pseud.). *See* Bloch, I.

Dunn, C. W. 1940. Stilbestrol-induced gynecomastia in the male. J. Amer. Med. Assoc. 115:2263–2264.

1941. Stilbestrol induced testicular degeneration in hypersexual males. J. Clin. Endocrinol. 1:643–648.

Durand-Wever, A.-M. 1952. The influence of the nervous system on the structure and functions of human genital organs. Int. J. Sexol. 5:209–211.

Durfee, T., Lerner, M., and Kaplan, N. 1940. The artificial production of seminal ejaculation. Anat. Rec. 76:65–68.

Duvall, E. M. 1947. Courtship and engagement. *In:* Fishbein, M., and Burgess, E. W., ed. Successful marriage, pp. 32–43.

1950. Facts of life and love for teenagers. New York, Association Press, xx + 360p. (incl. xiv + 16 pl.).

Duvall, E. M., and Hill, R. 1945. When you marry. Boston, D. C. Heath & Co., xiv + 450p.

Duvall, S. M. 1952. Men, women, and morals. New York, Association Press, xvi + 336p.

East, N. 1951. Society and the criminal. Springfield, Ill., Charles C Thomas, x + 437p.

Eberhard, E. F. W. 1924. Die Frauenemanzipation und ihre erotischen Grundlagen. Wien und Leipzig, Wilhelm Braumüller, x + 916p.

1927. Feminismus und Kulturuntergang. Die erotischen Grundlagen der Frauenemanzipation. Wien und Leipzig, Wilhelm Braumüller, xv + 654p.

Eberth, C. J. 1904. Die männlichen Geschlechtsorgane. Jena, Gustav Fischer, xi + 310p.

Edson, N. W. 1936. Love, courtship and marriage. New York, American Social Hygiene Association, Publication No. 932, 18p.

Ehrmann, W. W. 1952. [Correspondence and tabulated data from a study entitled Premarital dating behavior.] Ms, n.pag.

1953. Premarital dating behavior. New York, Dryden Press (in preparation).

Eissler, K. R., et al., ed. 1949. Searchlights on delinquency. New psychoanalytic studies . . . New York, International Universities Press, xviii + 456p.

Elkisch, P. 1945. Children's drawings in a projective technique. Psychol. Monogr., v. 58, no. 1, 31p.

Elliott, G. L., and Bone, H. 1929. The sex life of youth. New York, Association Press, xi + 142p.

Ellis, Albert. 1947. Questionnaire versus interview methods in the study of human love relationships. Amer. Sociol. Rev. 12:541–553.

1948. Questionnaire versus interview methods in the study of human love relationships. II. Uncategorized responses. Amer. Sociol. Rev. 13:61–65.

1949. A study of human love relationships. J. Genetic Psychol. 75:61–71.

1951. The folklore of sex. New York, Charles Boni, 313p.

Ellis, Havelock. 1905. Studies in the psychology of sex. Volume IV. Sexual selection in man. I. Touch. II. Smell. III. Hearing. IV. Vision. Philadelphia, F. A. Davis Co., xi + 270p.

1906. Studies in the psychology of sex. Volume V. Erotic symbolism. The mechanism of detumescence. The psychic state in pregnancy. Philadelphia, F. A. Davis Co., x + 285p.

1910. Ed. 3 (ed. 1, 1900). Studies in the psychology of sex. Volume I. The evolution of modesty. The phenomena of sexual periodicity. Autoerotism. Philadelphia, F. A. Davis Co., xv + 352p.

1910. Studies in the psychology of sex. Volume VI. Sex in relation to society. Philadelphia, F. A. Davis Co., xvi + 656p.

1913. Ed. 2 (ed. 1, 1903). Studies in the psychology of sex. Volume III.

Analysis of the sexual impulse. Love and pain. The sexual impulse in women. Philadelphia, F. A. Davis Co., xii + 353p.
1915. Ed. 3 (ed. 1, 1901). Studies in the psychology of sex. Volume II. Sexual inversion. Philadelphia, F. A. Davis Co., xi + 391p.
1921 (orig. 1897). (Gennep, A. van, trans.). Études de psychologie sexuelle. II. L'inversion sexuelle, 338p. III. L'impulsion sexuelle, viii + 439p. Paris, Mercure de France.
1925. (Gennep, A. van, trans.). Études de psychologie sexuelle. V. Le symbolisme érotique. Le mecanisme de la détumescence. Paris, Mercure de France, 284p.
1926. (Gennep, A. van, trans.). Études de psychologie sexuelle. VI. L'état psychique pendant la grossesse. La mère et l'enfant. Paris, Mercure de France, 212p.
1927. (Gennep, A. van, trans.). Études de psychologie sexuelle. VII. L'éducation sexuelle. Paris, Mercure de France, 220p.
1928. (Gennep, A. van, trans.). Études de psychologie sexuelle. VIII. L'évaluation de l'amour. La chasteté. L'abstinence sexuelle. Paris, Mercure de France, 219p.
1929. Man and woman. A study of secondary and tertiary sexual characters. Boston and New York, Houghton Mifflin Company, vi + 495p.
1934. (Gennep, A. van, trans.). Études de psychologie sexuelle. XVII. Les charactères sexuels physiques secondaires et tertiaires. Paris, Mercure de France, 252p.
1936. Studies in the psychology of sex. New York, Random House, 6v. in 2. v. 1 pt. 1, xxxix + 339p. pt. 2, xii + 353p. pt. 3, xi + 391p. v. 2 pt. 1, x + 285p. pt. 2, xvi + 539p. pt. 3, xvi + 750p.
1952. (Gawsworth, J., ed.). Sex and marriage. Eros in contemporary life. New York, Random House, xiii + 219p.
Ellis, H., and Moll, A. 1911, 1921, ed. 2. 3. Autoerotische Äusserungen des Geschlechtstriebes. B. Erotisches Nachtträumen. *In:* Moll, Albert, ed. Handbuch der Sexualwissenschaften, pp. 612–616.
Elsberg, C. A. 1925. Tumors of the spinal cord. New York, Paul B. Hoeber, viii + 421p.
Elwin, V. 1947. The Muria and their ghotul. Bombay, Geoffrey Cumberlege, Oxford University Press, xxix + 730p., 1 col. pl.
Emmens, C. W., and Parkes, A. S. 1947. Effect of exogenous estrogens on the male mammal. Vitamins and Hormones 5:233–272.
Enders, R. K. 1945. Induced changes in the breeding habits of foxes. Sociometry 8:53–55.
Engel, G. L. 1950. Fainting; physiological and psychological considerations. Springfield, Ill., Charles C Thomas, xii + 141p.
England, L. R. 1949–1950. Little Kinsey: An outline of sex attitudes in Britain. Pub. Opin. Quart. 13:587–600.
England, L. R. 1950. A British sex survey. Int. J. Sexol. 3:148–154.
Engle, E. T. 1942. The testis and hormones. *In:* Cowdry, E. V., ed. Problems of ageing, pp. 475–494.
1952. The male reproductive system. *In:* Cowdry, E. V. Problems of ageing, pp. 708–729.
Engle, E. T., and Shelesnyak, M. C. 1934. First menstruation and subsequent menstrual cycles of pubertal girls. Human Biol. 6:431–453.
Englisch, P. 1931. Sittengeschichte Europas. Berlin, Gustav Kiepenheuer; Wien, Phaidon-Verlag, 442p., 139 pl.
English, O. S. 1947. Sexual adjustment in marriage. *In:* Fishbein, M., and Burgess, E. W., ed. Successful marriage, pp. 102–116.
English, O. S., and Pearson, G. H. J. 1945. Emotional problems of living. Avoiding the neurotic pattern. New York, W. W. Norton & Co., 438p.
Epstein, L. M. 1948. Sex laws and customs in Judaism. New York, Bloch Publishing Co., x + 251p.

Escamilla, R. F., and Lisser, H. 1941. Testosterone therapy of eunuchoids. II. Clinical comparison of parenteral implantation, and oral administration of testosterone compounds in male eunuchoidism. J. Clin. Endocrinol. 1:633–642.

Eulenburg, A. 1902. Sadismus und Masochismus. Wiesbaden, J. F. Bergmann, 89p.
    1934. (Kent, H., trans.). Algolagnia. The psychology, neurology and physiology of sadistic love and masochism. New York, New Era Press, 200p., 10 pl.

Evans, C. B. S. 1941 (orig. 1935). Sex practice in marriage. New York, Emerson Books, 128p.

Evans-Pritchard, E. E. 1940. The Nuer . . . The modes of livelihood and political institutions of a Nilotic people. Oxford, Clarendon Press, xii + 271p., 30 pl.

Everett, M. S. 1948. Preparation for marriage. Chicago, Roosevelt College, 41p.

Exner, M. J. 1915. Problems and principles of sex instruction. A study of 948 college men. New York, Association Press, 39p.
    1933. The question of petting. New York, American Social Hygiene Association (Publication No. 853), 14p.

Faegre, M. L. 1943. Understanding ourselves. A discussion of social hygiene for older boys and girls. Minneapolis, Minn., University of Minnesota Press, 44p.

Faller, A. 1950. Befruchtung, Vererbung, Geschlechtsbestimmung, eheliche Vereinigung. *In:* Horstein, X. von, and Faller, A., ed. Gesundes Geschlechts Leben, pp. 221–240.

Famin, César. 1832. Peintures, bronzes et statues erotiques, formant la collection du cabinet secret du Musée Royal de Naples. Paris, Typographie Éverat, 116p., xli pl.

Famin, César (Colonel Fanin, pseud.). **1871** (orig. 1832). The Royal Museum at Naples, being some account of the erotic paintings, bronzes, and statues contained in that famous "Cabinet Secret." London, priv. print., xviii + 122p., 60 col. pl.

Farmer, J. S., and Henley, W. E., comp. and ed. **1890–1904.** Slang and its analogues past and present . . . with synonyms in English, French, German, Italian, etc. . . . [London], priv. print., 7v. v. 1, x + 405p. v. 2, 406p. v. 3, 387p. v. 4, 399p. v. 5, 381p. v. 6, 378p. v. 7, 380p.

Farnham, M. F. 1951. The adolescent. New York, Harper & Brothers, xi + 243p.

Farris, E. J., ed. **1950a.** The care and breeding of laboratory animals. New York, John Wiley & Sons, etc., xvi + 515p.
    1950b. Human fertility and problems of the male. White Plains, N. Y., Author's Press, xvi + 211p.

Fauconney, J. *See* Caufeynon, Dr.

Federal Reserve Bulletin. 1946–1953. Annual surveys of consumer finances. Volumes 32–39.

Federn, P. 1927–1928. Die Wiener Diskussion aus dem Jahre 1912. Ztschr. f. psa. Pädagogik 2:106–112.

Fehlinger, H. 1945 (orig. 1921). (Herbert, S. and Mrs. S., trans.). Sexual life of primitive people. New York, United Book Guild, 133p.

Feinier, L., and Rothman, T. 1939. Study of a male castrate. J. Amer. Med. Assoc. 113:2144–2146.

Feldhusen, F. 1909. Die Sexualenquete unter der Moskauer Studentenschaft. [Chlenov Study, 1904.] Ztschr. f. Bekämpfung der Geschlechtskrankh. 8:211–224, 245–255.

Fenichel, O. 1945. Ed. 2 (ed. 1, 1924). The psychoanalytic theory of neurosis. New York, W. W. Norton & Company, x + 703p.

Féré, C. S. 1904. The evolution and dissolution of the sexual instinct. Paris, Charles Carrington, xxiv + 358p.

1932. Scientific and esoteric studies in sexual degeneration in mankind and in animals. New York, Anthropological Press, priv. print., xiv + 325p.

Ferenczi, S. 1936. (Bunker, H. A., trans.). Male and female: Psychoanalytic reflections on the "theory of genitality," and on secondary and tertiary sex differences. Psychoanal. Quart. 5:249–260.

Ferguson, L. W. 1938. Correlates of woman's orgasm. J. Psychol. 6:295–302.

Fetscher, R. 1928. Der Geschlechtstrieb. Einführung in die Sexualbiologie unter besonderer Berücksichtigung der Ehe. München, Ernst Reinhardt, 156p.

Fifteen plagues of a maiden-head. See Anon. The fifteen plagues . . .

Filler, W., and Drezner, N. 1944. The results of surgical castration in women under forty. Amer. J. Obstet. & Gynec. 47:122–124.

Finger, F. W. 1947. Sex beliefs and practices among male college students. J. Abnorm. & Soc. Psychol. 42:57–67.

Fink, L. A. 1950. Premarital sex experience of girls in Sidney: A survey of 100 girls. Int. J. Sexol. 4:33–35.

Firth, R. W. 1936. We, the Tikopia; a sociological study of kinship in primitive Polynesia. New York, American Book Co., xxv + 605p.

Fishbein, M., and Burgess, E. W., ed. 1947. Successful marriage. An authoritative guide to problems related to marriage from the beginning of sexual attraction to matrimony and the successful rearing of a family. Garden City, N. Y., Doubleday & Co., xxi + 547p.

Fisher, Mary S. 1938. Romance and realism in love and marriage. In: Folsom, J. K., ed. Plan for marriage, pp. 1–28.

Fletcher, J. [1849]. Summary of the moral statistics of England and Wales. London, priv. print., xi + 228p., xii pl.

Fleuret, F. See Hernandez, L.

Flood, [E.]. [?1901]. Emasculation in twenty-six cases. n.impr., 19p. [Probably address to Mass. Society for Mental Hygiene.]

Folsom, J. K. 1937. Changing values in sex and family relations. Amer. Sociol. Rev. 2:717–726.

1942. Love and courtship. In: Becker, H., and Hill, R., ed. Marriage and the family, pp. 153–189.

Folsom, J. K., ed. 1938. Plan for marriage. An intelligent approach to marriage and parenthood proposed by the staff of Vassar College. New York and London, Harper & Brothers, xii + 305p.

Foras, A. de. 1886. Le droit du seigneur au moyen-age. Étude critique et historique. Chambery, André Perrin, xix + 281p.

Forberg, F. K. 1884 (orig. 1824). (Smithson, J., trans.). Manual of classical erotology (De figuris Veneris). "Manchester, Julian Smithson, priv. print . . ." [Brussels, Charles Carrington], 2v. v. 1, xviii + 261p., 10 pl. v. 2, 250p., 8 pl.

Ford, C. S. 1945. A comparative study of human reproduction. New Haven, Conn., Yale University Press, 111p.

Ford, C. S., and Beach, F. A. 1951. Patterns of sexual behavior. New York, Harper & Brothers, and Paul B. Hoeber, ix + 307p.

Forel, A. 1905. Die sexuelle Frage. Eine naturwissenschaftliche, psychologische, hygienische und soziologische Studie für Gebildete. München, Ernst Reinhardt, viii + 587p.

1922 (ex German ed. 2, 1906). (Marshall, C. F., trans.). The sexual question. A scientific, psychological, hygienic and sociological study. Brooklyn, N. Y., Physicians and Surgeons Book Co., xv + 536p.

Fortune, R. F. 1932. Sorcerers of Dobu. The social anthropology of the Dobu Islanders of the western Pacific. New York, E. P. Dutton & Co., xxviii + 318p., viii pl.

Fortune Quarterly Survey: VII. 1937. Fortune (Jan.) 15:86–87, 150–156, 162–168.

Fortune Survey. 1943. Fortune (Aug.) 28:10–30.

Foster, R. G. 1950 (ed. 1, 1944). Marriage and family relationships. New York, The Macmillan Co., xvi + 316p.

Fowler, O. S. 1875. Sexual science; including manhood, womanhood, and their mutual interrelations; love its laws, power etc. . . . as taught by phrenology. Philadelphia, National Publishing Co., xxx + 930p.

Frank, A. 1952 (orig. 1947). (Mooyaart, B. M., trans.). Anne Frank: The diary of a young girl. Garden City, N. Y., Doubleday & Company, 285p.

Frank, R. T. 1940. The sex hormones: Their physiologic significance and use in practice. J. Amer. Med. Assoc. 114:1504–1512.

Franz, S. I. 1913. The accuracy of localization of touch stimuli on different bodily segments. Psychol. Rev. 20:107–128.

Fraser, R. W., et al. 1941. Colorimetric assay of 17-ketosteroids in urine. A survey of the use of this test in endocrine investigation, diagnosis, and therapy. J. Clin. Endocrinol. 1:234–256.

Freeman, G. L. 1948. The energetics of human behavior. Ithaca, N. Y., Cornell University Press, vii + 344p.

Freeman, K. 1949. The pre-socratic philosophers. A companion to Diels, Fragmente der Vorsokratiker. Oxford, Basil Blackwell, xiii + 486p.

Freud, Anna. 1923. The relation of beating-phantasies to a daydream. Int. J. Psychoanal. 4:89–102.

    1951. Observations on child development. Psychoanalytic study of the child 6:18–30.

Freud, Sigmund. 1910. [Letter to Dr. F. S. Krauss]. Anthropophyteia 7:472–474.

    1910. (Brill, A. A., trans.). Three contributions to the sexual theory. New York, Journal of Nervous & Mental Disease Publishing Co., x + 91p.

    1912. In: Wiener psychoanalytische Vereinigung. Die Onanie, pp. 133–140.

    1921. (Paul, E. and C., trans.). Preface to: A young girl's diary. New York, Thomas Seltzer, xvi + 284p.

    1922 (orig. 1910). (Brill, A. A., trans.). Leonardo da Vinci. A psychosexual study of an infantile reminiscence. London, Kegan Paul, Trench, Trubner & Co., 130p., 4 pl.

    1924. (Riviere, J., and Strachey, A. and J., trans.). Collected Papers. London, Hogarth Press and Institute of Psychoanalysis, 4v. v. 1, 359p. v. 2, 404p. v. 3, 607p. v. 4, 508p.

    1924 (orig. 1919). 'A child is being beaten.' A contribution to the study of the origin of sexual perversions. In: Collected papers, v. 2, pp. 172–201.

    1933. (Sprott, W. J. H., trans.). New introductory lectures on psychoanalysis. New York, W. W. Norton & Co., 257p.

    1935. (Riviere, J., trans.). A general introduction to psychoanalysis. New York, Perma Giants, 412p.

    1938. (Brill, A. A., ed.). The basic writings of Sigmund Freud. New York, Modern Library, vi + 1001p.

    1949 (orig. 1940). (Strachey, J., trans.). An outline of psychoanalysis. New York, W. W. Norton & Co., 127p.

    1950. (Strachey, J., trans. and ed.). Collected papers. Volume V. London, Hogarth Press and Institute of Psycho-Analysis, 396p.

    1950 (orig. 1931). Female homosexuality. In: Collected papers, v. 5, pp. 252–272.

Friedeburg, L. von. [1950]. Ein Versuch ueber Meinung und Verhalten im Bereich der Zwischengeschlechtlichen Beziehungen in Deutschland . . . Allensbach, Institut für Demoskopie etc., 40p. (mimeo.).

Friedjung, J. K. 1923. Die kindliche Sexualität und ihre Bedeutung für Erziehung und ärztliche Praxis. Berlin, Julius Springer, 37p.

Friedlaender, K. F. 1921. (Hirschfeld, Magnus, ed.). Die Impotenz des Weibes. Leipzig, Ernst Bircher (Sexus Monographien no. 2), xii + 88p.

Fromme, A. 1950. The psychologist looks at sex and marriage. New York, Prentice-Hall, xv + 248p.

Fuchs, E. 1909–1912. Illustrierte Sittengeschichte vom Mittelalter bis zur Gegenwart. 1. Renaissance. 2. Die galante Zeit. 3. Das bürgerliche Zeitalter.

München, Albert Langen, 3v. v. 1, 1909, x + 500p., 59 pl. v. 2, 1910, x + 484p., 65 pl. v. 3, 1912, x + 496p., 63 pl.

Fuller, J. K. 1950. Communication from Justin K. Fuller, M.D., Medical Consultant to Department of Corrections. *In:* Kilpatrick, V., et al. Partial report of Assembly Interim Committee on crime and corrections (California), pp. 68–72.

Fulton, J. F. 1949. Ed. 3. Physiology of the nervous system. New York, Oxford University Press, xii + 667p.

Fulton, J. F., ed. 1949. *See* Howell, W. H.

Fulton, J. F., Ranson, S. W., and Frantz, A. M., ed. 1940. The hypothalamus and central levels of autonomic function. Proceedings of the Association for Research in Nervous and Mental Disease . . . 1939. Baltimore, Williams & Wilkins Co. (Research Publication Vol. XX), xxx + 980p.

Furtwängler, A., and Reichhold, K. 1904. Griechische Vasenmalerei. Auswahl hervorragender Vasenbilder. Tafeln, Serie 1 & 2. München, F. Bruckmann.

Gallagher, T. F., et al. 1937. The daily urinary excretion of estrogenic and androgenic substances by normal men and women. J. Clin. Invest. 16:695–703.

Gallonio, A. 1591. Historia delle sante vergini Romane . . . E de' gloriosi martiri papia e mauro soldati . . . Roma, Presso Ascanio, e Girolamo Donangeli, [6] + 350 + [24]p. *Also:* ibid. Trattato de gli instrumenti de martirio e delle varie maniere di martoriare usate da' gentili contro Christiani . . . , [4] + 159 + [9]p., incl. 47 pl.

Gantt, W. H. 1944. Experimental basis for neurotic behavior. Origin and development of artificially produced disturbances of behavior in dogs. New York, London, Paul B. Hoeber, xv + 211p.

1949. Psychosexuality in animals. *In:* Hoch, P. H., and Zubin, J., ed. Psychosexual development in health and disease, pp. 33–51.

1950. Disturbances in sexual functions during periods of stress. *In:* Wolff, H. G., et al., ed. Life stress and bodily disease, pp. 1030–1050.

Ganzfried, S. 1927. (Goldin, H. E., trans.). Code of Jewish law (Kitzur schulchan aruch). A compilation of Jewish laws and customs. New York, Hebrew Publishing Co., 4v. in 1. [ix]p. index. v. 1, 154p. v. 2, 150p. v. 3, 121p. v. 4, 137p.

Gardner, W. U. 1949. Reproduction in the female. *In:* Howell, W. H. (Fulton, J. F., ed.). A textbook of physiology, pp. 1162–1188.

Garnier, P. 1921 (orig. 1889). Anomalies sexuelles, apparents et cachés. Paris, Garnier Frères, 544p.

Geist, S. H. 1941. Androgen therapy in the human female. J. Clin. Endocrinol. 1:154–161.

Gerling, R. 1928. Der Geschlechtsverkehr der Ledigen. Berlin, Orania Verlag, 115p.

Gichner, L. E. 1949. Erotic aspects of Hindu sculpture. U.S.A., priv. print., 1 pl. + 56p.

Gilbert Youth Research. 1951. How wild are college students? [Unsigned article on a survey by Gilbert Youth Research.] Pageant (Nov.), pp. 10–21.

Glass, S. J., Deuel, H. J., and Wright, C. A. 1940. Sex hormone studies in male homosexuality. Endocrinol. 26:590–594.

Glass, S. J., and Johnson, R. H. 1944. Limitations and complications of organotherapy in male homosexuality. J. Clin. Endocrinol. 4:540–544.

Glueck, S., and Glueck, E. T. 1934. Five hundred delinquent women. New York, Alfred A. Knopf, xxiv + 539 + x p.

Goldstein, K., and Steinfeld, J. I. 1942. The conditioning of sexual behavior by visual agnosia. Bull. Forest Sanitarium 1:37–45.

Goldzieher, M. A., and Goldzieher, J. W. 1949. Toxic effects of percutaneously absorbed estrogens. J. Amer. Med. Assoc. 140:1156.

Golla, F. L., and Hodge, R. S. 1949. Hormone treatment of the sexual offender. Lancet 256:1006–1007.

Gollancz, V. 1952. My dear Timothy. An autobiographical letter to his grandson. New York, Simon and Schuster, 439p., [4] pl.

Golossowker. See Weissenberg, S.

Goode, W. J. 1949. Problems in postdivorce adjustment. Amer. Sociol. Rev. 14: 394–401.

Goodenough, W. H. 1949. Premarital freedom on Truk: Theory and practice. Amer. Anthropologist 51:615–620.

Goodland, R. 1931. A bibliography of sex rites and customs. An annotated record of books, articles, and illustrations in all languages. London, George Routledge & Sons, v + 752p.

Gorer, G. 1934. The Marquis de Sade. A short account of his life and work. New York, Liveright Publishing Corp., 264p.

——— 1938. Himalayan village; an account of the Lepchas of Sikkim. London, M. Joseph, 2 pl., 7–510p., 32 pl.

——— 1948. The American people. A study in national character. New York, W. W. Norton & Co., 246p.

Gould, H. N., and Gould, M. R. 1932. Age of first menstruation in mothers and daughters. J. Amer. Med. Assoc. 98:1349–1352.

Grafenberg, E. 1950. The role of urethra in female orgasm. Int. J. Sexol. 3:145–148.

Gray, M. 1951. The changing years. What to do about menopause. New York, Doubleday & Co., 224p.

Greek anthology. See Anon. The Greek anthology.

Greenblatt, R. B. 1947. Office endocrinology. Springfield, Ill., Charles C Thomas, xiv + 303p.

Greulich, W. W. 1944. Physical changes in adolescence. In: Henry, N. B., ed. The forty-third yearbook of the National Society for the Study of Education. Part I, Adolescence, pp. 8–32.

Greulich, W. W., et al. 1938. A handbook of methods for the study of adolescent children. Washington, D. C., Society for Research in Child Development, (Monogr. for the Soc. . . . Vol. 3, No. 1, Ser. No. 15), xvii + 406p.

Griffith, E. F. 1947 (ed. 1, 1935). Modern marriage. London, Methuen & Co., xi + 303p.

Groves, E. R. 1933. Marriage. New York, Henry Holt & Co., xvi + 552p.

Groves, E. R., Groves, G. H., and Groves, C. 1943. Sex fulfillment in marriage. New York, Emerson Books, 319p.

Groves, Gladys H. 1942. Marriage and family life. Boston, Houghton Mifflin Co., x + 564 + iv p.

Gruenberg, B. C., ed. 1922. High schools and sex education. Washington, D. C., U. S. Public Health Service, vii + 98p.

Gruenberg, B. C. 1932. Ed. 3 (ed. 1, 1923). Parents and sex education. For parents of young children. New York, Viking Press, viii + 112p.

Gruenberg, B. C., and Kaukonen, J. L. 1940. Ed. 2. High schools and sex education. Washington, D. C., U. S. Public Health Service (Educational Publication No. 7), xix + 110p.

Gruenberg, Sidonie M., et al. 1943. When children ask about sex. New York, Child Study Association of America, 16p.

Gudden, H. 1911. Pubertät und Schule. München, Aerztliche Rundschau, Otto Gmelin, 31p.

Gurewitsch, Z. A., and Grosser, F. J. 1929. Das Geschlechtsleben der Gegenwart. Ztschr. f. Sexualwiss. 15:513–546.

Gurewitsch, Z. A., and Woroschbit, A. J. 1931. Das Sexualleben der Bäuerin in Russland. Ztschr. f. Sexualwiss. u. Sexualpolit. 18:51–74, 81–110.

Guttmacher, M. S. 1951. Sex offenses. The problem, causes and prevention. New York, W. W. Norton & Co., 159p.

Guttmacher, M. S., and Weihofen, H. 1952. Sex offenses. J. Crim. Law, Criminol. & Police Sci. 43:153–175.

Guyon, R. 1929, 1933, 1934, 1936, 1937, 1938. Études d'éthique sexuelle. I. La légitimité des actes sexuels, 399p. II. La liberté sexuelle, 364p. III. Révision des institutions classiques (Mariage. Famille.), 392p. IV. Politique rationelle de sexualité. La reproduction humaine, 378p. V. Politique rationelle de sexualité. Le plaisir sexuel, 325p. VI. La persécution des actes sexuels. 1.—Les courtisanes, 432p. Saint-Denis, Dardaillon (and) Dardaillon et Dagniaux.

    1933. (Flugel, J. C. and Ingeborg, trans., Haire, N., ed.). Sex life & sex ethics. London, John Lane, xxii + 386p.

    1948. (Flugel, J. C. and Ingeborg, trans.). The ethics of sexual acts. New York, Alfred A. Knopf, [25] + 383 + xxvii p.

Guze, H. 1951. Sexual attitudes in the scientific medical literature. Int. J. Sexol. 5:97–100.

Hafez, E. S. E. 1951. Mating behaviour in sheep. Nature 167:777–778.

Haire, Norman, ed. 1937. Encyclopaedia of sexual knowledge. New York, Eugenics Publishing Co., xx + 567p.

Haire, Norman. 1948. Everyday sex problems. London, Frederick Muller, 268p.

    1951. The encyclopedia of sex practice. London, Encyclopaedic Press, 836p. (incl. 80 pl., 32 col. pl.)

Hall, G. S. 1904. Adolescence. Its psychology and its relations to physiology, anthropology, sociology, sex, crime, religion and education. New York, D. Appleton and Co., 2v. v. 1, xx + 589p. v. 2, vi + 784p.

Hall, W. S. 1920. Sex training in the home. Plain talks on sex life covering all periods and relationships from childhood to old age. Chicago, Midland Press, 128p.

Hallowell, A. I. 1949. Psychosexual adjustment, personality and the good life in a non-literate culture. In: Hoch, P. H., and Zubin, J., ed. Psychosexual development in health and disease, pp. 102–123.

Halverson, H. M. 1940. Genital and sphincter behavior of the male infant. Pedag. Sem. & J. Genet. Psychol. 56:95–136.

Hamblen, E. C. 1945. Endocrinology of woman. Springfield, Ill., Charles C Thomas, xii + 574p.

Hamburger, C. 1948. Normal urinary excretion of neutral 17-ketosteroids with special reference to age and sex variations. Acta Endocrinol. 1:19–37.

Hamilton, G. V. 1914. A study of sexual tendencies in monkeys and baboons. J. Anim. Behav. 4:295–318.

    1929. A research in marriage. New York, Albert & Charles Boni, xiii + 570p.

    1936. Homosexuality, defensive. [Homosexuality as a defense against incest.] In: Robinson, Victor, ed. Encyclopaedia sexualis, pp. 334–342.

Hamilton, G. V., and Macowan, K. 1929. What is wrong with marriage. New York, Albert & Charles Boni, xxi + 319p.

Hamilton, H. B., and Hamilton, J. B. 1948. Ageing in apparently normal men. I. Urinary titers of ketosteroids and of alpha-hydroxy and beta-hydroxy ketosteroids. J. Clin. Endocrinol. 8:433–452.

Hamilton, J. B. 1937a. Treatment of sexual underdevelopment with synthetic male hormone substance. Endocrinol. 21:649–654.

    1937b. Induction of penile erection by male hormone substances. Endocrinol. 21:744–749.

    1941. Therapeutics of testicular dysfunction. J. Amer. Med. Assoc. 116:1903–1908.

    1943. Demonstrated ability of penile erection in castrate men with markedly low titers of urinary androgens. Proc. Soc. Exper. Biol. & Med. 54:309–312.

    1948. The role of testicular secretions as indicated by the effects of castration in man and by studies of pathological conditions and the short lifespan associated with maleness. Rec. Prog. in Hormone Res. 3:257–322.

Hamilton, J. B., and Hubert, G. R. 1940. Vocal changes in eunuchoidal and castrated men upon administration of male hormone substance. Amer. J. Physiol. 129:P372–P373.

Hamilton, W. J., Boyd, J. D., and Mossman, H. W. 1947. Human embryology. (Prenatal development of form and function.) Baltimore, Williams & Wilkins Co., viii + 366p.

Hammond, W. A. 1887. Sexual impotence in the male and female. Detroit, George S. Davis, 305p.

Hara, Kōzan. 1938. Nihon Kōshoku bijutsu shi. [History of Japanese erotic art.] Tokyo, Kinryūdō, 384p., 16 pl., 1 col. pl.

Hardenbergh, E. W. 1949. The psychology of feminine sex experience. Int. J. Sexol. 2:224–228.

Harper, R. A. 1949. Marriage. New York, Appleton Century Crofts, xi + 308p.

Harper, R. F. 1904. Ed. 2. The code of Hammurabi, King of Babylon. Chicago, University of Chicago Press, xv + 192 + 103 pl.

Hartman, C. G. 1931. On the relative sterility of the adolescent organism. Science 74:226–227.

1936. Time of ovulation in women. Baltimore, Williams & Wilkins Co., x + 226p.

1940. The effect of testosterone on the monkey uterus and the administration of steroidal hormones in the form of Deanesly-Parkes pellets. Endocrinol. 26:449–471.

1945. The mating of mammals. Ann. N. Y. Acad. Sci. 46:23–44.

Hartman, F. A., and Brownell, K. A. 1949. The adrenal gland. Philadelphia, Lea & Febiger, 581p.

Hartwich, A. H., Kaus, G., and Kind, A. 1931. Die Brautnacht . . . Eine Morphologie ihrer Erscheinungsformen. Wien, Verlag für Kulturforschung, 192p., 7 pl., 8 col. pl., + 6 photog. pl.

Harvey, O. L. 1932a. Some statistics derived from recent questionnaire studies relative to human sexual behavior. J. Soc. Psychol. 3:97–100.

1932b. A note on the frequency of human coitus. Amer. J. Sociol. 38:64–70.

Hattwick, LaB. A. 1937. Sex differences in behavior of nursery school children. Child Develop. 8:343–355.

Hawke, C. C. 1950. Castration and sex crimes. [Address delivered to Illinois Academy of Criminology], 8p. mimeo.

Heckel, N., and McDonald, J. H. 1952. The effects of testosterone propionate upon the spermatogenic function of the human testis. Ann. N. Y. Acad. Sci. 55:725–733.

Hediger, H. 1950. (Sircom, G., trans.). Wild animals in captivity. London, Butterworth, ix + 207p., [17] pl.

Hegar, A. 1878. Die Castration der Frauen vom physiologischen und chirurgischen Standpunkte aus. Leipzig, Breitkopf und Härtel, iv + 144p.

Heidel, A. 1946. The Gilgamesh epic and Old Testament parallels. Chicago, University of Chicago Press, ix + 269p.

Heller, C. G., and Maddock, W. O. 1947. The clinical uses of testosterone in the male. Vitamins and Hormones 5:393–432.

Heller, C. G., and Nelson, W. O. 1948. The testis-pituitary relationship in man. Rec. Prog. in Hormone Res. 3:229–255.

Heller, C. G., Nelson, W. O., and Roth, A. A. 1943. Functional prepuberal castration in males. J. Clin. Endocrinol. 3:573–588.

Hellmann. See Weissenberg, S.

Hemphill, F. M. 1952. A sample survey of home injuries. Publ. Health Rept. 67:1026–1034.

Hemphill, R. E. 1944. Return of virility after prefrontal leucotomy with enlargement of gonads. Lancet 247:345–346.

Heneman, H. G., Jr., and Paterson, D. G. 1949. Refusal rates and interview quality. Int. J. Opin. & Attit. Res. 3:392–398.

Henninger, J. M. 1941. Exhibitionism. J. Crim. Psychopath. 2:357–366.

Henry, G. W. 1938. Ed. 3 (ed. 1, 1925). Essentials of psychiatry. Baltimore, Williams & Wilkins Co., xii + 465p.

1941. Sex variants. A study of homosexual patterns. New York, London, Paul B. Hoeber, 2v. v. 1, xxi + 546p. v. 2, vii + 547–1179p.

Henry, Jules. 1949. The social function of child sexuality in Pilaga Indian culture. *In:* Hoch, P. H., and Zubin, J., ed. Psychosexual development in health and disease, pp. 91–101.

Henry, N. B., ed. 1944. The forty-third yearbook of the National Society for the Study of Education. Part I, Adolescence. Chicago, University of Chicago Press, x + 358p.

Henz, W. 1910. Probenächte. Sexual-Probleme 6:740–750.

Hernandez, L., pseud. [Fleuret, F., and Perceau, L.]. **1920.** Les procès de bestialité au XVIe et XVIIe siècles. Paris, Bibliotheque des Curieux, 238p., 4 pl.

Herondas. 1921 (orig. 3rd cent. B.C.). (Buck, M. S., trans.). The mimes of Herondas. New York, priv. print., 119p.

Hesnard, A. 1933. Strange lust: The psychology of homosexuality. New York, Amethnol Press, 256p.

Heyn, A. 1921. Studien zur Physiologie des Geschlechtslebens der Frau. Geschlecht und Gesellschaft 10:405–408.

1924. Über sexuelle Träume (Pollutionen) bei Frauen. Archiv f. Frauenkunde 10:60–69.

Hikatsu-sho. *See* Anon. Hikatsu-sho.

Hill, W. W. 1935. The status of the hermaphrodite and transvestite in Navaho culture. Amer. Anthropologist 37:273–279.

Himes, N. E. 1940. Your marriage. A guide to happiness. New York, Toronto, Farrar & Rinehart, xiv + 434p.

Hirning, L. C. 1947. Genital exhibitionism, an interpretive study. J. Clin. Psychopath. 8:557–564.

Hirsch, E. W. 1949. Modern sex life. With case histories. New York, Permabooks, xv + 236p.

Hirschfeld, M., ed. 1899–1921. Jahrbuch für sexuelle Zwischenstufen. Unter besonderer Berücksichtigung der Homosexualität. Leipzig, Max Spohr, Volumes 1–21.

Hirschfeld, Magnus (Praetorius, Numa, pseud.). 1911. Homosexuelle Pissoirinschriften aus Paris. Anthropophyteia 8:410–422.

Hirschfeld, M. 1916–1921. Sexualpathologie. Ein Lehrbuch für Ärzte und Studierende. Part 1: Geschlechtliche Entwicklungsstörungen mit besonderer Berücksichtigung der Onanie. Part 2: Sexuelle Zwischenstufen. Das männliche Weib und der weibliche Mann. Part 3: Störungen im Sexualstoffwechsel mit besonderer Berücksichtigung der Impotenz. Bonn, A. Marcus & E. Webers Verlag, 3v. in 1. v. 1, 1916, xv + 211p. v. 2, 1921, x + 279p. v. 3, 1920, xi + 340p.

1920 (ex. 1914 ed.). Die Homosexualität des Mannes und des Weibes. Berlin, Louis Marcus, xvii + 1067p.

1926, 1928, 1930. Geschlechtskunde auf Grund dreiszigjähriger Forschung und Erfahrung bearbeitet. I. Band: Die körperseelischen Grundlagen, xv + 638p. II. Band: Folgen und Folgerungen, 659p. III. Band: Einblicke und Ausblicke, 780p. IV. Band: Bilderteil, 904p. Stuttgart, Julius Püttmann.

1928. Kastration bei Sittlichkeitsverbrechern. Ztschr. f. Sexualwiss. 15:54–55.

1935. (Rodker, J., trans.). Sex in human relationships. London, John Lane, xxii + 218p.

1936. Homosexuality. *In:* Robinson, Victor, ed. Encyclopaedia sexualis, pp. 321–334.

1940, rev. ed. Sexual pathology. A study of derangements of the sexual instinct. New York, Emerson Books, xii + 368p.

1944. Sexual anomalies and perversions. Physical and psychological development and treatment. A summary of the works of the late . . . Dr. Magnus Hirschfeld. London, Francis Aldor, 630p.

1948, rev. ed. Sexual anomalies. The origins, nature, and treatment of sexual disorders. New York, Emerson Books, 538p.

Hirschfeld, M., and Bohm, E. 1930. Sexualerziehung. Der Weg durch Natürlichkeit zur neuen Moral. Berlin, Universitas Deutsche Verlags-Aktiengesellschaft, 234p.

Hirschfeld, M., and Linsert, R. 1930. Liebesmittel. Eine Darstellung der geschlechtlichen Reizmittel (Aphrodisiaca). Berlin, Man Verlag, x + 395p.

Histoire et psycho-physiologie du vice . . . See Anon. Histoire et psycho-physiologie du vice . . .

Hitschmann, E., and Bergler, E. 1936. (Weil, P. L., trans.). Frigidity in women. Its characteristics and treatment. Washington and New York, Nervous and Mental Disease Publishing Co., v + 76p.

Hoch, P. H., and Zubin, J., ed. 1949. Psychosexual development in health and disease. Proceedings of the thirty-eighth annual meeting of the American Psychopathological Association . . . June 1948. New York, Grune & Stratton, viii + 283p.

Hodann, M. 1929. Onanie. Weder Laster noch Krankheit. Berlin, Universitas-Deutsche Verlags-Aktiengesellschaft, 91p.

1932. Geschlecht und Liebe. Berlin, Büchergilde Gutenberg, 264p.

1937. (Browne, S., trans.). History of modern morals. London, William Heinemann, xv + 338p.

Hoff, H. E. 1949. Cardiac output: Regulation and estimation. In: Howell, W. H. (Fulton, J. F., ed.). A textbook of physiology, pp. 660–680.

Hoffman, J. 1944. Female endocrinology. Including sections on the male. Philadelphia and London, W. B. Saunders Co., xv + 788p.

Hohman, L. B., and Schaffner, B. 1947. The sex lives of unmarried men. Amer. J. Sociol. 52:501–507.

Hollingshead, A. B. 1949. Elmtown's youth. The impact of social classes on adolescents. New York, John Wiley & Sons, London, Chapman & Hall, xi + 453p.

Hooker, C. W. 1949. Reproduction in the male. In: Howell, W. H. (Fulton, J. F., ed.). A textbook of physiology, pp. 1189–1206.

Hooton, E. A. 1942. Man's poor relations. Garden City, N. Y., Doubleday, Doran & Co., xl + 412p., 74 pl., 11 fig.

1946, rev. ed. (Ed. 1, 1931). Up from the ape. New York, The Macmillan Co., 1 pl. + xxii + 788p., 40 pl.

Hoover, J. E. 1947. How safe is your daughter? Amer. Mag. (July), pp. 32–33, 102–104.

Horace. 1914 (orig. 1st cent. B.C.). (Bennett, C. E., trans.). Odes and epodes. Cambridge, Mass., Harvard University Press, xx + 431p.

1926. (Fairclough, H. R., trans.). Satires, epistles and ars poetica. Cambridge, Mass., Harvard University Press, xxx + 509p.

Hornstein, F. X. von, and Faller, A., ed. 1950. Gesundes Geschlechts Leben. Handbuch für Ehefragen. Olten, Switzerland, Otto Walter, 452p.

Horrocks, J. E. 1951. (Carmichael, L., ed.). Psychology of adolescence. Behavior and development. Boston, Houghton Mifflin Co., xxvi + 614p.

Horvitz, D. G. 1952. Sampling and field procedures of the Pittsburgh morbidity survey. Pub. Health Rpt. 67:1003–1012.

Hoskins, R. G. 1941. Endocrinology. The glands and their functions. New York, W. W. Norton & Co., 388p.

Hotchkiss, R. S. 1944. Fertility in men. Philadelphia, J. B. Lippincott Co., xiii + 216p.

Houssay, B. A., et al. 1951. (Lewis, J. T. and Olive T., trans.). Human physiology. New York, McGraw-Hill Book Co., xvi + 1118p.

Howard, J. E., and Scott, W. W. 1950. The testes. *In:* Williams, R. H., ed. Textbook of endocrinology, pp. 316–348.

Howard, P., et al. 1950. Testicular deficiency: a clinical and pathological study. J. Clin. Endocrinol. 10:121–186.

Howell, W. H. 1949. Ed. 16 (ed. 1, 1905). (Fulton, J. F., ed.). A textbook of physiology. Philadelphia and London, W. B. Saunders Co., xl + 1258p.

Hoyer, E. 1929. Das lüsterne Weib. Sexualpsychologie der begehrenden unbefriedigten und schamlosen Frau. Wien, Leipzig, Verlag für Kulturforschung, 256p., 30 pl.

Huffman, J. W. 1950. The effect of gynecologic surgery on sexual reactions. Amer. J. Obstet. & Gynec. 59:915–917.

Huggins, C., Stevens, R. E., and Hodges, C. V. 1941. Studies on prostatic cancer. II. The effects of castration on advanced carcinoma of the prostate gland. Arch. Surgery 43:209–223.

Hughes, W. L. 1926. (Edwards, M. S., ed.). Sex experiences of boyhood. J. Soc. Hyg. 12:262–273.

Huhner, M. 1945. Ed. 3. The diagnosis and treatment of sexual disorders in the male and female including sterility and impotence. Philadelphia, F. A. Davis Co., xiii + 516p.

Hunter, J. 1786. A treatise on the venereal diseases. 12 + 398 + 19p., vii pl.

Hurxthal, L. M. 1943. Sublingual use of testosterone in 7 cases of hypogonadism: Report of 3 congenital eunuchoids occurring in one family. J. Clin. Endocrinol. 3:551–556.

Huschka, M. 1938. The incidence and character of masturbation threats in a group of problem children. Psychoanal. Quart. 7:338–356.

Hutton, I. E. 1942. (Ed. 1, 1932). The sex technique in marriage. New York, Emerson Books, 160p.

Hutton, L. 1935, 1937. The single woman and her emotional problems. Baltimore, William Wood and Co., xv + 173p.

     1950. The unmarried. *In:* Neville-Rolfe, S., ed. Sex in social life, pp. 414–434.

Hyman, H. T. 1946. An integrated practice of medicine. Philadelphia and London, W. B. Saunders Co., 5v. v. 1, xxvii + 1032 + lxiii p. v. 2, xvi + 1033–2010 + xliv p. v. 3, xv + 2011–3095 + xlvii p. v. 4, xvii + 3097–4131 + xlvi p. Index, 4133–4336p.

Hyndman, O. R., and Wolkin, J. 1943. Anterior chordotomy. Further observations on physiologic results and optimum manner of performance. Arch. Neurol. & Psychiat. 50:129–148.

Indiana State Board of Health. n.d. Parents part. [Revised from orig. by U. S. Pub. Health Serv.] Indianapolis, Ind., 13p.

Inge, W. R. 1930. Christian ethics and modern problems. New York, London, G. P. Putnam's Sons, ix + 427p.

Iovetz-Tereshchenko, N. M. 1936. Friendship-love in adolescence. London, George Allen & Unwin, xvi + 367p.

Isaacs, Susan S. F. 1939. Social development in young children; a study of beginnings. New York, Harcourt, Brace and Co., xii + 480p.

Ito, P. K. 1942. Comparative biometrical study of physique of Japanese women born and reared under different environments. Human Biol. 14:279–351.

Jackson, L. 1949. A study of sado-masochistic attitudes in a group of delinquent girls by means of a specially designed projective test. Brit. J. Med. Psychol. 22:53–65.

Jacobus X. *See* Jacolliot, Louis.

Jacolliot, L. (Jacobus X. . . . French army surgeon). 1900. Medico-legal examination of the abuses, aberrations, and dementia of the genital sense. Paris, Charles Carrington, 19 + 543p.

Jefferis, B. G., and Nichols, J. L. 1912 (ed. 1, 1894). Search lights on health. Light on dark corners. A complete sexual science and a guide to purity and physical manhood. Advice to maiden, wife, and mother. Love, courtship, and marriage. Naperville, Ill., J. L. Nichols, 487p.

Jeffress, L. A., ed. 1951. Cerebral mechanisms in behavior. The Hixon symposium. New York, John Wiley & Sons; London, Chapman & Hall, xiv + 311p., 1 col. pl.

Jenkins, I. 1944. The legal basis of literary censorship. Va. Law Rev. 31:83–118.

Jenkins, M. 1931. The effect of segregation on the sex behavior of the white rat as measured by the obstruction method. *In:* Warden, C. J., ed. Animal motivation. Experimental studies on the white rat, pp. 179–261.

Jensen, M. B. 1947. A case of sadism expressed through pictorial mutilations. Amer. Psychologist 2:277.

Jiiro haya-shinan. *See* Anon. Jiiro haya-shinan.

Jonsson, G. 1951. Sexualvanor hos svensk ungdom. *In:* Wangson, Otto, et al. Ungdomen möter samhället, (Bilaga A), pp. 160–204.

Jordan, H. E. 1934. A textbook of histology. New York and London, Appleton-Century Co., xxvii + 738p.

Justinian, (pseud.), ed. [1939]. Americana sexualis. Chicago, priv. print., 40p.

Juvenal and Persius. 1789 (orig. 1st–2nd cent. A.D.). (Madan, M., trans.). New and literal translation of Juvenal and Persius; with copious explanatory notes, by which these difficult satirists are rendered easy and familiar to the reader. London, priv. print., 2v. v. 1, x + 448p. v. 2, 476 + 20p.

Juvenal and Flaccus. 1817 (orig. 1st–2nd cent. A.D.). (Gifford, W., trans.). The satires of Decimus Junius Juvenalis, and of Aulus Persius Flaccus, translated into English verse. London, W. Bulmer & Co., 2v. v. 1, xi + lxxxii + 384p. v. 2, 163p.

Kadis, A. L. 1952. Latency period. *In:* Brower, D., and Abt, L. E., ed. Progress in clinical psychology, v. 1, sec. 2, pp. 361–368.

Kahn, F. 1937. Unser Geschlechtsleben. Ein Führer und Berater für jedermann. Zürich und Leipzig, Albert Müller, 393p., 32 pl.

1939. (Rosen, G., trans.). Our sex life. A guide and counsellor for everyone. New York, Alfred A. Knopf, xxxvi + 459p.

Kahn, S. 1937. Mentality and homosexuality. Boston, Meador Publishing Co., 249p.

Kallmann, F. J. 1952. Comparative twin study on the genetic aspects of male homosexuality. J. Nerv. & Ment. Dis. 115:283–298.

Kantor, J. R. 1924. Principles of psychology. New York, Alfred A. Knopf, 2v. v. 1, xix + 473p. v. 2, xii + 524p.

Karsch, F. 1900. Päderastie und Tribadie bei den Tieren auf Grund der Literatur. Jahrb. f. Sex. Zwisch. 2:126–160.

Karsch-Haack, F. 1906. Forschungen über gleichgeschlechtliche Liebe. 1. Die Mongoloiden. München, Seitz & Schauer, xvi + 134p.

1911. Das gleichgeschlechtliche Leben der Naturvölker. München, Ernst Reinhardt, xvi + 668p., xiii pl.

Käser, J. 1830. Bemerkungen über die Unzucht und die unehelichen Geburten . . . München, Mich. Lindauer, xii + 74p. + 1 pl.

Katz, D., and Allport, F. H. 1931. Students' attitudes, a report of the Syracuse University reaction study. Syracuse, N. Y., The Craftsman Press, xxviii + 408p.

Kearns, W. M. 1941. Oral therapy of testicular deficiency. Methyl testosterone administered orally to patients with marked testicular deficiency. J. Clin. Endocrinol. 1:126–130.

Keiser, S., and Schaffer, D. 1949. Environmental factors in homosexuality in adolescent girls. Psychoanal. Rev. 36:283–295.

Kelleher, E. J. 1952. The role of psychiatry in programs for the control and treatment of sex offenders. [Speech delivered before the Illinois Academy of Criminology, May, 1952.] Ms, (mimeo.), 10p.

Keller, D. H. 1942. The truth about "self-abuse." New York, Sexology Magazine (Personal Problem Library, v. 2), 16p.

Kellogg, Rhoda. 1953. Babies need fathers, too. New York, Comet Press Books, 256p.

Kelly, G. L. 1930. Sexual feeling in woman. Augusta, Ga., Elkay Co., xviii + 270p.
    1947. Technic of marriage relations. *In:* Fishbein, M., and Burgess, E. W., ed. Successful marriage, pp. 92–101.

Kelly, P. C. M. 1951. The Catholic book of marriage. New York, Farrar, Straus & Young, 299p.

Kempf, E. J. 1917. The social and sexual behavior of infra-human primates with some comparable facts in human behavior. Psychoanal. Rev. 4:127–154.

Kenigsberg, S., Pearson, S., and McGavack, T. H. 1949. The excretion of 17-ketosteroids. I. Normal values in relation to age and sex. J. Clin. Endocrinol. 9:426–429.

Kenyon, A. T. 1938. The effect of testosterone propionate on the genitalia, prostate, secondary sex characters, and body weight in eunuchoidism. Endocrinol. 23:121–134.

Kenyon, A. T., et al. 1940. A comparative study of the metabolic effects of testosterone propionate in normal men and women and in eunuchoidism. Endocrinol. 26:26–45.

Kepler, E. J., and Locke, W. 1950. The adrenals. Part I. Chronic adrenal hyperfunction. *In:* Williams, R. H., ed. Textbook of endocrinology, pp. 180–248.

Kilpatrick, V., et al. 1950. Partial report of Assembly Interim Committee on Crime and Corrections. California State Assembly, House Resolution no. 243, 83p., 16 pl.

Kimbrough, R. A., and Israel, S. L. 1948. The use and abuse of estrogen. J. Amer. Med. Assoc. 138:1216–1220.

Kind, A., and Moreck, C. 1930. Gefilde der Lust. Morphologie, Physiologie und sexual-psychologische Bedeutung der sekundären Geschlechtsmerkmale des Weibes. Wien, Leipzig, Verlag für Kulturforschung, 351p., 14 pl., 26 col. pl.

Kinsey, A. C. 1941. Criteria for a hormonal explanation of the homosexual. J. Clin. Endocrinol. 1:424–428.
    1947. Sex behavior in the human animal. Ann. N. Y. Acad. Sci. 47:635–637.

Kinsey, A. C., Pomeroy, W. B., and Martin, C. E. 1948. Sexual behavior in the human male. Philadelphia and London, W. B. Saunders Co., xv + 804p.

Kinsey, A. C., et al. 1949. Concepts of normality and abnormality in sexual behavior. *In:* Hoch, P. H., and Zubin, J., ed. Psychosexual development in health and disease, pp. 11–32.

Kirkendall, L. A. 1947. Understanding sex. Chicago, Science Research Associates, 48p.

Kisch, E. H. 1907. Das Geschlechtsleben des Weibes in physiologischer, pathologischer und hygienischer Beziehung. Berlin, Urban & Schwarzenberg, viii + 728p.
    1918. Ed. 3 (ed. 1, 1916). Die sexuelle Untreue der Frau. 1. Die Ehebrecherin. Bonn, A. Marcus & E. Weber, viii + 206p.
    1926 (orig. 1910). (Paul, N. E., trans.). The sexual life of woman in its physiological, pathological and hygienic aspects. New York, Allied Book Company, xi + 686p.

Klumbies, G., and Kleinsorge, H. 1950a. Das Herz in Orgasmus. Medizinische Klinik 45:952–958.
    1950b. Circulatory dangers and prophylaxis during orgasm. Int. J. Sexol. 4:61–66.

Klüver, H., and Bucy, P. C. 1938. An analysis of certain effects of bilateral temporal lobectomy in the rhesus monkey, with special reference to "psychic blindness." J. Psychol. 5:33–54.

1939. Preliminary analysis of functions of the temporal lobes in monkeys. Arch. Neurol. & Psychiat. 42:979–1000.

Knight, R. P. 1943. Functional disturbances in the sexual life of women. Frigidity and related disorders. Bull. Menninger Clin. 7:25–35.

Koch, F. C. 1937. The male sex hormones. Physiol. Rev. 17:153–238.

1938. The chemistry and biology of male sex hormones. Baltimore, Williams & Wilkins Co., from the Harvey Society Lecture Series 33:205–236.

Koch, H. L. 1935. An analysis of certain forms of so-called "nervous habits" in young children. J. Genetic Psychol. 46:139–170.

Kogon, E. 1950. (Norden, H., trans.). The theory and practice of hell. The German concentration camps and the system behind them. New York, Farrar, Straus & Co., 307p.

Kolb, L. C. 1949. An evaluation of lobotomy and its potentialities for future research in psychiatry and the basic sciences. J. Nerv. & Ment. Dis. 110: 112–148.

Kopp, Marie E. 1933. Birth control in practice. Analysis of ten thousand case histories of the Birth Control Clinical Research Bureau. New York, Robt. M. McBride & Co., vii + 290p.

1938. Surgical treatment as sex crime prevention measure. J. Crim. Law & Criminol. 28:692–706.

Koran. 1935. Ed. 3 (ed. 1, 1916). (Maulvi, Muhammad Ali, trans. and ed.). The holy Qur-án. Containing the Arabic text with English translation . . . Lahore, India, Ahmadiyya Anjuman-i-Isháat-i-Islam, cxii + 1275p.

Krafft-Ebing, R. von. 1901. Neue Studien auf dem Gebiete der Homosexualität. Jahrb. f. Sex. Zwisch. 3:1–36.

1922 (orig. 1906). (Rebman, F. J., trans.). Psychopathia sexualis, with especial reference to the antipathic sexual instinct. A medico-forensic study. Brooklyn, N. Y., Physicians and Surgeons Book Co., xiii + 617p.

1924. Ed. 16 (orig. 1886). (Moll, Albert, ed.). Psychopathia sexualis. Mit besonderer Berücksichtigung der konträren Sexualempfindung. Stuttgart, Ferdinand Enke, v + 832p.

Krauss, F. S. 1911. Ed. 2 (ed. 1, 1907). Das Geschlechtsleben in Glauben, Sitte, Brauch und Gewohnheitsrecht der Japaner. Leipzig, Ethnolog. Verlag, viii + 226p. + [68] pl.

1931. Japanisches Geschlechtsleben in zwei Bänden. Erster Band: Das Geschlechtsleben in Sitte, Brauch, Glauben und Gewohnheitsrecht des Japanischen Volkes. Leipzig, "Anthropophyteia" Verlag für Urtriebkunde, 432p. + 100 pl. on 95p.

Krauss, F. S., and Satow, Tamio. 1931. (Ihm, Hermann, ed. and trans.). Japanisches Geschlechtsleben, in zwei Bänden. Zweiter Band: Abhandlungen und Erhebungen über das Geschlechtsleben des Japanischen Volkes. Leipzig, "Anthropophyteia" Verlag für Urtriebkunde, 654p.

Krauss, F. S., ed. Anthropophyteia. See Anthropophyteia.

Kroger, W. S., and Freed, S. C. 1950. Psychosomatic aspects of frigidity. J. Amer. Med. Assoc. 143:526–532.

1951. Psychosomatic gynecology: including problems of obstetrical care. Philadelphia and London, W. B. Saunders Co., xvii + 503p.

Kronfeld, A. 1923. Exhibitionismus. In: Marcuse, M., ed. Handwörterbuch der Sexualwissenschaft, pp. 121–122.

1923. Masochismus. In: Marcuse, M., ed. Handwörterbuch der Sexualwissenschaft, pp. 313–314.

1923. Sexualpsychopathologie. Leipzig und Wien, Franz Deuticke, viii + 134p.

Kuhn, M. H. 1942. The engagement. In: Becker, H., and Hill, R., ed. Marriage and the family, pp. 211–233.

Kühn, R. 1932. Die Frau bei den Kulturvölkern. Berlin, Neufeld & Henius, 256p., 77 photog. pl.

Kuntz, A. 1945. Ed. 3. The autonomic nervous system. Philadelphia, Lea & Febiger, 1 pl. + 687p.
    1951. Visceral innervation and its relation to personality. Springfield, Ill., Charles C Thomas, viii + 152p.

Lampl-DeGroot, J. 1950. On masturbation and its influence on general development. Psychoanalytic study of the child 5:153–174.
Landes, R. 1938. The Ojibwa woman. New York, Columbia University Press, viii + 247p.
Landis, C., et al. 1940. Sex in development. A study of the growth and development of the emotional and sexual aspects of personality together with physiological, anatomical, and medical information on a group of 153 normal women and 142 female psychiatric patients. New York, London, Paul B. Hoeber, xx + 329p.
Landis, C., and Bolles, M. M. 1942. Personality and sexuality of the physically handicapped woman. New York, London, Paul B. Hoeber, xii + 171p.
Landis, J. T., and Landis, M. G. 1948. Building a successful marriage. New York, Prentice-Hall, xii + 559p.
Landis, P. H. 1945. Adolescence and youth. The process of maturing. New York and London, McGraw-Hill Book Co., xiii + 470p.
Lange, J. 1934. Die Folgen der Entmannung Erwachsener. An der Hand der Kriegserfahrungen dargestellt. Arbeit und Gesundheit 24:51–102 + append.
Langworthy, O. R. 1944. Behavior disturbances related to decomposition of reflex activity caused by cerebral injury. An experimental study of the cat. J. Neuropath. & Exper. Neurol. 3:87–100.
Lanval, M. 1946. L'amour sous le masque. Bruxelles, Le Laurier, 208p.
    1950. (Gibault, P., trans.). An inquiry into the intimate lives of women. (L'amour sous le masque.) New York, Cadillac Publishing Co., xii + 243p.
Lashley, K. S. 1916. Reflex secretion of the human parotid gland. J. Exper. Psychol. 1:461–493.
    1938. The thalamus and emotion. Psychol. Rev. 45:42–61.
    1951. The problem of serial order in behavior. In: Jeffress, L. A., ed. Cerebral mechanisms in behavior, pp. 112–146.
Lasker, G. W., and Thieme, F. P. 1949. Yearbook of physical anthropology, 1948, Volume 4. New York, Viking Fund, v + 217p.
Laubscher, B. J. F. 1938. Sex, custom and psychopathology. A study of South African pagan natives. New York, Robert M. McBride & Co., xv + 347p., 16 pl.
Laughlin, H. H. 1922. Eugenical sterilization in the United States. Chicago, Psychopathic Laboratory of the Municipal Court, xiii + 502p.
Laurent, E. 1904. Sexuelle Verirrungen. Sadismus und Masochismus. Berlin, H. Barsdorf, iv + 271p.
Lazarsfeld, S. 1931. Wie die Frau den Mann erlebt. Fremde Bekenntnisse und eigene Betrachtungen. Leipzig, Wien, Schneider & Co., 331p., 24 pl.
Leach, M., and Fried, J., ed. 1949–1950. Dictionary of folklore, mythology and legend. New York, Funk & Wagnalls Co., 2v. v. 1, xii + 531p. v. 2, 532–1196p.
Lecky, W. E. H. 1881 (orig. c. 1877). History of European morals from Augustus to Charlemagne. New York, D. Appleton and Co., 2v. v. 1, xxiv + 468p. v. 2, xi + 407p.
Lees, H. 1944. The word you can't say. Hygeia 22:336–337, 388–390.
Lenormant, M. F. 1867. Chefs-d'oeuvre de l'art antique. Ser. 2. Monuments de la peinture et de la sculpture. Paris, A. Lévy, 4v. v. 1, 85p. + 122 pl. v. 2, 123p. + 102 pl. v. 3, 107p. + 135 pl. v. 4, 214p. + 168 pl.
Leuba, C. 1948. Ethics in sex conduct. New York, Association Press, 164p.

Levine, M. I. 1951. Pediatric observations on masturbation in children. Psychoanalytic study of the child 6:117–124.

Levy, D. M. 1928. Fingersucking and accessory movements in early infancy. An etiologic study. Amer. J. Psychiat. 7:881–918.

Levy, J., and Munroe, R. 1938. The happy family. New York, Alfred A. Knopf, 319p.

Levy, S. S. 1951. Interaction of institutions and policy groups: The origin of sex crime legislation. The Lawyer and Law Notes 5:3–12.

Lewinsky, H. 1944. On some aspects of masochism. Int. J. Psychoanal. 25:150–155.

Lewis, T. 1942. Pain. New York, The Macmillan Co., xiii + 192p.

Licht, H. 1925–1928. Sittengeschichte Griechenlands. Die griechische Gesellschaft. Das Liebesleben der Griechen. Die Erotik in der griechischen Kunst (Ergänzungen zu Band 1 und 2). Dresden und Zürich, Paul Aretz, 3v. v. 1, 1925, 319p., 16 pl. v. 2, 1926, 263p., 16 pl. v. 3, 1928, 279p., 16 pl.
  1932. (Freese, J. H., trans.). Sexual life in ancient Greece. London, George Routledge & Sons, xv + 557p.

Liepmann, W. 1922. Psychologie der Frau. Berlin, Wien, Urban & Schwarzenberg, [12] + 322p.

Limborch, P. 1692. Historia inquisitionis. Cui subjungitur liber sententiarum inquisitionis tholosanae. Amstelodam, Henricum Westenium, 2pts. Pt. 1, [16] + 384 + [12]p. Pt. 2, [8] + 397 + [19]p.
  1816. The history of the Inquisition . . . with a particular description of its secret prisons, modes of torture . . . etc. Abridged (and trans.). London, W. Simpkin & R. Marshall, xvi + 542p.

Lindner, R. M., and Seliger, R. V., ed. 1947. Handbook of correctional psychology. New York, Philosophical Library, 691p.

Lindsey, B. B., and Evans, W. 1929. The companionate marriage. New York, Garden City Publishing Co., xxxiv + 396p.

Linton, R. 1936. The study of man; an introduction. New York, London, D. Appleton-Century Co. (Century Social Science Series), viii + 503p.

Lion, E. G., et al. 1945. An experiment in the psychiatric treatment of promiscuous girls. San Francisco, City and County Dept. of Public Health, 68p.

Lipschütz, A. 1924. The internal secretions of the sex glands. The problem of the "puberty gland." Cambridge, Heffer & Sons; Baltimore, Williams and Wilkins Co., xviii + 513p.

Lisser, H. 1942. Sexual infantilism of hypothyroid origin. J. Clin. Endocrinol. 2:29–32.

Lisser, H., Escamilla, R. F., and Curtis, L. E. 1942. Testosterone therapy of male eunuchoids. III. Sublingual administration of testosterone compounds. J. Clin. Endocrinol. 2:351–360.

Lloyd, J. W. 1931. The Karezza method, or magnetation the art of connubial love. Roscoe, Calif., priv. print., 64p.

Locke, H. J. 1951. Predicting adjustment in marriage: a comparison of a divorced and a happily married group. New York, Henry Holt and Co., xx + 407p.

Loewenfeld, L. 1908. Über sexuelle Träume. Sexual-Probleme 4:588–601.
  1911a. Über die Sexualität im Kindesalter. Sexual-Probleme 7:444–454, 516–534.
  1911b. Über die sexuelle Konstitution und andere Sexualprobleme. Wiesbaden, J. F. Bergmann, 231p.

London, L. S. 1952. Dynamic psychiatry. Vol. 1. Basic principles. Vol. 2. Transvestism—desire for crippled women. Vol. 3. Frustrated women. New York, Corinthian Publications, 3v. v. 1, vi + 98p. v. 2, v + 129p. [incl. 50 pl.]. v. 3, vii + 132p.

London, L. S., and Caprio, F. S. 1950. Sexual deviations. Washington, D. C., Linacre Press, xviii + 702p.

Long, H. W., pseud. (Smith, William Hawley). 1922 (orig. 1919). Sane sex life and sane sex living. New York, Eugenics Publishing Co., 151p.

Longus. 1896 (orig. 3rd cent. A.D.). Pastorals of Longus. Athens, Athenian Society, priv. print., xxxii + 227p. Gr. opp. 227p. Eng.

1916 (orig. 3rd cent. A.D.). (Thornley, G., and Edmonds, J. M., trans.). Daphnis & Chloe. London, William Heinemann, xxiii + 247p.

Lorie, J. H., and Roberts, H. V. 1951. Basic methods of marketing research. New York, McGraw-Hill Book Co., xii + 453p.

Louttit, C. M. 1927. Reproductive behavior of the guinea pig. I. The normal mating behavior. J. Comp. Psychol. 7:247–263.

1929. Reproductive behavior of the guinea pig. II. The ontogenesis of the reproductive behavior pattern. J. Comp. Psychol. 9:293–304.

Lowry, O. 1938. A virtuous woman. Sex life in relation to the Christian life. Grand Rapids, Mich., Zondervan Publishing House, 160p.

Lucian. 1895 (orig. 2nd cent. A.D.). (Jacobitz, C., trans.). Lucian. The ass, the Dialogues of courtesans, and the Amores. Athens, Athenian Society, priv. print., 225p. Gr. opp. 225p. Eng.

Ludovici, A. M. 1948. Untapped reserves of sadism in modern men and women. J. Sex Educ. 1:95–100.

Ludwig, D. J. 1950. The effect of androgen on spermatogenesis. Endocrinol. 46: 453–481.

Lundberg, F., and Farnham, Marynia F. 1947. Modern woman, the lost sex. New York and London, Harper & Brothers, vii + 497p.

Luquet, G. H. 1910. Sur la survivance des caractères du dessin enfantin dans des graffiti à indications sexuelles. Anthropophyteia 7:196–202.

1911. Représentation de la vulve dans les graffiti contemporains. Anthropophyteia 8:210–214.

Lynch, K. M., Jr. 1952. Recovery of the rat testis following estrogen therapy. Ann. N. Y. Acad. Sci. 55:734–741.

Lynd, R. S., and Lynd, H. M. 1929. Middletown. A study in contemporary American culture. New York, Harcourt, Brace and Co., xi + 550p.

1937. Middletown in transition. A study in cultural conflicts. New York, Harcourt, Brace and Co., xviii + 604p.

Macandrew, R. 1946. Friendship, love affairs and marriage. An explanation of men to women and of women to men. London, Wales Publishing Co., 150p.

Maccoby, E. E. 1951. Television: Its impact on school children. Pub. Opin. Quart. 15:421–444.

Macfadden, B. 1922. Sex talks to boys. New York, Macfadden Publications (Sex Education Series No. 1), 36p.

Maddock, W. O., et al. 1952. The assay of urinary estrogens as a test of human Leydig cell function. Ann. N. Y. Acad. Sci. 55:657–673.

Maes, J. P. 1939. Neural mechanism of sexual behavior in the female cat. Nature 144:598–599.

Magaldi, E. 1931. Le iscrizioni parietali pompeiane. Naples, Cimmaruta, 148p.

Magoun, F. A. 1948. Love and marriage. New York, Harper & Brothers, xvii + 369p.

Makurabunko. See Anon. Makurabunko.

Malamud, W., and Palmer, G. 1932. The role played by masturbation in the causation of mental disturbances. J. Nerv. & Ment. Dis. 76:220–233, 366–379.

Malchow, C. W. 1923. The sexual life, embracing the natural sexual impulse, normal sexual habits and propagation, together with sexual physiology and hygiene. St. Louis, C. V. Mosby Co., 317p.

Malinowski, B. 1928. The anthropological study of sex. In: Marcuse, M., ed. Verhandlungen des I International Kongresses für Sexualforschung 5:92–108.

1929. The sexual life of savages in Northwestern Melanesia . . . New York, Halcyon House, xxviii + 603p., 96 pl.

Mallett, D. T. 1935. Mallett's index of artists. International-biographical. New York, Peter Smith, xxxiv + 493p.

1940. Supplement to Mallett's index of artists. New York, Peter Smith, xxxviii + 319p.

Mantegazza, P. n.d. Ger. ed. 3 (orig. 1885). Anthropologisch-kulturhistorische Studien über die Geschlechtsverhältnisse des Menschen. Jena, Hermann Costenoble, x + 434p.

Marañón, G. 1929. Ed. 5. Tres ensayos sobre la vida sexual. Sexo, trabajo y deporte maternidad y feminismo, educación sexual y diferenciación sexual. Madrid, Biblioteca Neuva, 250p.

1932. (Span. ed. 1, 1930). (Wells, W. B., trans.). The evolution of sex and intersexual conditions. London, George Allen & Unwin, 344p.

Marcuse, M., ed. 1923. Handwörterbuch der Sexualwissenschaft. Enzyklopädie der natur- und kulturwissenschaftlichen Sexualkunde des Menschen. Bonn, A. Marcus & E. Weber, iv + 481p.

Marcuse, M. 1924. Die demi-vierge. Ztschr. f. Sexualwiss. 11:143–153.

1926. Neuropathia sexualis. In: Moll, Albert, ed. Handbuch der Sexualwissenschaften, v. 2, pp. 843–902.

Marcuse, M., ed. 1927. Verhandlungen des I. Internationalen Kongresses für Sexualforschung . . . 1926. Berlin und Köln, A. Marcus & E. Weber, 5v. in 1, 225 + 249 + 217 + 230 + 183p.

Marinello, G. 1563. Le medicine partenenti alle infermitá. Venitio, Francesco de Franceschi Senese, [16]p. + 258 lvs. (i.e. 516p.) + [16]p.

Marriage, love and woman amongst the Arabs. See Anon. Marriage, love and woman amongst the Arabs.

Marro, A. 1922 (orig. 1900). (Medici, J. P., trans., Marie, A., ed.). La puberté chez l'homme et chez la femme . . . Paris, Alfred Costes, xvi + 536p., 4 pl.

Marshall, F. H. A. 1922. Ed. 2. The physiology of reproduction. London and New York, Longmans, Green, and Co., xvi + 770p.

1936. The Croonian Lecture. Sexual periodicity and the causes which determine it. Philos. Trans. Roy. Soc. London 226 (Ser. B): 423–456.

Marshall, F. H. A., and Hammond, J. 1944. Fertility and animal-breeding. London, H.M.S.O. (Ministry of Agriculture . . . Bull. no. 39), 42p.

Martial. 1919 (orig. 1st cent. A.D.). (Ker, W. C. A., trans.). Epigrams. I. London, William Heinemann, v. 1, xxii + 491p.

1920. (Ker, W. C. A., trans.). Epigrams. II. London, William Heinemann, v. 2, 568p.

1921. (Buck, M. S., trans.). Epigrams. In fifteen books. Priv. print., x + 423p.

Martineau, L. 1886. Ed. 2. Leçons sur les déformations vulvaires et anales produites par la masturbation, le saphisme, la défloration et la sodomie. Paris, Adrien Delahaye et Émile Lecrosnier, iii + 190p.

Martinez, J. A. 1947. El homosexualismo y su tratamiento. Una serie de tres conferencias dictadas en el Tribunal Supremo de la República, bajo los auspicios de la "Asociacion Nacional de Funcionarios del Poder Judicial." Mexico City, Ediciones Botas, 153p.

Maslow, A. H. 1942. Self-esteem (dominance-feeling) and sexuality in women. J. Soc. Psychol. 16:259–294.

Masters, W. H. 1952. The female reproductive system. In: Cowdry, E. V. (Lansing, A. I., ed.). Problems of ageing, pp. 651–685.

Maximow, A. A., and Bloom, W. 1952. Ed. 6. A textbook of histology. Philadelphia and London, W. B. Saunders Co., x + 616p.

May, Geoffrey. 1931. Social control of sex expression. New York, William Morrow & Co., xi + 307p.

Mazer, C., and Israel, S. L. 1946. Ed. 2. Diagnosis and treatment of menstrual disorders and sterility. New York and London, Paul B. Hoeber, xii + 570p.

1951. Ed. 3. Diagnosis and treatment of menstrual disorders and sterility. New York, Paul B. Hoeber, xiv + 583p., 2 col. pl.

McCance, R. A., Luff, M. C., and Widdowson, E. E. 1937. Physical and emotional periodicity in women. J. Hyg. 37:571–611.

McCartney, J. L. 1929. Dementia praecox as an endocrinopathy with clinical and autopsy reports. Endocrinol. 13:73–87.

McCullagh, E. P. 1939. Treatment of testicular deficiency with testosterone propionate. J. Amer. Med. Assoc. 112:1037–1044.

McCullagh, E. P., and Renshaw, J. F. 1934. Effects of castration in adult male. J. Amer. Med. Assoc. 103:1140–1143.

McDermott, J. F., and Taft, K. B., ed. 1932. Sex in the arts. A symposium. New York, Harper & Brothers, xviii + 328p.

McDonald, H. C. 1952. Playtime with Patty and Wilbur. Culver City, Calif., Murray & Gee, 33p.

McKenzie, F. F., and Berliner, V. 1937. The reproductive capacity of rams. Columbia, Mo., University of Missouri (Agricultural Experiment Station Research Bull. 265), 143p. [incl. 20 pl.].

McKenzie, F. F., Miller, J. C., and Bauguess, L. C. 1938. The reproductive organs and semen of the boar. Columbia, Mo., University of Missouri (Agricultural Experiment Station Research Bull. 279), 122p.

McKenzie, F. F., and Terrill, C. E. 1937. Estrus, ovulation, and related phenomena in the ewe. Columbia, Mo., University of Missouri (Agricultural Experiment Station Research Bull. 264), 1 col. pl. + 88p. [incl. 10 pl.].

McKenzie, K. G., and Proctor, L. D. 1946. Bilateral frontal lobe leucotomy in the treatment of mental disease. Canada Med. Assoc. 55:433–441.

McPartland, John. 1947. Sex in our changing world. New York, Rinehart & Co., 280p.

Mead, M. 1939. From the South Seas. Studies of adolescence and sex in primitive societies. New York, William Morrow & Co., xxxv + [3] + 304 + 384 + xiv + 335p.

1949. Male and female. A study of the sexes in a changing world. New York, William Morrow & Co., xii + 477p.

Meagher, J. F. W. 1929. Ed. 2. A study of masturbation and the psychosexual life. New York, William Wood and Co., 130p.

Meagher, J. F. W., and Jelliffe, S. E. 1936. Ed. 3. A study of masturbation and the psychosexual life. Baltimore, William Wood and Co., xii + 149p.

Meibomius, J. H. n.d. (orig. 1761). A treatise of the use of flogging in veneral affairs . . . Also of the office of the loins and reins. Written to the famous Christianus Cassius, Bishop of Lubeck, and Privy-Councillor to the Duke of Holstein. London, n.publ. (facs. repr.), 83p.

Mendelsohn, M. 1896. Ist das Radfahren als eine gesundheitsgemässe Uebung anzusehen und aus ärtzlichen Gesichtspunkten zu empfehlen? Dtsch. med. Wchnschr. 22:381–384.

Menninger, K. A. 1938. Man against himself. New York, Harcourt, Brace and Co., xii + 485p.

Menzies, K. 1921. Ed. 2. Autoerotic phenomena in adolescence. An analytical study of the psychology and psychopathology of onanism. New York, Paul B. Hoeber, viii + 100p.

Merrill, L. 1918. A summary of findings in a study of sexualism among a group of one hundred delinquent boys. J. Delinq. 3:255–267.

Merry Andrew. See Anon., comp. The merry Andrew.

Meyer, F. 1929. Helps to purity. A frank, yet reverent instruction on the intimate matters of personal life for adolescent girls. Cincinnati, O., St. Francis Book Shop (Father Fulgence's Book No. 8), [viii] + 91p., incl. 1 pl.

Michels, R. 1914. (Paul, E. and C., trans.). Sexual ethics. A study of borderland questions. London, George Allen & Unwin, New York, Charles Scribner's Sons, xv + 296p.

Michigan. Governor's Study Commission. (Richards, R. M., chairman.) **1951.** Report of the . . . Commission on the deviated criminal sex offender. Lansing, State of Michigan, xi + 245p.

Miller, G. S., Jr. 1931. The primate basis of human sexual behavior. Quart. Rev. Biol. 6:379–410.

Miller, M. M., and Robinson, D. M., trans. and ed. **1925.** The songs of Sappho, including the recent Egyptian discoveries. The poems of Erinna. Greek poems about Sappho. Ovid's epistle of Sappho to Phaon. Lexington, Ky., Maxwelton Co., xiv + 436p., 10 pl.

Miller, N. E., Hubert, G., and Hamilton, J. B. 1938. Mental and behavioral changes following male hormone treatment of adult castration, hypogonadism, and psychic impotence. Proc. Soc. Exper. Biol. & Med. 38:538–540.

Mills, C. A. 1937. Geographic and time variations in body growth and age at menarche. Human Biol. 9:43–56.

Mills, C. A., and Ogle, C. 1936. Physiologic sterility of adolescence. Human Biol. 8:607–615.

Mirbeau, O. 1931 (orig. 1899). (Bessie, A. C., trans.). Torture Garden. (Le jardin des supplices). New York, Claude Kendall, 284p., 9 pl.

Möbius, P. J. 1903. Über die Wirkungen der Castration. Halle a. d. S., Carl Marhold, 99p.

Moldau. [1911]. Die Onanie oder Selbstbefleckung. Ihre Ursachen, Folgen und Heilung. Radebeul-Dresden, M. Wolf, 68p.

Moll, A. 1899. Ed. 3 (ed. 1, 1891). Die konträre Sexualempfindung. Berlin, H. Kornfeld, xvi + 651p.

[1909]. Das Sexualleben des Kindes. Leipzig, F.C.W.Vogel, viii + 313p.

1912. The sexual life of the child. New York, The Macmillan Co., xv + 339p.

1931. (Popkin, M., trans.). Perversions of the sex instinct. A study of sexual inversion based on clinical data and official documents. Newark, N. J., Julian Press, 237p.

Moll, A., ed. **1911.** Handbuch der Sexualwissenschaften mit besonderer Berücksichtigung der kulturgeschichtlichen Beziehungen. Leipzig, F.C.W.Vogel, xxiv + 1029p.

1921. Ed. 2. Handbuch der Sexualwissenschaften mit besonderer Berücksichtigung der kulturgeschichtlichen Beziehungen. Leipzig, F.C.W.Vogel, xxiv + 1046p.

1926. Handbuch der Sexualwissenschaften. Leipzig, F.C.W.Vogel, v. 1, xxviii + 736p., x pl. v. 2, 737–1280p.

Montagu, M. F. Ashley. 1937. Coming into being among the Australian Aborigines. London, George Routledge & Sons, xxxv + 362p., iv photog. pl.

1946. Adolescent sterility. A study in the comparative physiology of the infecundity of the adolescent organism in mammals and man. Springfield, Ill., Charles C Thomas, ix + 148p.

Moore, C. R. 1939. Biology of the testes. In: Allen, Edgar, et al., ed. Sex and internal secretions, pp. 353–451.

1942. The physiology of the testis and application of male sex hormone. J. Urol. 47:31–44.

Moore, T. V. 1945. The pathogenesis and treatment of homosexual disorders: a digest of some pertinent evidence. J. Personality 14:47–83.

Moraglia, G. B. 1897. Die Onanie beim normalen Weibe und bei den Prostituierten. Berlin, Priber & Lammers, 21p.

Moreck, C. 1929. Kultur- und Sittengeschichte der neuesten Zeit. Das Genussleben des modernen Menschen. Dresden, Paul Aretz, 451p., 32 pl.

Müller, E. 1929. Ein Beitrag zur Sexualforschung in der Volksschule. Ztschr. f. Päd. Psych. 30:467–477.

Munro, D., Horne, H., and Paull, D. 1948. The effect of injury to the spinal cord and cauda equina on the sexual potency of men. N. Eng. J. Med. 239:903–911.

Murdock, G. P. 1949. The social regulation of sexual behavior. In: Hoch, P. H., and Zubin, J., ed. Psychosexual development in health and disease, pp. 256–266.
1949. Social structure. New York, The Macmillan Co., xix + 387p.

Nathanson, I. T., Towne, L. E., and Aub, J. C. 1941. Normal excretion of sex hormones in childhood. Endocrinol. 28:851–865.
Naumburg, M. 1947. Studies in the "free" art expression of behavior problem children and adolescents as a means of diagnosis and therapy. New York, Coolidge Foundation, Nervous and Mental Disease Monographs No. 71, xii + 225p.
1950. Schizophrenic art: its meaning in psychotherapy. New York, Grune & Stratton, viii + 247p., viii col. pl.
Neely, W. C. 1940. Family attitudes of denominational college and university students, 1929 and 1936. Amer. Sociol. Rev. 5:512–522.
Nefzawi. n.d. The perfumed garden. A manual of Arabian erotology. Paris, Librairie "Astra," priv. print., 167p.
Negri, V. 1949. Psychoanalysis of sexual life. Los Angeles, Western Institute of Psychoanalysis, 274p.
Nelson, J. 1888. A study of dreams. Amer. J. Psychol. 1:367–401.
Neumann, H. 1936. Marriage and morals. J. Soc. Hyg. 22:102–114.
Neville-Rolfe, S., ed. 1950. Sex in social life. New York, W. W. Norton & Co., 504p.
New York. Law Revision Commission. 1937. Study relating to rape, abduction, seduction, corrupting morals of minors and related sexual offenses contained in the New York Penal Law. Albany, N. Y., J. B. Lyon Co., 116p.
New York. Mayor's Committee for the Study of Sex Offenders. 1944. Report. City of New York, 100p.
Newcomb, T. 1937. Recent changes in attitudes toward sex and marriage. Amer. Sociol. Rev. 2:659–667.
Niedermeyer, A. 1950. Anomalien des Geschlechtstriebes. In: Hornstein, X. von, and Faller, A., ed. Gesundes Geschlechts Leben, pp. 143–159.
Niemoeller, A. F. 1935. American encyclopedia of sex. New York, Panurge Press, xiv + 277p.
Nimuendajú, C. 1939. The Apinayé. Washington, D. C., Catholic University of America Press (Anthropological Series No. 8), vi + 189p.
Noldin, H. 1904. Summa theologiae moralis. Volume 3: De sacramentis. Innsbruck, Fel. Rauch (C. Pustet), 798p.
Northcote, H. 1916 (orig. 1906). Christianity and sex problems. Philadelphia, F. A. Davis Co., xvi + 478p.
Norton, H. 1949. The third sex. Portland, Ore., Facts Publishing Co., 112p.
Nowlis, V. 1941. Companionship preference and dominance in the social interaction of young chimpanzees. Compar. Psychol. Monogr. v. 17, no. 1, 57p.
Nystrom, A. 1908. (Swed. ed. 1, 1904). (Sandzen, C., trans.). The natural laws of sexual life. Kansas City, Mo., Burton Co., 260p.

O'Brien, P. 1950. Emotions and morals. Their place and purpose in harmonious living. New York, Grune & Stratton, xiii + 241p.
Ohlson, W. E. 1937. Adultery: A review. Boston Univer. Law Rev. 17:328–368, 533–622.
Orsi, A. 1913. Lussuria e castitá. Seguito alla "Donna nuda." Saggio di psicologia. Milano, "La Broderie," 263p.
Ovid. 1921 (orig. 1st cent. B.C.—1st cent. A.D.). (Showerman, G., trans.). Heroides and Amores. London, William Heinemann; New York, G. P. Putnam's Sons, 524p.
1929. (Mozley, J. H., trans.). The art of love, and other poems. London, William Heinemann; New York, G. P. Putnam's Sons, xiv + 382p.
1930. (May, J. L., trans.). The love books of Ovid. Being the Amores,

Ars amatoria, Remedia amoris, and Medicamina faciei femineae of Publius Ovidius Naso. New York, Rarity Press, priv. print., xxxiii + 216p.
1930. (Young, C. D., and Marlowe, Christopher, trans.). The love books of Ovid. A completely unexpurgated . . . edition. Together with the Elegie. Priv. print., 302p., 18 col. pl.

Ozaki, Hisaya. 1928. Edo nan-bungaku kō-i. [Unorthodox thoughts on the light or erotic literature of the Tokugawa period.] Tokyo, Nakanishi Shobō, 578p., 61 pl.

Parke, J. R. 1906. Human sexuality. A medico-literary treatise on the laws, anomalies, and relations of sex with especial reference to contrary sexual desire. Philadelphia, Professional Publishing Company, x + 476p.

Parker, R. A. 1935. A yankee saint. John Humphrey Noyes and the Oneida community. New York, G. P. Putnam's Sons, 322p., 15 pl.

Parker, V. H. n.d. Sex education for parent groups. Outline of four lectures for popular presentation. New York, American Social Hygiene Association (Publication No. A-163), 14p.

Parkes, A. S. 1950. Androgenic activity of the ovary. Rec. Prog. in Hormone Res. 5:101–114.

Parshley, H. M. 1933. The science of human reproduction. Biological aspects of sex. New York, W. W. Norton, xv + 319p.

Parsons, E. C. 1916. The Zuni La' Mana. Amer. Anthropologist 18:521–528.

Paschkis, K. E., and Rakoff, A. E. 1950. Some aspects of the physiology of estrogenic hormones. Rec. Prog. in Hormone Res. 5:115–149.

Patten, B. M. 1946. Human embryology. Philadelphia, Toronto, The Blakiston Co., xv + 776p.

Payne, F. 1949. Changes in the endocrine glands of the fowl with age. J. Gerontol. 4:193–199.
1952. Cytological changes in the cells of the pituitary, thyroids, adrenals and sex glands of ageing fowl. In: Cowdry, E. V. (Lansing, A. I., ed.). Problems of ageing, pp. 381–402.

Pearl, R. 1930. The biology of population growth. New York, Alfred A. Knopf, xiv + 260p.

Peck, M. W., and Wells, F. L. 1923. On the psycho-sexuality of college graduate men. Ment. Hyg. 7:697–714.
1925. Further studies in the psycho-sexuality of college graduate men. Ment. Hyg. 9:502–520.

Penfield, W., and Erickson, T. C. 1941. Epilepsy and cerebral localization; a study of the mechanism, treatment and prevention of epileptic seizures. Springfield, Ill., Charles C Thomas, x + 623p.

Penfield, W., and Rasmussen, T. 1950. The cerebral cortex of man. A clinical study of localization of function. New York, The Macmillan Co., xv + 248p.

Pepys, S. 1942 (orig. 1659–1669). (Wheatley, H. B., ed., Bright, M., transcr.). The diary of Samuel Pepys. Transcribed . . . from the shorthand manuscript . . . New York, Heritage Press, 2v. var. pag.

Perceau, L. See Hernandez, L.

Perloff, W. H. 1949. Role of the hormones in human sexuality. Psychosom. Med. 11:133–139.
1951. The hormonal balance of the normal menstrual cycle. In: Mazer, C., and Israel, S. L. Diagnosis and treatment of menstrual disorders and sterility, pp. 118–128.

Peterson, K. M. 1938. Early sex information and its influence on later sex concepts. [Thesis . . . Master of Arts.] Boulder, Colo., University of Colorado, Ms, vii + 136p.

Petronius. 1913 (orig. 1st cent. A.D.). (Heseltine, M., trans.). Petronius (Satyricon, fragments, poems). London, William Heinemann, 363p.
1922. (Firebaugh, W. C., trans.). The satyricon . . . unexpurgated translation in which are incorporated the forgeries of Nodot and Marchena,

and the readings introduced . . . by De Salas. New York, Boni & Liveright, priv. print., 2v. v. 1, xxxi + 258p., 36 pl. v. 2, 259–516p., 11 pl.

1927. (?Wilde, Oscar, trans.). The satyricon of Petronius Arbiter . . . Chicago, Pascal Covici, 2v. v. 1, lxxxv + 206p. v. 2, 207–497p.

Pfuhl, E. 1923. Malerei und Zeichnung der Griechen. v. 1, xv + 503p. v. 2, 504–918p. v. 3. Verzeichnisse und Abbildungen, [919]–981p. + 805 fig. on 361 pl. München, F. Bruckmann.

Pillay, A. P. 1950. Premarital sex activities of Indian males. A survey of 381 patients of a sexological clinic. Int. J. Sexol. 4:80–84.

Pilpel, H. F., and Zavin, T. 1952. Your marriage and the law. New York, Toronto, Rinehart & Co., xv + 358p.

Pincus, G., and Thimann, K. V., ed. 1948, 1950. The hormones. Physiology, chemistry and applications. New York, Academic Press, 2v. v. 1, xi + 886p. v. 2, ix + 782p.

Plautus. 1916 (orig. 3rd–2nd cent. B.C.). (Nixon, P., trans.). Plautus. In five volumes. I. Amphitryon. The comedy of asses. The pot of gold. The two Bacchises. The captives. Cambridge, Mass., Harvard University Press, xix + 570p.

Pliny. 1942 (orig. 1st cent. A.D.). (Rackham, H., trans.). Natural history. In ten volumes. II. Cambridge, Mass., Harvard University Press, ix + 664p.

Ploscowe, Morris. 1951. Sex and the law. New York, Prentice-Hall, ix + 310p.

Plutarch. 1905 (orig. 1st–2nd cent. B.C.). (Goodwin, W. W., et al., trans.). Essays and miscellanies. Boston and New York, Little, Brown and Co., 5v. var. pag.

1914, 1916. (Perrin, B., trans.). Lives. In eleven volumes. I, II, IV. London, William Heinemann, v. 1, xix + 582p. v. 2, ix + 631p. v. 4, ix + 467p.

Podolsky, Edward. 1942. Sex today in wedded life. New York, Simon Publications, Section I, xx + 240p.

Poirier, L. J. 1952. Anatomical and experimental studies on the temporal pole of the macaque. J. Comp. Neurol. 96:209–248.

Polatin, P., and Douglas, D. B. 1951. Spontaneous orgasm in a case of schizophrenia. Unpubl. Ms, 13p.

Pomeroy, H. S. 1888. The ethics of marriage. New York and London, Funk & Wagnalls Co., 197p.

Popenoe, P. 1938. Preparing for marriage. Los Angeles, Calif., American Institute of Family Relations, 23p.

1943. Marriage. Before and after. New York, Wilfred Funk, xiv + 246p.

1952. Love: The American way is best. Pageant v. 8 (Sept.), pp. 4–8.

Potter, La F. 1933. Strange loves. A study in sexual abnormalities. New York, National Library Press, ix + 243p.

Pouillet, T. 1897. Ed. 7 (ed. 1, 1876). L'onanisme chez la femme. Ses formes, ses causes, ses signes, ses consequences, et son traitment. Paris, Vigot Frères, 216p.

Powdermaker, H. 1933. Life in Lesu. The study of a Melanesian society in New Ireland. New York, W. W. Norton & Co., 1 pl. + 352p., 12 pl. + 2 folded pp.

Praetorius, Numa. See Hirschfeld, M.

Pratt, J. P. 1939. Sex functions in man. In: Allen, E., et al., ed. Sex and internal secretions, pp. 1263–1334.

Praz, M. 1951. (Ed. 1, 1933). (Davidson, A., trans.). The romantic agony. Translated from the Italian. London, Oxford University Press, xix + 502p.

Priapeia. See Anon. Priapeia.

Priesel, R., and Wagner, R. 1930. Gesetzmässigkeiten im Auftreten der extragenitalen sekundären Geschlechtsmerkmale bei Mädchen. Ztschr. f. d. ges. Anat. 15:333–352.

Priester, H. M. 1941. The reported dating practices of one hundred and six high school seniors in an urban community. Ithaca, N. Y., Cornell University Master's Thesis, 115 + xvii + 5p.

Prince, M. 1914. The unconscious. New York, The Macmillan Co., xii + 549p.

Pringle, H. F. 1938. What do the women of America think about morals? Ladies' Home Journal, v. 55 (May), pp. 14–15, 49, 51–52.

Pritchard, J. B., ed. 1950. Ancient Near Eastern texts relating to the Old Testament. Princeton, N. J., Princeton University Press, xxi + 526p.

Propertius. 1895 (orig. 1st cent. B.C.). (Gantillon, P. J. F., trans.). The elegies of Propertius . . . with metrical versions of select elegies by Nott and Elton. London, George Bell & Sons, viii + 187p.

Pryor, H. B. 1936. Certain physical and physiologic aspects of adolescent development in girls. J. Pediat. 8:52–62.

Public Opinion and Sociological Research Division, SCAP, Japan. 1950. [Abstract of survey on attitude toward foreign countries, etc.]. Int. J. Opin. & Attit. Res. 4:452–453.

Pussep, L. M. 1922. Der Blutkreislauf im Gehirn beim Koitus. In: Weil, Arthur, ed. Sexualreform und Sexualwissenschaft, pp. 61–85.

Quanter, R. 1901. Die Leibes- und Lebensstrafen bei allen Völkern und zu allen Zeiten. Eine kriminalhistorische Studie. Dresden, H. R. Dohrn, 467p.

Rado, S. 1933. Fear of castration in women. Psychoanal. Quart. 2:425–475.

Radvanyi, L. 1951. Measurement of the effectiveness of basic education. Int. J. Opin. & Attit. Res. 5:347–366.

Ramsey, G. V. 1943. The sex information of younger boys. Amer. J. Orthopsychiat. 13:347–352.

——— 1943. The sexual development of boys. Amer. J. Psych. 56:217–234.

Rank, Otto. 1912. [no title]. In: Wiener psychoanalytische Vereinigung. Die Onanie . . . , pp. 107–129.

Rasmussen, A. 1934. Die Bedeutung sexueller Attentate auf Kinder unter 14 Jahren für die Entwicklung von Geisteskrankheiten und Charakteranomalien. Acta Psychiat. et Neurol. 9:351–434.

Rau, H. 1903. Die Grausamkeit, mit besonderer Bezugnahme auf sexuelle Faktoren. Berlin, H. Barsdorf, iv + 248p.

Read, A. W. 1935. Lexical evidence from folk epigraphy in Western North America: a glossorial study of the low element in the English vocabulary. Paris, priv. print., 83p.

Read, J. M. 1941. Atrocity propaganda, 1914–1919. New Haven, Conn., Yale University Press, xiii + 319p.

Reed, C. A. 1946. The copulatory behavior of small mammals. J. Comp. Psychol. 39:185–206.

Reich, Annie. 1951. The discussion of 1912 on masturbation and our present-day views. Psychoanalytic study of the child 6:80–94.

Reich, W. 1942. The discovery of the orgone. Vol. 1: The function of the orgasm. New York, Orgone Institute Press, xxxv + 368p.

——— 1945. The sexual revolution. Toward a self-governing character structure. New York, Orgone Institute Press, xxvii + 273p.

Reichard, G. A. 1938. Social life. In: Boas, F., ed. General anthropology, pp. 409–486.

Reik, T. 1941. Masochism in modern man. New York, Toronto, Farrar & Rinehart, vi + 439p.

Reiman, C. F., and Schroeter, L. W. 1951. Implications of the Kinsey Report upon criminal laws concerning sexual behavior. Unpublished thesis, Harvard Law School, var. pag.

Reisinger, L. 1916–1917. Einige Bemerkungen zur Spezifität des männlichen und weiblichen Geschlechtstriebes. Ztschr. f. Sexualwiss. 3:343–345.

Reiskel, K. 1906. Skatologische Inscriften. Anthropophyteia 3:244–246.

Remplein, H. 1950 (ed. 1, 1948). Die seelische Entwicklung in der Kindheit und Reifezeit. München, Basel, Ernst Reinhardt, 430p.

Reymert, M. L., ed. 1950. Feelings and emotions. The Mooseheart symposium . . . New York, McGraw-Hill Book Co., xviii + 603p.

Reynolds, E. L. 1946. Sexual maturation and the growth of fat, muscle and bone in girls. Child Develop. 17:121–144.

Reynolds, E. L., and Wines, J. V. 1949. Individual differences in physical changes associated with adolescence. In: Lasker, G. W., and Thieme, F. P., ed. Yearbook of physical anthropology, 1948, v. 4, pp. 89–110.

Rice, T. B. 1933. In training for boys of high school age. Chicago, American Medical Association, Bureau of Health . . . , 48p.

Rice, V. A., and Andrews, F. N. 1951 (ed. 1, 1926). Breeding and improvement of farm animals. New York, McGraw-Hill Book Co., xiv + 787p.

Richter, G. M. A. 1936. Red-figured Athenian vases in the Metropolitan Museum of Art. New Haven, Conn., Yale University Press, 2v. v. 1, xlvii + 249p. v. 2, viii p. + 181 pl.

1942. Kouroi. A study of the development of the Greek kouros from the late seventh to the early fifth century B.C. New York, Oxford University Press, xxi + 428p., incl. 483 fig. on 135 pl.

Rickles, N. K. 1950. Exhibitionism. Philadelphia, J. B. Lippincott Co., 198p.

Rioch, D. McK. 1938. Certain aspects of the behavior of decorticate cats. Psychiat. 1:339–345.

Riolan, Dr. (pseud.). 1927. La masturbation dans les deux sexes. Paris, Librairie Artistique et Médicale, 91p.

Roark, D. B., and Herman, H. A. 1950. Physiological and histological phenomena of the bovine estrual cycle with special reference to vaginal-cervical secretions. Columbia, Mo., University of Missouri (Agricultural Experiment Station Research Bull. 455), 70p. [incl. 7 pl.].

Roberts, A., Donaldson, J., and Coxe, A. C., ed. 1885. The Ante-Nicene fathers. Translations of the writings of the fathers down to A.D. 325. Volume II: Fathers of the second century. Buffalo, N. Y., Christian Literature Publishing Co., vii + 629p.

Robie, W. F. 1925. The art of love. London, Medical Research Society, 386p.

Robinson, S. 1938. Experimental studies of physical fitness in relation to age. Arbeitsphysiologie 10:251–323.

Robinson, V., ed. 1936. Encyclopaedia sexualis. New York, Dingwall-Rock, xx + 819p.

Robinson, W. J. 1931. Woman. Her sex and love life. New York, Eugenics Publishing Co., 415p.

Rockwood, L. D., and Ford, M. E. N. 1945. Youth, marriage, and parenthood. New York, John Wiley & Sons, xiii + 298p.

Rohleder, H. 1902. Ed. 2. Die Masturbation. Eine Monographie für Ärzte, Pädagogen und gebildete Eltern. Berlin, Fischer's medicin. Buchhandlung, xxiii + 336p.

1907. Vorlesungen über Geschlechtstrieb und gesamtes Geschlechtsleben des Menschen. Band 1. Das normale, anormale und paradoxe Geschlechtsleben, xvi + 600p. Band 2. Das perverse Geschlechtsleben des Menschen, auch vom Standpunkte der lex lata und der lex ferenda, xvi + 545p. Berlin, Fischer's medicin. Buchhandlung.

1918. Ed. 2. Normale, pathologische und künstliche Zeugung beim Menschen. (Monographien über die Zeugung beim Menschen. Band I.) Leipzig, G. Thieme, xvi + 317p.

1921. Die Masturbation. Eine Monographie für Ärzte, Pädagogen und gebildete Eltern. Berlin, Fischer's medicin. Buchhandlung, xxvii + 384p.

1923, 1925. Vorlesungen über das gesamte Geschlechtsleben des Menschen. Band II. Die normale und anormale Kohabitation und Konzeption (Befruchtung), 357p. Band IV. Die homosexuellen Perversionen des Menschen, auch vom Standpunkt der lex lata und lex ferenda, 403p. Berlin, Fischer's medicin. Buchhandlung.

1927. Die Masturbation. *In:* Stern, E. ed. Die Erziehung und die sexuelle Frage, pp. 279–300.

Roland de la Platière, M. J. P. 1864. Mémoires de Madame Roland. Paris, Henri Plon, [2] + 443p.

Root, W. S., and **Bard, P.** 1947. The mediation of feline erection through sympathetic pathways with some remarks on sexual behavior after deafferentiation of the genitalia. Amer. J. Physiol. 151:80–90.

Rosanoff, A. J. 1938. Manual of psychiatry and mental hygiene. New York, John Wiley & Sons, xviii + 1091p.

Rosenbaum, J. 1845. Die Onanie oder Selbstbefleckung, nicht sowohl Laster oder Sünde, sondern eine wirkliche Krankheit. Leipzig, Gebauer, iv + 267p.

Rosenthal, H. C. 1951. Sex habits of: European women vs. American women. A digest of two important new surveys . . . Pageant Mag. (March), pp. 52–59.

Rosenzweig, S., and **Hoskins, R. G.** 1941. A note on the ineffectualness of sex-hormone medication in a case of pronounced homosexuality. Psychosom. Med. 3:87–89.

Ross, R. T. 1950. Measures of the sex behavior of college males compared with Kinsey's results. J. Abnorm. & Soc. Psychol. 45:753–755.

Rossen, R., Kabat, H., and **Anderson, J. P.** 1943. Acute arrest of cerebral circulation in man. Arch. Neurol. & Psychiat. 50:510–528.

Roubaud, F. 1876. (Ed. 1, 1855). Traité de l'impuissance et de la stérilité chez l'homme et chez la femme . . . Paris, J. B. Baillière et Fils, xvi + 804p.

Rowlands, I. W., and **Parkes, F. R. S.** 1935. The reproductive processes of certain mammals. VIII. Reproduction in foxes. Proc. Zool. Soc. London 1935: 823–841.

Ruch, T. C. 1949. Somatic sensation. *In:* Howell, W. H. (Fulton, J. F., ed.). A textbook of physiology, pp. 292–315.

Ruland, L. 1934. (Rattler, T. A., trans.). Pastoral medicine. St. Louis and London, B. Herder Book Co., viii + 344p.

Russell, B. 1929. Marriage and morals. New York, Horace Liveright, 320p.

Rutgers, J. 1934. (Haire, N., trans.). The sexual life. Dresden, R. A. Giesecke, 448p.

Sacher-Masoch, L. von. [190–]. Grausame Frauen (Sphinxe). Leipzig, Leipziger Verlag, 112p.

1902. (Beaufort, L. de, trans.). La Vénus a la fourrure. Roman sur la flagellation. Paris, Charles Carrington, xxxvii + 216p.

Sade, D. A. F. de. 1797. Histoire de Justine, ou les malheurs de vertu. Hollande, 4v. v. 1, 347p. v. 2, 351p. v. 3, 356p. v. 4, 366p.

1904. Les 120 journées de Sodome, ou l'École du libertinage. Publié pour la première fois d'après de manuscrit original. Paris, Club des Bibliophiles, 8 + 543p.

Sadger, J. 1921. Die Lehre von den Geschlechtsverirrungen (Psychopathia sexualis) auf psychoanalytischer Grundlage. Leipzig und Wien, Franz Deuticke, 458p.

Sadler, W. S. 1948. Adolescence problems. A handbook for physicians, parents, and teachers. St. Louis, C. V. Mosby Co., 466p.

Sadler, W. S., and **Sadler, L. K.** 1944. Living a sane sex life. Chicago, New York, Wilcox & Follett Co., xii + 344p.

Salmon, U. J. 1941. Rationale for androgen therapy in gynecology. J. Clin. Endocrinol. 1:162–179.

Salmon, U. J., and **Geist, S. H.** 1943. Effect of androgens upon libido in women. J. Clin. Endocrinol. 3:235–238.

Salter, W. T., Humm, F. D., and **Oesterling, M. J.** 1948. Analogies between urinary 17-ketosteroids and urinary "estroid," as determined microchemically. J. Clin. Endocrinol. 8:295–314.

Sanchez, T. 1637. . . . De sancto matrimonio sacramento disputationum . . . Tomus primus (Liber IX). Lugduni, Societas Typographorum, var. pag.

Sand, K., and Okkels, H. 1938. The histological variability of the testis from normal and sexual-abnormal, castrated men. Endokrinologie 19:369–374.

Sandblom, P. 1948. Precocious sexual development produced by an interstitial cell tumor of the testis. Acta Endocrinol. 1:107–120.

Sanger, Margaret H. n.d. What every girl should know. Reading, Pa., Sentinel Printing Co., 91p.

Sappho. See Miller, M. M., and Robinson, D. M.; Weigall, A.; and Wharton, H. T.

Schapera, I. 1941. Married life in an African tribe. New York, Sheridan House, xvii + 364p., 8 pl.

Schbankov. See Weissenberg, S.

Scheinfeld, A. 1944. Women and men. New York, Harcourt, Brace and Co., xv + 453p.

Scheuer, O. F. 1923. [Discussion on] Jungfernschaft [virginity]. In: Marcuse, M., ed. Handwörterbuch der Sexualwissenschaft, pp. 242–244.

Schlichtegroll, C. F. von. 1901. Sacher-Masoch und der Masochismus. Dresden, H. R. Dohrn, 96p.

Schmidt, Karl. 1881. Jus prima noctis. Eine geschichtliche Untersuchung. Freiburg im Breisgau, Herder, xliii + 397p.

Schonfeld, W. A., and Beebe, G. W. 1942. Normal growth and variation in the male genitalia from birth to maturity. J. Urol. 48:759–777 + [2 tables].

Scott, J. C. 1930. Systolic blood-pressure fluctuations with sex, anger and fear. J. Comp. Psychol. 10:97–114.

Sears, R. R. 1943. Survey of objective studies of psychoanalytic concepts. New York, Social Science Research Council, xiv + 156p.

Selling, L. S. 1938. The endocrine glands and the sex offender. Med. Record (May 18), 9p.
    1947. The extra-institutional treatment of sex offenders. In: Lindner, R. M., and Seliger, R. V., ed. Handbook of correctional psychology, pp. 226–232.

Selye, Hans. 1949. Textbook of endocrinology. Montreal, Acta Endocrinologica, xxxii + 914p.

Semans, J. H., and Langworthy, O. R. 1938. Observations on the neurophysiology of sexual function in the male cat. J. Urol. 40:836–846.

Shadle, A. R. 1946. Copulation in the porcupine. J. Wildlife Management 10:159–162 + 1 pl.

Shadle, A. R., Smelzer, M., and Metz, M. 1946. The sex reactions of porcupines (Erethizon d. dorsatum) before and after copulation. J. Mammal. 27:116–121.

Shelden, C. H., and Bors, E. 1948. Subarachnoid alcohol block in paraplegia. Its beneficial effect on mass reflexes and bladder dysfunction. J. Neurosurgery 5:385–391.

Sherwin, R. V. 1949. Sex and statutory law (in all 48 states). New York, Oceana Publications, 2 pts. in 1 v. pt. 1, [6] + 90p. pt. 2, [6] + 74p.
    1950. Some legal aspects of homosexuality. Int. J. Sexol. 4:22–26.
    1951. Sex expression and the law. II. Sodomy: a medico-legal enigma. Int. J. Sexol. 5:10–13.

Shinozaki, Nobuo. 1951. Report on sexual life of Japanese. Tokyo, Research Institute of Population Problems, Ministry of Welfare, 38p.

Shock, N. W. 1950. Physiological manifestations of chronic emotional states. In: Reymert, M. L., ed. Feelings and emotions, pp. 277–283.

Shultz, Gladys D. 1949. Widows: wise and otherwise. A practical guide for the woman who has lost her husband. Philadelphia and New York, J. B. Lippincott Co., 285p.

Shuttleworth, F. K. 1937. Sexual maturation and the physical growth of girls age six to nineteen. Washington, D. C., Nat. Res. Council (Soc. Res. Child Devel. Mono. 12), xx + 253p.

1938a. The adolescent period, a graphic and pictorial atlas. Washington, D. C., Nat. Res. Council (Soc. Res. Child Devel. Mono. 16), v + 246p.

1938b. Sexual maturation and the skeletal growth of girls age six to nineteen. Washington, D. C., Nat. Res. Council (Soc. Res. Child Devel. Mono. 18), vii + 56p.

1939. The physical and mental growth of girls and boys age six to nineteen in relation to age at maximum growth. Washington, D. C., Nat. Res. Council (Soc. Res. Child Devel. Mono. 22), vi + 291p.

1951. The adolescent period: a graphic atlas. Evanston, Ill., Society for Research in Child Development (Monog. Ser. 49, no. 1), n. pag.

Siegler, S. L. 1944. Fertility in women. Causes, diagnosis and treatment of impaired fertility. Philadelphia, J. B. Lippincott Co., vii + 438p.

Simon, C. 1947. Homosexualists and sex crimes. Int. Assoc. Chiefs Police, 8p.

Singer, M. R. D. 1940. Behavior and development in puberty praecox. Charlottesville, Va., University of Virginia (M.A. Thesis), [4] + 79p.

Slater, E., and Woodside, M. 1951. Patterns of marriage. A study of marriage relationships in the urban working classes. London, Cassell & Co., 311p.

Smith, G. F. 1924. Certain aspects of the sex life of the adolescent girl. J. Applied Psychol. 8:347–349.

Smith, G. V. S. 1950. The ovaries. *In:* Williams, R. H., ed. Textbook of endocrinology, pp. 349–449.

Smith, M. B. 1951. The single woman of today. Her problems and adjustment. London, Watts & Co., xiv + 130p.

Smith, P. E., and Engle, E. T. 1927. Experimental evidence regarding the role of the anterior pituitary in the development and regulation of the genital system. Amer. J. Anat. 40:159–217.

Smitt, J. W., ed. 1951. Hvorfor er de sådan? En studie over homosexualitetens problemer. København, Hans Reitzel, 223p.

Snow, W. F. 1941. Women and their health. New York, American Social Hygiene Association (Publication No. A-328), 15p.

Soule, S. D., and Bortnick, A. R. 1941. Stilbestrol. A clinical study in estrogenic therapy. J. Clin. Endocrinol. 1:53–57.

Sparrow, W. S., ed. 1905. Women painters of the world, from . . . Caterina Vigri . . . to Rosa Bonheur . . . London, Hodder & Stoughton, 332p. (incl. 38 pl., 7 col. pl., + unlisted pl.)

Speigel, L. A. 1951. A review of contributions to a psychoanalytic theory of adolescence. Psychoanalytic study of the child 6:375–393.

Spitz, R. A. 1949. Autoerotism. Some empirical findings and hypotheses of three of its manifestations in the first year of life. Psychoanalytic study of the child 3–4:85–120.

1952. Authority and masturbation. Some remarks on a bibliographical investigation. Psychoanal. Quart. 21:490–527.

Spragg, S. D. S. 1940. Morphine addiction in chimpanzees. Comp. Psychol. Monogr. v. 15, no. 7, 132p.

Squier, R. 1938. The medical basis of intelligent sexual practice. *In:* Folsom, J. K., ed. Plan for marriage, pp. 113–137.

Stall, S. 1897. What a young husband ought to know. Philadelphia, Vir Publishing Co., 1 pl. + 300p.

Steinach, E. 1940. (Loebel, Josef, ed.). Sex and life. Forty years of biological and medical experiments. New York, Viking Press, x + 252p.

Steiner, M. 1912. *In:* Wiener psychoanalytische Vereinigung. Die Onanie, pp. 129–132.

Steinhårdt, I. D. 1938. Sex talks to girls (twelve years and older). Philadelphia, J. B. Lippincott Co., 221p.

Stekel, W. 1895. Ueber Coitus im Kindesalter. Wiener medizinische Blätter 18 (Apr.18):247–249.

1912. *In:* Wiener psychoanalytische Vereinigung. Die Onanie, pp. 29–45.

1920. Onanie und Homosexualität. (Die homosexuelle Neurose.) Vol. 2 of:

Störungen des Trieb- und Affektlebens. (Die parapathischen Erkrankungen.) Berlin, Wien, Urban & Schwarzenberg, xii + 527p.

1920. Die Geschlechtskälte der Frau. (Eine Psychopathologie des weiblichen Liebeslebens.) Vol. 3 of: Störungen des Trieb- und Affektlebens. (Die parapathischen Erkrankungen.) Berlin, Wien, Urban & Schwarzenberg, xii + 402p.

1922. (Van Teslaar, J. S., trans.). Bi-sexual love, the homosexual neurosis. Boston, Richard G. Badger, viii + 359p.

1923. Ed. 3. Onanie und Homosexualität. (Die homosexuelle Parapathie.) Berlin und Wien, Urban & Schwarzenberg, xii + 600p.

1925. Sadismus und Masochismus, für Ärzte und Kriminalogen dargestellt. Leipzig, Verlag der Psychotherapeutischen Praxis, vi + 765p.

1926. (Van Teslaar, J. S., trans.). Frigidity in woman. New York, Liveright, 2v. v. 1, 304p. v. 2, 314p.

1929. (Brink, L., trans.). Sadism and masochism. New York, Liveright Publishing Corp. 2v. v. 1, 441p. v. 2, 473p.

1930. (Parker, S., trans.). Sexual aberrations. New York, Liveright Publishing Corp., 2v. v. 1, 369p. v. 2, 355p.

1950. (Van Teslaar, J. S., trans.). Auto-erotism. A psychiatric study of onanism and neurosis. New York, Liveright Publishing Corp., vii + 289p.

Stephens, A. O. 1947. Premarital sex relationships. *In:* Fishbein, M., and Burgess, E. W., ed. Successful marriage, pp. 44–54.

Stern, B. 1933. (Berger, D., trans.). The scented garden. Anthropology of the sex life in the Levant. New York, American Ethnological Press, 443p.

Stern, E., ed. 1927. Die Erziehung und die sexuelle Frage. Berlin, Union deutsche Verlagsgesellschaft, 382p. + xxx pl.

Stevens, S. S., ed. 1951. Handbook of experimental psychology. New York, John Wiley & Sons, xi + 1436p.

Stier, E. 1938. Schädigung der sexuellen Funktionen durch Kopftrauma. Deutsch. med. Wchnschr. 64:145–147.

Stiles, H. R. 1869. Bundling; its origin, progress and decline in America. Albany, Joel Munsell, 139p.

Stockham, A. B. 1901? Karezza, ethics of marriage. Chicago, Stockham Publishing Co., viii + 144p.

Stokes, W. R. 1948. Modern pattern for marriage. The newer understanding of married love. New York, Rinehart and Co., xiv + 143p.

Stone, C. P. 1922. The congenital sexual behavior of the young male albino rat. J. Comp. Psychol. 2:95–153.

1923a. Further study of sensory functions in the activation of sexual behavior in the young male albino rat. J. Comp. Psychol. 3:469–473.

1923b. Experimental studies of two important factors underlying masculine sexual behavior: the nervous system and the internal secretion of the testis. J. Exper. Psychol. 6:85–106.

1927. The retention of copulatory ability in male rats following castration. J. Comp. Psychol. 7:369–387.

1939. Copulatory activity in adult male rats following castration and injections of testosterone propionate. Endocrinol. 24:165–174.

1939. Sex drive. *In:* Allen, E., et al., ed. Sex and internal secretions, pp. 1213–1262.

Stone, C. P., and Barker, R. G. 1937. On the relationship between menarcheal age and certain measurements of physique in girls of the ages 9 to 16 years. Human Biol. 9:1–28.

Stone, H. M., and Stone, A. S. 1937. A marriage manual. A practical guide-book to sex and marriage. New York, Simon and Schuster, xi + 334p.

1952. Rev. ed. A marriage manual. A practical guidebook to sex and marriage. New York, Simon and Schuster, xiv + 301p.

Stone, L. A. 1924. Sex searchlights and sane sex ethics. Chicago, Science Publishing Co., 1 pl. + xxvii + 606p.

Stookey, B. 1943. The management of intractable pain by chordotomy. *In:* Wolff, H. G., et al., ed. Pain, pp. 416–433.

Stopes, M. C. 1931. Married love. A new contribution to the solution of sex difficulties. New York, Eugenics Publishing Co., xxiii + 165p.

Stotijn, C. P. J. 1946. Pubertas praecox. 's-Gravenhage, Martinus Nijhoff, 62p.

Strain, F. B. 1948. The normal sex interests of children from infancy to childhood. New York, Appleton-Century-Crofts, vii + 210p.

Strakosch, F. M. 1934. Factors in the sex life of seven hundred psychopathic women. Utica, N. Y., State Hospitals Press, 102p.

Stratz, C. H. 1909. Ed. 3 (ed. 1, 1903). Der Körper des Kindes und seine Pflege. Für Eltern, Erzieher, Ärzte und Künstler. Stuttgart, Ferdinand Enke, xviii + 386p., 4 pl.

1926. Lebensalter und Geschlechter. Stuttgart, Ferdinand Enke, x + 194p., 1 pl.

Sturgis, F. R. 1907. The comparative prevalence of masturbation in males and females. Amer. J. Dermatology (Sept.), pp. 396–400.

ca. 1908. Notes and reflections on the causes which induce marital infelicity due to the relations of the sexes. n.impr., 32p.

Stürup, G. K. 1946a. A psychiatric establishment for investigation, training and treatment of psychologically abnormal criminals. Acta Psychiat. et Neurol. 21:781–793.

[1946]b. . . . Treatment of criminal psychopaths. Scandinavian Psychiatrists, Congress 8, Report, 21–33p.

Sullivan, Mark. 1935. Our times; the United States, 1900–1925. Volume 6, The twenties. New York, Charles Scribner's Sons, xx + 674p.

Sundt, E. 1857. Om Saedeligheds-Tilstanden i Norge. Christiania, Norway, J. Chr. Abelsted, iv + 326 + xx p.

[Swinburne, A. C.]. 1907. Sadopaideia. Being the experiences of Cecil Prendergast, undergraduate of the University of Oxford. Showing how he was led through the pleasant paths of masochism to the supreme joys of sadism. Edinburg, G. Ashantee & Co., 2v. v. 1, 124p. v. 2, 167p.

[?Swinburne, A. C.] et al. 1888. The Whippingham papers; a collection of contributions in prose and verse, chiefly by the author of the "Romance of Chastisement." London, n.publ., var. pag.

Takara bunko. *See* Anon. Takara bunko.

Talbot, H. S. 1949. A report on sexual function in paraplegics. J. Urol. 61:265–270.

1952. The sexual function in paraplegics. J. Nerv. & Ment. Dis. 115:360–361.

Talbot, N. B., et al. 1943. Excretion of 17-ketosteroids by normal and by abnormal children. Amer. J. Dis. Child. 65:364–375.

1952. Functional endocrinology from birth through adolescence. Cambridge, Mass., Harvard University Press for The Commonwealth Fund, xxx + 638p.

Talmey, B. S. 1910. Ed. 6. Woman. A treatise on the normal and pathological emotions of feminine love. New York, Practitioners' Publishing Co., xii + 262p.

1912. Neurasthenia sexualis. A treatise on sexual impotence in men and in women. New York, Practitioners' Publishing Co., xi + 196p.

1915. Love. A treatise on the science of sex-attraction. For the use of physicians and students of medical jurisprudence. New York, Practitioners' Publishing Co., viii + 438p.

Talmud. 1935–1948. (Epstein, I., and Simon, M., trans. and ed.). The Babylonian Talmud . . . (Volume 1–28.) London, Soncino Press, var. pag.

Tandler, J., and Grosz, S. 1913. Die biologischen Grundlagen der sekundären Geschlechtscharaktere. Berlin, Julius Springer, [4] + 169p.

Tappan, P. W. 1949. Juvenile delinquency. New York, McGraw-Hill Book Co., x + 613p.

1950. The habitual sex-offender. Report and recommendations. New Jersey Commission on the Habitual Sex Offender, 65 + xxi p. + 1 pl.

1951. Treatment of the sex offender in Denmark. Amer. J. Psychiat. 108: 241–249.

Tauber, E. S. 1940. Effects of castration upon the sexuality of the adult male. Psychosom. Med. 2:74–87.

Taylor, W. S. 1933. A critique of sublimation in males: a study of forty superior single men. Genetic Psychol. Monogr. 13, no. 1, 115p.

Terman, L. M. 1938. Psychological factors in marital happiness. New York and London, McGraw-Hill Book Co., xiv + 474p.

1951. Correlates of orgasm adequacy in a group of 556 wives. J. Psychol. 32:115–172.

Terman, L. M., and Miles, C. C. 1936. Sex and personality. New York and London, McGraw-Hill Book Co., xi + 600p.

Theocritus, Bion, Moschus, et al. 1912 (orig. 3rd cent. B.C.). (Edmonds, J. M., trans.). The Greek bucolic poets. London, William Heinemann, xxviii + 527p.

Thieme, U., and Becker, F. 1907–1947. Allgemeines Lexikon der bildenden Künstler, von der Antike bis zur Gegenwart. Leipzig, Wilhelm Engle-mann, v. 1–4; E. A. Seemann, v. 5–34, 36.

Thorn, G. W., and Forsham, P. H. 1950. The pancreas and diabetes mellitus. In: Williams, R. H., ed. Textbook of endocrinology, pp. 450–561.

Thornton, H., and Thornton, F. 1939. How to achieve sex happiness in marriage. New York, Vanguard Press, 155p.

Thornton, N. 1946. Problems in abnormal behavior. Philadelphia, Toronto, The Blakiston Co., x + 244p.

Thorpe, L. P., and Katz, B. 1948. The psychology of abnormal behavior. A dynamic approach. New York, Ronald Press, xvi + 877p.

Tibullus. See Catullus, et al.

Tinklepaugh, O. L. 1928. The self-mutilation of a male macacus rhesus monkey. J. Mammalogy 9:293–300.

1933. The nature of periods of sex desire in woman and their relation to ovulation. Amer. J. Obstet. & Gynec. 26:333–345.

Tissot, S. A. 1764. Ed. 3 (ed. 1, 1760). L'onanisme. Dissertation sur les maladies produites par la masturbation. Lausanne, Marc Chapius et Cie, xxiv + 264p.

1773–1774. Die Erzeugung der Menschen und Heimlichkeiten der Frauen-zimmer. Frankfurt, n.publ., 4v. in 1. v. 1, (1773), 100p. v. 2, (1774), 96p. v. 3, (1774), 95p. v. 4, (1774), 96p.

1775. Ed. 10. L'onanisme. Dissertation sur les maladies produites par la masturbation. Toulouse, Laporte, xvi + 210p.

1777. Ed. 11. L'onanisme. Dissertation sur les maladies produites par la masturbation. Lausanne, Grasset & Cie, xvi + 210p.

1785 (Ex ed. 8). L'onanisme, dissertation sur les maladies produites par la masturbation. Lausanne, Franc. Grasset & Comp., xx + 268p.

Townsend, C. W. 1896. Thigh friction in children under one year of age. Trans. Amer. Pediat. Soc. 8:186–189.

Turner, C. D. 1948. General endocrinology. Philadelphia and London, W. B. Saunders Co., xii + 604p.

Turner, H. H. 1950. The clinical use of testosterone. Springfield, Ill., Charles C Thomas, vii + 69p.

Über die Behandlung der Unarten, Fehler und Vergehungen der Jugend. See Anon. Über die Behandlung der Unarten . . .

Undeutsch, U. 1950. Die Sexualität im Jugendalter. Studium Generale 3:433–454.

U. S. Bureau of the Census. 1943. Sixteenth census . . . 1940. Population. Vol. IV: Characteristics by age, marital status, relationship, education, and citizenship. Pt. 1: U. S. Summary. Washington, D. C., U.S.G.P.O., xii + 183p.

1945. Sixteenth census . . . 1940. Population. Differential fertility, 1940 and 1910. Women by number of children ever born. Washington, D. C., U.S.G.P.O., ix + 410p.

U. S. Bureau of the Census. Sampling staff. [?1947]. A chapter in population sampling. Washington, D. C., U.S.G.P.O., vi + 141p.

U. S. Bureau of the Census. 1952. Current population reports. Population characteristics. Washington, D. C., U.S.G.P.O., (Series P-20, No. 38), 20p.

U. S. Department of Labor. 1949. Family income, expenditures and savings in 1945. Washington, D. C., U.S.G.P.O. (Bull. No. 956), v + 41p.

1952. Family income, expenditures, and savings in ten cities. Washington, D. C., U.S.G.P.O. (Bull. No. 1065), v + 110p.

U. S. Public Health Service. 1930. Sex education in the home. Washington, D. C., U.S.G.P.O., 7p.

1934. Keeping fit. A pamphlet for adolescent boys. Washington, D. C., (V. D. Bulletin No. 55), 15p.

1937. Manpower. Washington, D. C., (V. D. Pamphlet No. 6), 15p.

Urbach, K. 1921. Über die zeitliche Gefühlsdifferenz der Geschlechter während der Kohabitation. Ztschr. f. Sexualwiss. 8:124-138.

Valentine, C. W. 1942. The psychology of early childhood. London, Methuen & Co., xiv + 522p.

Van der Hoog, P. H. 1934. De sexueele revolutie. Amsterdam, Nederlandsche Keurboekerij, 350p.

Van de Velde, T. H. 1926. Het volkomen huwelijk. Een studie omtrent zijn physiologie en zijn techniek. Amsterdam, N. V. Em. Querido, xxiv + 334p. + viii pl.

1930. Ideal Marriage. New York, Covici Friede, xxvi + 323p. + 8 pl.

Van Gulik, R. H. 1952. Erotic colour prints of the Ming Period, with an essay on Chinese sex life from the Han to the Ch'ing Dynasty, B.C. 206–A.D. 1644. Tokyo, priv. print., 3v. v. 1, 242p. v. 2, 210p. v. 3, 50p. + illus.

Vatsyayana. 1883–1925 (orig. betw. 1st and 6th cent. A.D.). (Burton, R. F., and Arbuthnot, F. F., trans.). The Kama Sutra of Vatsyayana. Translated from the Sanscrit by The Hindoo Kama Shastra Society. Benares, New York, Society of the Friends of India, [Guy d'Isère], priv. print., xxi + 175p., 8 pl.

Vecki, V. G. 1920. Ed. 6. Sexual impotence. Philadelphia, W. B. Saunders Co., viii + 424p.

Vest, S. A., and Howard, J. E. 1938. Clinical experiments with the use of male sex hormones. I. Use of testosterone propionate in hypogonadism. J. Urol. 40:154-183.

Vigman, F. K. 1952. Sexual precocity of young girls in the United States. Int. J. Sexol. 6:90–91.

Virgil. 1918 (orig. 1st cent. B.C.). (Fairclough, H. R., trans.). Virgil. I. Eclogues, Georgics, Aeneid I–VI. II. Aeneid VII–XII. Minor poems. Cambridge, Mass., Harvard University Press, 2v. v. 1, xvi + 593p. v. 2, 583p. ea. v., 1 pl.

Virgin's dream. See Anon. The virgin's dream.

Vorberg, G. 1921. Die Erotik der Antike in Kleinkunst und Keramik. München, Georg Müller, priv. print., 34p. + 113 pl.

Vorberg, G., ed. 1926. Ars erotica veterum. Ein Beitrag zum Geschlechtsleben des Altertums. Stuttgart, Julius Püttmann, [3]p., 47 pl.

Waehner, T. S. 1946. Interpretation of spontaneous drawings and paintings. Genetic Psychol. Monogr. v. 33:3–70.

Walker, K., and Strauss, E. B. 1939. Sexual disorders in the male. Baltimore, Williams & Wilkins Co., xiv + 248p.

Wallace, V. H. 1948. Sex in art. Inter. J. Sexol. 2:20–23.

Waller, W. 1937. The rating and dating complex. Amer. Sociol. Rev. 2:727–734.

1951. (Ed. 1, 1938). The family. A dynamic interpretation. New York, Dryden Press, xviii + 637p.

Walsh, W. T..1940. Characters of the Inquisition. New York, P. J. Kenedy & Sons, xiii + 301p.

Wangson, O., et al. 1951. Ungdomen möter sämhallet. Ungdomsvårdskommittens slutbetänkande. Stockholm, Justitiedepartementet, Statens Offentliga Wredningar 1951:41, 221p.

Ward, E. 1938. The Yoruba husband-wife code. Washington, D. C., Catholic University of America Press (Anthropological Series No. 6), 1 pl. +[v]– viii + 178p.

Warden, C. J. 1931. Animal motivation. Experimental studies on the albino rat. New York, Columbia University Press, xii + 502p.

Warner, M. P. 1943. The premarital medical consultation. Clinical premarital procedures as an aid to biologic and emotional adjustments of marriage. Med. Woman's J. 50:293–300.

Warner, W. L., and Lunt, P. S. 1941. The social life of a modern community. New Haven, Conn., Yale University Press, xx + 460p.

1942. The status system of a modern community. New Haven, Conn., Yale University Press, xx + 246p.

Washington Public Opinion Laboratory. (Dodd, S. C., and Bachelder, J. E., director.) 1950. [Abstract of survey on marriage and divorce, etc.] Int. J. Opin. & Attit. Res. 4:467–470.

Wassermann, B. J., ed. 1938. Ceramics del antiguo Peru, de la coleccion Wassermann-San Blas. Buenos Aires, Jacobo Pueser, xxxi pl. + 365 pl.

Weatherhead, L. D., assisted by Greaves, M. 1932. The mastery of sex through psychology and religion. New York, The Macmillan Co., xxv + 246p.

Webster, R. C., and Young, W. C. 1951. Adolescent sterility in the male guinea pig. Fertil. & Steril. 2:175–181.

Weigall, A. 1932. Sappho of Lesbos: her life and times. London, Thornton Butterworth, 319p., 14 pl.

Weil, A., ed. 1922. Sexualreform und Sexualwissenschaft. Vorträge gehalten auf der I. internationalen Tagung für Sexualreform auf sexualwissenschaftlicher Grundlage in Berlin. Stuttgart, J. Püttmann, 288p.

Weil, A. 1943. The chemical growth of the brain of the white rat and its relation to sex. Growth 7:257–264.

1944. The influence of sex hormones upon the chemical growth of the brain of white rats. Growth 8:107–115.

Weil, A., and Liebert, E. 1943. The correlation between sex and chemical constitution of the human brain. Quart. Bull. Northwestern U. Med. Schl. 17:117–120.

Weisman, A. I. 1941. Spermatozoa and sterility. A clinical manual. New York, London, Paul B. Hoeber, xvi + 314p.

1948. The engaged couple has a right to know. (A modern guide to happy marriage.) New York, Renbayle House, 256p.

Weiss, E., and English, O. S. 1949. Éd. 2. Psychosomatic medicine. The clinical application of psychopathology to general medical problems. Philadelphia and London, W. B. Saunders Co., xxx + 803p.

Weissenberg, S. 1924a. Das Geschlechtsleben der russischen Studentinnen. [Schbankov Study, 1908. Reported 1922.] Ztschr. f. Sexualwiss. 11:7–14.

1924b. Das Geschlechtsleben des russischen Studententums der Revolutionszeit. [Hellmann Study, 1923.] Ztschr. f. Sexualwiss. 11:209–216.

1925. Weiteres über das Geschlechtsleben der russischen Studentinnen. [Golossowker Study, 1922–1923.] Ztschr. f. Sexualwiss. 12:174–176.

Welander, E. 1908. Några ord om de veneriska sjukdomarnas bekämpande. Hygiea N:r 12:1–32.

West, James. See Withers, Carl.

Westbrook, C. H., Lai, D. G., and Hsiao, S. D. 1934. Some physical aspects of adolescence in Chinese students. Chinese Med. J. 48:37–46.

Westermarck, E. 1912, 1917. (Ed. 1, 1906). The origin and development of the moral ideas. London, Macmillan and Co., 2v. v. 1, xxi + 716p. v. 2, xix + 865p.

    1922. (Ed. 1, 1891). The history of human marriage. New York, Allerton Book Co., 3v. v. 1, x + 571p. v. 2, xi + 595p. v. 3, viii + 587p.

Wexberg, E. 1931. (Wolfe, W. B., trans.). The psychology of sex: an introduction. New York, Blue Ribbon Books, xxvi + 215p.

Wharton, F. 1932. Ed. 12. (Ruppenthal, J. C., ed.). Wharton's criminal law. Rochester, N. Y., Lawyers Co-operative Publishing Co., 3v. v. 1, viii + 1157p. v. 2, iii + 1159–2446p. v. 3, 2447–3358p.

Wharton, H. T., trans. and ed. 1895 (ed. 1, 1885). Sappho. Memoir, text, selected renderings, and a literal translation. London, John Lane, xx + 217p., 3 pl.

Wharton, L. R. 1947. Ed. 2. Gynecology, with a section on female urology. Philadelphia and London, W. B. Saunders Co., xxi + 1027p.

Whitelaw, G. P., and Smithwick, R. H. 1951. Some secondary effects of sympathectomy. With particular reference to disturbance of sexual function. New Eng. J. Med. 245:121–130.

Whiting, J. W. M. 1941. Becoming a Kwoma. Teaching and learning in a New Guinea tribe. New Haven, Conn., Yale University Press, xix + 226p., 8 pl.

Whitney, L. F. 1950. The dog. In: Farris, E. J., ed. The care and breeding of laboratory animals, pp. 182–201.

Wiener psychoanalytische Vereinigung. 1912. Die Onanie. Vierzehn Beiträge zu einer Diskussion der Wiener psychoanalytischen Vereinigung. Wiesbaden, J. F. Bergmann, iv + 140p.

Wikman, K. R. van. 1937. Die Einleitung der Ehe. Abö, Sweden, Institut für Nordische Ethnologie, xliv + 395p.

Williams, R. H., ed. 1950. Textbook of endocrinology. Philadelphia and London, W. B. Saunders Co., xii + 793p.

Williams, W. L. 1943. The diseases of the genital organs of domestic animals. Worcester, Mass., Ethel W. Plimpton, xvii + 650p., 3 col. pl. incl. 196 fig.

Willoughby, R. R. 1937. Sexuality in the second decade. Washington, D. C., Society for Research in Child Development (Monograph for the Society . . . , Vol. 2, No. 3, Serial No. 10), iii + 57p.

Wilson, J. G., and Young, W. C. 1941. Sensitivity to estrogen studied by means of experimentally induced mating responses in the female guinea pig and rat. Endocrinol. 29:779–783.

Windsor, Edward, pseud. (Malkin, S.). 1937. The Hindu art of love. New York, Falstaff Press, xiv + 276p.

Wirz, P. 1950. Hygiene der Geschlechtsorgane. In: Hornstein, X. von, and Faller, A., ed. Gesundes Geschlechtsleben, pp. 115–132.

Withers, Carl. (West, James, pseud.). 1945. Plainville, U. S. A. New York, Columbia University Press, xviii + 238p.

Wittels, D. G. 1948. What can we do about sex crimes? Sat. Eve. Post. v. 221 (Dec. 11), pp. 30–31, 47–69.

Wolf, C. 1934. Die Kastration bei sexuellen Perversionen und Sittlichkeitsverbrechen des Mannes. Basel, B. Schwabe & Co., xii + 300p.

Wolff, H. G., Gasser, H. S., and Hinsey, J. C., ed. 1943. Pain. Proceedings of the Association for Research in Nervous and Mental Diseases, 1942. Baltimore, Williams & Wilkins Co. (Research Publication Vol. XXIII), xii + 468p.

Wolff, H. G., Wolf, S. G., and Hare, C. C., ed. 1950. Life stress and bodily disease. Proceedings of the Association for Research in Nervous and Mental Diseases, 1949. Baltimore, Williams & Wilkins Co. (Research Publication Vol. XXIX), xxiii + 1135p.

Wolman, B. 1951. Sexual development in Israeli adolescents. Amer. J. Psychother. 5:531–559.

Wood-Allen, M. 1905. (Ed. 1, 1897). What a young girl ought to know. Philadelphia, Vir Publishing Co., 1 pl. + 194p.

Woodside, M. 1950. Sterilization in North Carolina. A sociological and psychological study. Chapel Hill, N. C., University of North Carolina Press, xv + 219p.

Woodworth, R. S., and Marquis, D. G. 1947. Ed. 5 (ed. 1, 1921). Psychology. New York, Henry Holt and Co., x + 677p.

Wright, H. 1937. The sex factor in marriage. New York, Vanguard Press, 172p.

Wulffen, E. 1913. Das Kind. Sein Wesen und seine Entartung. Berlin, P. Langenscheidt, xxiv + 542p.

Wulffen, E., et al. 1931. Die Erotik in der Photographie. Die geschichtliche Entwicklung der Aktphotographie und des erotischen Lichtbildes und seine Beziehungen zur Psychopathia Sexualis. Wien, Verlag für Kulturforschung, 254p., 31 pl.

Yarros, R. S. 1933. Modern woman and sex. A feminist physician speaks. New York, Vanguard Press, 218p.

Yates, F. 1949. Sampling methods for censuses and surveys. New York, Hafner Publishing Co., xiv + 318p.

Yerkes, R. M. 1935. A second-generation captive-born chimpanzee. Science 81: 542–543.

     1939. Sexual behavior in the chimpanzee. Human Biol. 11:78–111.

     1943. Chimpanzees. A laboratory colony. New Haven, Conn., Yale University Press, xv + 321p.

Yerkes, R. M., and Elder, J. H. 1936a. Oestrus, receptivity, and mating in chimpanzee. Comp. Psychol. Monogr. 13, no. 5, 39p.

     1936b. The sexual and reproductive cycles of chimpanzees. Proc. Nat. Acad. Sci. 22:276–283.

Yerkes, R. M., and Yerkes, A. W. 1929. The great apes. A study of anthropoid life. New Haven, Conn., Yale University Press, xix + 652p.

Young, P. V. 1949. Ed. 2. Scientific social surveys and research. New York, Prentice-Hall, xxviii + 621p.

Young, W. C. 1941. Observations and experiments on mating behavior in female mammals. Quart. Rev. Biol. 16:(pt. 1) 135–156, (pt. 2) 311–335.

Young, W. C., et al. 1939. Sexual behavior and sexual receptivity in the female guinea pig. J. Comp. Psychol. 27:49–68.

Young, W. C., and Rundlett, B. 1939. The hormonal induction of homosexual behavior in the spayed female guinea pig. Psychosom. Med. 1:449–460.

Zeitschrift für psychoanalytische Pädagogik. 1927–1928. [Special issue on onanism.] v. 2, no. 4, 5, & 6.

Zell, T. 1921. Geheimpfade der Natur. 1. Die Diktatur der Liebe. 2. Neue Dokumente zur Diktatur der Liebe. Hamburg, Hoffmann & Campe, 2v. v. 1, 307p. v. 2, 288p.

Zikel, Heinz. [1909]. Die Kälte der Frauen. Aerztliche Ratschläge und Beobachtungen aus dem Leben. Berlin und Leipzig, Schweizer & Co., 94p.

Zitrin, A., and Beach, F. A. 1945. Discussion of the paper. In: Hartman, C. G. The mating of mammals, pp. 40–44.

Zuckerman, S. 1932. The social life of monkeys and apes. London, Kegan Paul, Trench, Trubner & Co., xii + 357p., 24 pl.

INDEX

All numbers refer to pages. **Bold face** entries refer to more extended treatment of each subject. Names of authors and titles cited are *italicized*. The letters T and F refer to material in tables and figures respectively. The abbreviation vs. means "in relation to."

CPSIA information can be obtained at www.ICGtesting.com
Printed in the USA
LVOW11s1538260215

428491LV00002B/402/P